W0080564

Concise Environmental Engineering: Principles and Applications

Concise Environmental Engineering: Principles and Applications

Contributors

Noelle Aarts, Anne Marike Lokhorst et al.

AURIS
Reference

www.aurisreference.com

Concise Environmental Engineering: Principles and Applications

Contributors: Noelle Aarts, Anne Marike Lokhorst et al.

Published by Auris Reference Limited

www.aurisreference.com

United Kingdom

Copyright 2016
Printed in 2017 for Sale in the Indian Subcontinent

The information in this book has been obtained from highly regarded resources. The copyrights for individual articles remain with the authors, as indicated. All chapters are distributed under the terms of the Creative Commons Attribution License, which permit unrestricted use, distribution, and reproduction in any medium, provided the original author and source are credited.

Notice

Contributors, whose names have been given on the book cover, are not associated with the Publisher. The editors and the Publisher have attempted to trace the copyright holders of all material reproduced in this publication and apologise to copyright holders if permission has not been obtained. If any copyright holder has not been acknowledged, please write to us so we may rectify.

Reasonable efforts have been made to publish reliable data. The views articulated in the chapters are those of the individual contributors, and not necessarily those of the editors or the Publisher. Editors and/or the Publisher are not responsible for the accuracy of the information in the published chapters or consequences from their use. The Publisher accepts no responsibility for any damage or grievance to individual(s) or property arising out of the use of any material(s), instruction(s), methods or thoughts in the book.

Concise Environmental Engineering: Principles and Applications

ISBN: 978-1-78154-969-8

British Library Cataloguing in Publication Data
A CIP record for this book is available from the British Library

Printed in the United Kingdom

Exclusively distributed by CBS Publishers & Distributors Pvt. Ltd.

Sales & Distribution Rights only for India, Pakistan, Bangladesh, Sri Lanka, Nepal and Bhutan. This book is not to be sold outside these territories.

Contents

List of Abbreviations

AVHRR	Advanced Very High Resolution Radiometer
AGW	Anthropogenic global warming
AMAP	Arctic Monitoring.Assessment Program
CCS	Carbon Capture and Storage
CDF	cation diffusion facilitators
CTCN	Climate Technology Centre and Network
CBO	Community Based Organization
CBOs	Community Based Organizations
DAD	Decide-announce-defend strategy
DIN	dissolved inorganic nitrogen
EO	Earth Observation
EO	Earth Observation
ECMWF	European Centre for Medium-Range Weather Forecasts
EUMETSAT	Exploitation of Meteorological Satellites
FAZ	Frankfurter Allgemeine Zeitung
DPG	German Physical Society
GEOSS	Global Earth Observation System of Systems
GEOSS	Global Earth Observation system of Systems
GEOSS	Global Earth Observation system of Systems
GEOSS	Global Earth Observation System of Systems
GEO	Group on Earth Observation
IMD	India Meteorological Department
IGCC	Integrated Gasification Combined Cycle
IPCC	Intergovernmental Panel on Climate Change
IPCC	Intergovernmental Panel on Climate Change
IPCC	Intergovernmental Panel on Climate Change
ILO	International Labour Organization
IMPACT	International Model for Policy Analysis of Agricultural Commodities and Trade
LVBC	Lake Victoria Basin Commission
LVRLAC	Lake Victoria Region Local Authorities Cooperation
LED	Light Emitting Diode
MDGs	Millennium Development Goals
MODIS	Moderate Imaging Spectroradiometer
MW	Molecular weights
NASA	National Aeronautics Space Administration
NOAA	National Oceanic and Atmospheric administration
NPP	Nature Policy Plan
NDVI	Normalized Difference Vegetation index
PFCAs	perfluorinated carboxylates

PFCs	Perfluorinated contaminants
FOSEs	perfluorinated sulfonamide alcohols
PFSAs	perfluorinated sulfonates
PFCs	perfluorinated.contaminants
PCS	phytochelatin synthase
PDM	Phytoremediation Dynamic Model
PE	potential evapotranspiration
QTL	quantitative trait loci
SAS	Story And Simulation
SPRU	Science and Technology Policy Research
SPD	Social democratic party
SDVI	Standardized Difference Vegetation Index
SOD	Superoxide dismutase
TEC	Technology Executive Committee
UNEP	United Nations Environmental Program

List of Contributors

Noelle Aarts
Wageningen University, The Netherlands
ASCoR (Amsterdam School for Communication Research),the Netherlands

Anne Marike Lokhorst
Wageningen University, The Netherlands

Rafael R. Canales-Pastrana
School of Science and Technology, Universidad del Turabo, Gurabo, Puerto Rico
Department of Natural Science and Mathematics, Inter American University, Bayamon, Puerto Rico

Marlio Paredes
School of Science and Technology, Universidad del Turabo, Gurabo, Puerto Rico

Wolfgang Römer
Department of Geography, RWTH (University) Aachen, Germany

Humberto Barbosa
Universidade Federal de Alagoas (UFAL), LAPIS, Brazil

Carolien Tote
Flemish Institute for Technological Research (VITO), Centre for Remote Sensing and Earth Observation, Boeretang, Mol, Belgium

Lakshmi Kumar
Atmospheric Science Research Laboratory, SRM University, India

Yazidhi Bamutaze
Department of Geography, Geo-Informatics and Climatic Sciences, Makerere University, Kampala, Uganda

Christopher Kipkoech Saina
Chepkoilel University College, School of Environmental Studies, Department of Applied Environmental Social Sciences, Eldoret, Kenya

Daniel Kipkosgei Murgor
Chepkoilel University College, School of Environmental Studies, Department of Applied Environmental Social Sciences, Eldoret, Kenya

Florence A.C Murgor
Chepkoilel University College, School of Environmental Studies, Department of Applied Environmental Social Sciences, Eldoret, Kenya

Paulo Roberto Ferreira Carneiro
Universidade Federal do Rio de Janeiro, Instituto Alberto Luiz Coimbra de Pós-Graduação e Pesquisa de Engenharia (COPPE/UFRJ), Laboratório de Hidrologia e Estudos do Meio Ambiente, Ilha do Fundão, Rio de Janeiro/RJ, Brazil

Marcelo Gomes Miguez
Universidade Federal do Rio de Janeiro, Escola Politécnica (POLI/UFRJ), Rio de Janeiro/RJ, Brazil

Dirk Loehr
Trier University of Applied Sciences, Environmental Campus Birkenfeld, Germany

Christine Majale-Liyala
Environmental Planning and Management, Kenyatta University, Nairobi, Kenya

Jennifer Koch
Center for Environmental Systems Research, University of Kassel, Germany

Florian Wimmer
Center for Environmental Systems Research, University of Kassel, Germany

Rüdiger Schaldach
Center for Environmental Systems Research, University of Kassel, Germany

Janina Onigkeit
Center for Environmental Systems Research, University of Kassel, Germany

José D. Carriquiry
Instituto de Investigaciones Oceanológicas, Universidad Autónoma de Baja California, Ensenada, Baja California, Mexico

Linda M. Barranco-Servin
Instituto de Investigaciones Oceanológicas, Universidad Autónoma de Baja California, Ensenada, Baja California, Mexico

Julio A. Villaescusa
Instituto de Investigaciones Oceanológicas, Universidad Autónoma de Baja California, Ensenada, Baja California, Mexico

Victor F. Camacho-Ibar
Instituto de Investigaciones Oceanológicas, Universidad Autónoma de Baja California, Ensenada, Baja California, Mexico

Hector Reyes-Bonilla
Universidad Autonoma de Baja California, Mexico

Amílcar L. Cupul-Magaña
Universidad de Guadalajara, Puerto Vallarta, Jal, Mexico

Melanie Mehes-Smith
Department of Biology, Nipissing University, North Bay, Ontario, Canada

Ewa Cholewa
Department of Biology, Nipissing University, North Bay, Ontario, Canada

Kabwe Nkongolo
Department of Biology, Laurentian University, Sudbury, Ontario, Canada

Humberto Barbosa
Universidade Federal de Alagoas (UFAL), LAPIS, Brazil

Carolien Tote
Flemish Institute for Technological Research (VITO), Centre for Remote Sensing and Earth Observation, Boeretang, Mol, Belgium

Lakshmi Kumar
Atmospheric Science Research Laboratory, SRM University, India

Yazidhi Bamutaze
Department of Geography, Geo-Informatics and Climatic Sciences, Makerere University, Kampala, Uganda

Sigmund Hågvar
Department of Ecology and Natural Resource Management, Norwegian University of Life Sciences, Norway

Volker Schneider
Department of Politics and Public Administration, University of Konstanz, Germany

Jana K. Ollmann
Department of Politics and Public Administration, University of Konstanz, Germany

Masachika Suzuki
Faculty of Commerce, Kansai University, Suita-shi, Osaka, Japan

Juan José Alava
School of Resource & Environmental Management, Faculty of Environment, Simon Fraser University, Burnaby, British Columbia, Canada
Facultad de Ingeniería Marítima, Ciencias Biológicas, Oceánicas y Recursos Naturales, Escuela Superior Politécnica del Litoral (ESPOL), Guayaquil, Ecuador

Mandy R.R. McDougall
School of Resource & Environmental Management, Faculty of Environment, Simon Fraser University, Burnaby, British Columbia, Canada

Mercy J. Borbor-Córdova
Facultad de Ingeniería Marítima, Ciencias Biológicas, Oceánicas y Recursos Naturales, Escuela Superior Politécnica del Litoral (ESPOL), Guayaquil, Ecuador

K. Paola Calle
Facultad de Ingeniería Marítima, Ciencias Biológicas, Oceánicas y Recursos Naturales, Escuela Superior Politécnica del Litoral (ESPOL), Guayaquil, Ecuador

Mónica Riofrio
Instituto Antártico Ecuatoriano (INAE), Guayaquil, Ecuador

Nastenka Calle
Pacific Institute for Climate Solution (PICS), Simon Fraser University, Burnaby, British Columbia, Canada

Michael G. Ikonomou
Institute of Ocean Sciences, Fisheries and Oceans Canada, Sidney, BC, Canada

Frank A.P.C. Gobas
School of Resource & Environmental Management, Faculty of Environment, Simon Fraser University, Burnaby, British Columbia, Canada

Preface

Environmental engineering is the application of science and engineering principles to protect and utilize natural resources, control environmental pollution, improve environmental quality to enable healthy ecosystems and comfortable habitation of humans. The text *Concise Environmental Engineering Principles and Applications* provides a concise and comprehensive coverage on environmental engineering. First chapter focuses on the role of government in environmental land use planning with a plea for valuing an integral perspective. Second chapter discusses previous knowledge on heavy metal cleanup techniques and mathematical model for their evaluation, then presents a novel model approach to characterize phytoremediation dynamic. The objective of third chapter is to demonstrate the role of the geomorphic response to environmental changes on a variety of temporal and spatial scales. Fourth chapter provides a variety of methodologies of processing chains over satellite data, allowing the monitoring of areas subject to or in risk of desertification and land degradation processes. Fifth chapter focuses on the importance of integrating indigenous knowledge systems and modern science based agricultural technologies to attain a food secure population in the face of climate change hence securing livelihoods and environmental sustainability. Sixth chapter deals with the need of integration of land use planning with water resource management, seeking to establish relations between the types of land use, urban settlements and the problems involving urban flooding. Seventh chapter describes the role of tradable planning permits in environmental land use planning. Eighth chapter focuses on policy arrangement of three urban centers in East Africa in order to conclude on which arrangement(s) presents the most flexible, robust and sustainable option for solid waste management. The objective of ninth chapter is to provide a comprehensive description of the integrated modeling system LandSHIFT. JR and of its validation. Tenth chapter describes conservation and sustainability of Mexican Caribbean coral reefs and the threats of a human-induced phase-shift. The importance of plants in the remediation of heavy metal polluted soil has been discussed in last chapter.

Chapter 1

THE ROLE OF GOVERNMENT IN ENVIRONMENTAL LAND USE PLANNING: TOWARDS AN INTEGRAL PERSPECTIVE

Noelle Aarts[1], [2] and Anne Marike Lokhorst[1]

[1] Wageningen University , The Netherlands

[2] ASCoR (Amsterdam School for Communication Research),the Netherlands

ABSTRACT

Dynamics and developments in the design and implementation of Dutch nature policies

In 1990 the Dutch parliament accepted the Nature Policy Plan (NPP), designed to conserve and develop nature over the next 30 years. The original plan was developed by ecologists and biologists with little involvement of the agricultural sector. Consequently the plan was about plants and animals, about biodiversity and valuable landscapes and about the realisation of a so-called ecological infrastructure that should connect isolated pieces of nature in the Netherlands. Nothing was said about farmers' behaviours related to nature whereas the implementation of the plan asked a lot from the farmers: they should start working in a nature-friendly way, or sell their land in case it had been located in the planned ecological infrastructure.

The NPP, in its original form, was the result of a centrally organized decide-announce-defend strategy (DAD): internally decided upon, publicly announced, and, because of a fierce public resistance, firmly defended. This strategy has clearly resulted in non-acceptance of the NPP by the majority of the Dutch farmers who did not immediately see advantages for them, but from whom cooperation was needed for a successful implementation. Therefore the government decided to realise the implementation of the NPP by means of participation.

A longitudinal study of such a participation process in the Drentsche Aa area in the north of the Netherlands, has shown that the ambition of Dutch nature conservation policymakers to involve multiple actors (farmers, citizens, recreationists) in nature policy processes has resulted in different patterns of citizen involvement (Van Bommel et al., 2008). A group of citizens appeared who wanted to be involved as stakeholders, but found that they had different views than the decision makers. Even though they were allowed to express their views in discussion meetings, it was clear that these views would not be taken into account. Roughly speaking, citizens who did agree upon the proposals were included, whereas those who did not agree were excluded from the participatory process. As a consequence, the process ended up just reproducing the government's dominant discourse. The citizens with different views and perspectives – not coincidently a group that mainly consisted of farmers - did not feel respected or represented because what was relevant to them was not part of the so-called formal perspective of the government and thus was ignored. Meaningful participation was no longer possible for them, and this resulted in active, self-organized powerful resistance to the policy.

Today more and more initiatives can be found involving farmers who organize themselves in interaction with other actors in the countryside, sharing similar problems or similar ideals, and explicitly avoiding the involvement of governments. The reason is that they no longer want to be confronted with policies that have been designed without their involvement and thus do not fit neither daily farming practices nor their identity as a farmer. In addition, they do not want to be dependent on the continuously changing rules and restrictions that they encounter when they, for instance, try to apply for a subsidy such as the Agri-Environment Scheme. Instead, they experiment and invest together, in collective management of nature at their farms, in collective meadow ownership, or in new co-operations for the production of biogas as an alternative energy source. These self-organizing initiatives are characterized by high commitment and responsibility, resulting in responsible and sustainable behaviours that go beyond the individual.

INTRODUCTION

In search of ways to influence citizens' decision making concerning land use, governments are continuously expanding their repertoire of strategic tools to steer people in the desired direction. These strategic tools range from policy instruments such as subsidies and agri-environmental schemes, that are used to reward desired nature related behaviours of farmers, to forms of participation trajectories in which citizens are invited to participate in policy processes related to the design and use of public spaces (Aarts and Leeuwis, 2010); a phenomenon also referred to as governance (Hajer & Wagenaar, 2003).

While such tools might appear promising and seem to offer a wide range of steering options for governments and policy makers, they each have some serious drawbacks that might hinder or even backfire on effective policy making. In this chapter we critically reflect on the downsides of the so called instrumental approach. Participative policy making is discussed as an alternative way of getting things done. We argue that both approaches have their unique fallbacks and share some features that lead them to be sub-optimal. Therefore the network perspective is added, which implies alternative conceptualisations of change, communication and planning. The chapter concludes with a plea for valuing an integral perspective, based on all three perspectives. Such an integral perspective does not offer a 100% guarantee for successful steering, but does enlarge the space for development of policies for land use planning and rural development that are the result of a process of co-creation between government and societal actors, and that therefore have gradually become acceptable, suitable and effective.

THE INSTRUMENTAL PERSPECTIVE

A common perspective on implementing governmental policies is the instrumental perspective: governments develop policies that are implemented with the help of a set of strategic tools, also referred to as policy instruments.

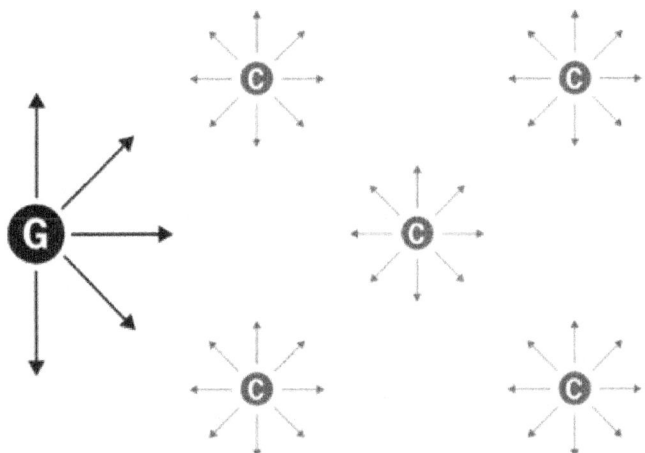

Figure 1: Interaction between government (G) and citizens (C): an instrumental perspective

Let us take a closer look to the instruments a government has available for steering citizens' behaviours.

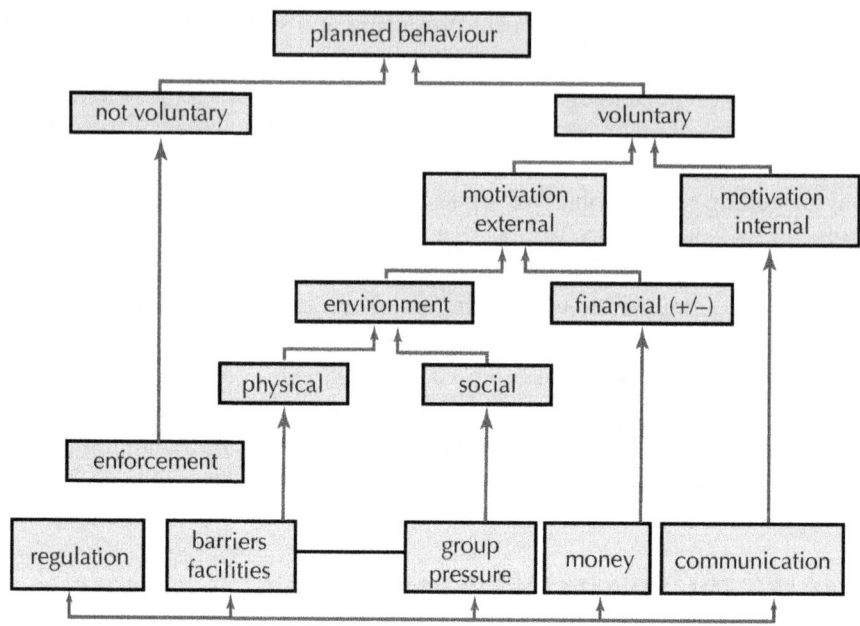

Figure 2: Policy instruments for behaviour change (Van Woerkum et al, 1999)

As can be seen in Figure 2, several policy instruments are available for governments looking to change behaviour. Within this perspective, regulation is not seen as a way to codify existing practices, but rather as an instrument to change the behaviour of people, and to facilitate that change. For several reasons it is not easy to develop and implement new rules. First, developing new regulation takes a lot of time as many different actors are involved in such a process. Second, especially in matters of nature conservation, many citizens, and certainly many farmers, are inclined to react negatively to regulation, for the very reason that such intervention takes place. Regulation undermines their feeling of freedom, ownership, and responsibility for looking after their own environment (Aarts and Van Woerkum, 1994). The physical environment can stimulate a certain behaviour, for example: playing grounds for children, recreation areas for tourists, museum for learning facilities. It can also hinder undesirable behaviour, for example: fences around valuable natural sites or roadblocks. Governments can try and (re-)design such environments as to promote the desired behaviour. Group pressure can be an important strategy to involve target groups in a policy program, but is hard to organize (for a successful example, see Lokhorst et al., 2010). More often it arises from an effective resonance between the results of other policy instruments. Money can be used as an incentive (subsidies) or as a disincentive (taxes). Finally, communication can be used when not much can be expected from other instruments. Different communication strategies are then applied to persuade people to change their behaviours (Petty and Caccioppo, 1986).

Most plans in the domain of spatial policies consist of a combination of instruments. They encompass regulation, facilities (like roads, water supply, etc.) and financial measures. As far as subsidies are involved, there is reason to be careful. Lokhorst et al. (in press a) critically review financial compensation in the domain of nature conservation. In Europe, these conservation measures are stimulated by agri-environment schemes (AES). However, as Lokhorst et al. (in press a) argue, financially rewarding conservation practices may create a dependency that is self-sustaining, costly, and therefore vulnerable. First of all, such financial policies are dependent on the current political climate and are thus susceptible to change. Indeed, many schemes on both the country and the European level have been altered or have even ceased to exist over the past years. Second, rewarding a behaviour can cause a decline in intrinsic motivation for this behaviour, a process called the "crowding out effect" (Frey, 1997). In this scenario, receiving a reward for performing a behaviour leads people to attribute their behaviour to this reward, causing a shift from intrinsic to extrinsic motivation. Should the reward then be taken away, people will no longer be motivated to perform the behaviour. Therefore, the vulnerability of financially dependent conservation practices are a threat to their continuity over longer time spans. Preliminary evidence for this idea was found in a study on the social psychological underpinnings of both subsidized and non-subsidized conservation (Lokhorst et al., 2011). In this study it was found that farmers' intention to engage in non-subsidized conservation was better explained by psychological aspects of their motivation than their intention to engage in subsidized conservation. While their motivation to perform subsidized conservation was driven mainly by their attitudes, when it came to non-subsidized conservation, feelings of moral obligation and self-identity played an important role. That is, farmers who saw themselves as the kind of people who conserve nature, and who felt that this was the right thing to do, were more likely to engage in non-subsidized conservation. For these farmers, conservation had become part of who they are, and their behavior is less likely to be affected by (changes in) financial policies.

Policies that have been developed without involving the people who are responsible for the implementation are not easy to implement. Such policies will only be accepted when people have the impression that they will benefit from the policy. This may be the case when subsidies are applied, however, as we have shown, this leads to a risky and unstable implementation. We should thus search for different ways of policymaking. Instead of the instrumental perspective in which the government is both the initiator and director of the policy process, the participation perspective, involving stakeholders in the process of policy-development for problem solution, has become a leading paradigm.

THE PARTICIPATION PERSPECTIVE

The participation perspective can be seen as a response to conceptualizations of steering that reflect great confidence in the malleability of society, but nevertheless do not often give the expected results (Glasbergen, 1995; Aarts and Van Woerkum, 2002; Arts and Van Tatenhove, 2005). Governments invite citizens with the aim to let them contribute to the development and implementation of policies.

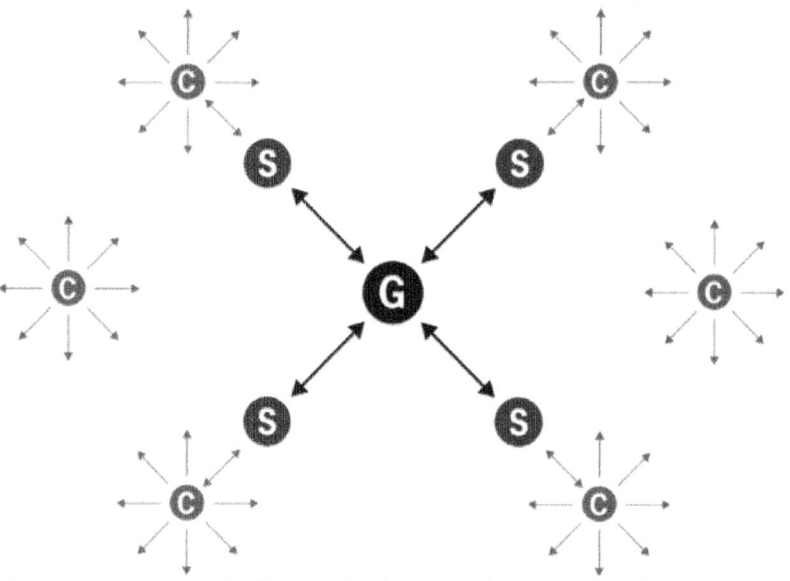

Figure 3: Interaction between government (G), statkeholders (S) and citizens (C): a participation perspective

The idea of public participation is not new. Since the late 1980s, public participation in land use issues has become the dominant discourse, in the Netherlands, but also outside the Netherlands (see Idrissou et al, 2011). Public participation may take place at (or across) various government levels. At the Dutch national level for instance, public meetings have been organized to discuss with citizens issues such as genetic modification of food products and nature conservation policies. In addition, citizen panels of the Ministry of Food and Nature are regularly consulted for the development of policies for sustainable food production. At the local level it is common that citizens are consulted for the design of public space in their neighbourhood.

Practices of public participation, however, do not always result in success. In the first place, we can notice a limited reach in society. Concrete efforts to organize participation in most cases do not involve more than a selected group of citizens, namely the so-called stakeholders that have an interest in the problem domain of which representatives are invited by the government to discuss and negotiate problem-definitions and possible solutions. There are many citizens who are not willing or not able to get their problem perceptions on the agenda of the government. Research of the way governments act in cases of conflict in public space shows that several preconditions have to be met before policy-windows are opened (Van Lieshout et al, 2008; see also Kingdon, 2002). The problem should be well-defined and solvable, and the political context should ask for dealing with the issue. We touch upon the phenomenon of self-referentiality as an inherent characteristic of governments, reducing reality to what is measurable and solvable (Wagemans, 2002).

Second, related to public participation, both in literature and in reality, a rather limited view on how people actually behave can be experienced. Public participation often starts from rather simplistic and prejudged ideas concerning people's motivations, possibilities and restrictions as well as how they behave in groups. Very limited attention is paid to what actually happens between people, including, for instance, processes of community development, processes of inclusion and exclusion, and processes of changing power-relationships, both within and between different groups in society. In addition, the reasonability and intelligence of people are systematically underestimated. Instead people are more often than not viewed as if they always behave in a completely selfish way, only concerned about money. For problems to be solved people only should change their behaviour: they should be open, honest, perfectly listening to each other and be ready to give in. In other words, it is not about understanding actual behaviours of people, but about striving after a sort of ideal behaviour that has to bring solutions.

Third, within the public participation model, instead of capitalizing on differences and diversity, we find a striving after consensus. In cases of complex land use issues this is in most cases neither realistic nor does it contribute to effective problem-solving because it takes away a lot of creativity and easily results in unsolvable impasses (Aarts and Leeuwis, 2010).

Finally, and maybe most important, efforts of public participation keep showing an obstinate illusion of central steering. It is still the government who defines the problem, including the direction for solution. The most important dilemmas of steering by participation concern matters of responsibility for the final result and of realizing ambitions, mainly expressed by the plea for the so-called 'primacy' of democratically elected bodies which is vested in the constitution. In the context of land use planning these dilemmas are often solved by simply denying people who bring in ideas that do not fit with the ambitions that have already been formulated by the government beforehand (Van Bommel et al, 2009; Turnhout et al, 2010). This may result in citizens that do not want to participate anymore, and instead start organizing themselves to realise their own ambitions in their own way. In other words, on the one hand, the government, in spite of their participation initiatives sticks to their power to decide what to do. On the other hand, by simply placing them outside the formal process, groups of citizens regain what Foucault has called productive power (Van der Arend, 2007: 53), referring to a certain amount of autonomy and freedom. The risk is that governments who apply public participation as an additional form of instrumental steering - which is the case when space for negotiation is lacking – create their own powerful antagonists (Turnhout et al, 2010).

It can be concluded that both the instrumental perspective and public participation that starts from a fixed policy tend to neglect the idea of citizens being active agents, interacting with each other and organizing themselves in order to attain their goals, for themselves and for others in their environment (Aarts & Leeuwis, 2010). A more advanced form of participation is the organizing of public commitment, in which an individual is asked to make a commitment to certain behaviour(s) in the presence of other people. In the social psychological literature, commitment-making is generally seen as a promising intervention technique (Abrahamse et al. 2005; De Young 1993; Dwyer et al. 1993; Katzev and Wang 1994) and has been shown to influence, a wide range of behaviours. Public commitment can influence behaviour in a number of ways (Lokhorst et al., in press b). First, it can change people's self-image such that the new behaviour becomes a part of their self- identity. Second, it can evoke a willingness to conform to either a societal or personal norm to engage in the behaviour in question. Third, it can set in motion a process generally referred to as cognitive elaboration (Petty and Cacioppo 1986), a process whereby the individual elaborates on the possible reasons to engage in the behaviour and strategies to accurately perform the behaviour, resulting in a strong positive attitude towards the behaviour.

In the domain of nature conservation, commitment making has been shown to affect people's intention to engage in conservation. In a study by Lokhorst et al. (2010), farmers were invited to participate in study groups in which they received feedback on their current conservation efforts. In this setting, farmers were asked to publicly commit to improving these efforts. A year later it was found that those who had made such a commitment showed actual improvement in terms of intention to conserve and area of (semi-) natural habitat. Commitment making seems like an effective tool, but requires working together with local groups and small-scale networks in order to truly be effective. It needs, in other words, to be realised in direct interaction with the people involved. This invites us to explore a third perspective: the network perspective.

THE NETWORK PERSPECTIVE

Already in the seventies the policy-scientist Scharpf (1978) argued that governments would lose their central and steering role. Scharpf referred to the tendency of increasing interweaving and organizational fragmentation within society. He predicted that governments would not be able to function without the co-operation of countless organizations and institutions. As a result, policy processes would have a 'network-like' structure (Scharpf, 1978).

Today, the idea that policies are shaped by 'pulling' and 'pushing' in complex interactions between different stakeholders has become commonly understood. Policy-processes are considered as on-going negotiation processes of which it is difficult to predict the results. Not only do circumstances change continuously, the figurations that people form on the base of their mutual dependences, continuously change as well (Elias, 1970). They shift, according to what is happening between them and in the world around them.

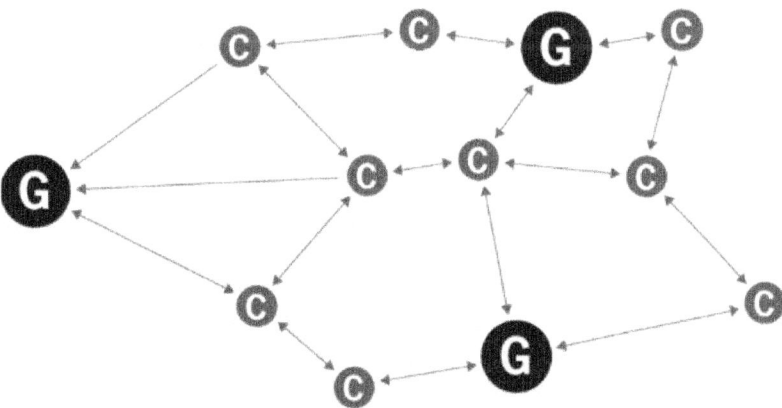

Figure 4: Interaction between government (G) and citizens (C): a network perspective

In view of unexpected developments decisions can (and will) be revised continuously. Thus, policies will eventually take shape in the interactions between different parties, involved in a network of a specific policy-domain and constantly trying to influence the process. This insight asks for further exploring the significance of a network perspective for the development and implementation of land use policies.

The network perspective refers to an endless collection of what Manual Castells (2004) calls 'interacting nodes', either people who have a special position or a combination of roles and functions that enable them to connect different networks, or specific technologies or policies that make different networks become active (Van Dijk et al, 2011). Such networks are neither centrally steered nor fixed, but instead constantly shaped through the pushing and pulling by different stakeholders who continuously do efforts to influence the situation. The network perspective thus starts from the assumption that people, instead of being passive and opportunistic, are active agents, interacting with each other and organizing themselves in order to get things done. In this perspective the role of governments is not to organize and manage a top-down or bottom-up process, but rather anticipate and make use of the self-organizing ability and initiatives of people. Operating in this mode has important implications for how a government interacts with society in order to get things done. It implies that governments must be alert, and constantly gather information about what happens in society, paying attention to informal networks. In doing so, it must develop a sensitivity for coinciding developments, and create room for experimentation especially 'where rules are not applied or are not yet developed' (Hüsken and De Jonge, 2005: 7). It may also need to redirect emergent developments at an early stage in case these are likely to go against the public interest and / or have unacceptable effects for specific groups of future generations. This requires a pro-active attitude in contacting societal actors and opening negotiations with them. In short, governments must, on the one hand, 'set free' and, on the other hand, 'stay connected'. This third mode of operating is in part an alternative to the instrumental and the participation perspective, in that there may be domains in which a government wants to delegate responsibility to societal forces without the deliberate organization of a participatory process. However, even when a government chooses for an instrumental mode or wants to organize a participation trajectory (e.g for reasons of creating legitimacy and / or formal commitment), there will be a need to make use of a network perspective. As we have seen, self-organizing dynamics will emerge, whether a government likes it or not. Hence, governments will have to develop the capacity to apply the three perspectives in an alternating way, and forge connections between them when needed.

In sum, integrating the three perspectives results in a set of relevant points of attention for governments when relating with society in order to develop and implement environmental land use policies:

- being constantly alert, by watching and being continuously informed about what is happening in society;
- providing the opportunity for coincidences to take place, by promoting and valuing diversity;
- creating room for experimentation, by leaving some space between rules and reality;
- connecting to what moves people, by co-constructing recognizable and understandable stories;
- problematizing the issue of societal accountability, instead of being focused only on rules and legislation;
- intervening if needed which implies a clear feeling of direction (see Aarts et al., 2007);
- working together with existing local groups and small-scale networks

ALTERNATIVE VIEWS ON CHANGE, COMMUNICATION AND PLANNING

In line with the dominance of the instrumental approach, we have become used to interpreting processes of change as goal-oriented activities where the use of a certain set of instruments will lead to the desired effect. However, most changes come about in a very different way. When looking to society from a network perspective we become aware of the fact that in many cases it is not as much the causality that determines the course of things, but rather the confluence of events at a certain point in time. In other words, it is the specific context that is the deciding factor. Moreover, whether it is a marriage, an economic crisis, the image of an organisation, or the design and use of land, structures and changes cannot be understood or explained by the behaviour of an involved individual (Elias, 1970, p. 148). Mutual interdependence between people and the way in which this is formed ultimately determine the course of things. People's activities and behaviours must therefore be understood and explained from the social bonds they have formed via the networks they are part of. In the words of Norbert Elias:

'From this intertwining, from the interdependence of people comes an order of a very specific nature, an order that is more compelling and stronger than the will and reason of each individual person that forms a part of the entwinement' (Elias, 1982, p. 240).

The focus on interdependence and interaction in relation to change also creates the necessity for a broader view of communication. Thinking in terms of individual senders and receivers, messages and channels, misses the target when we aim for an integral perspective on land use planning. In a broader view of communication, the interactions between people and groups of people are the unit of analysis. It is in such interactions that meanings are constructed, confirmed or contested. The dynamics brought about by communication are part of a whole variety of networks in which meanings are continuously negotiated.

The emphasis on interactions as the source and carrier of change stands in sharp contrast with the tendency to plan in terms of goal / means that characterises our society and in which many planners still seem to believe. It is high time that we start applying the alternative planning models that have been developed by now (Whittington, 2001; Stacey & Griffin, 2005; Stacey, 2001). The common essence of these models is that they encompass context and dynamics.

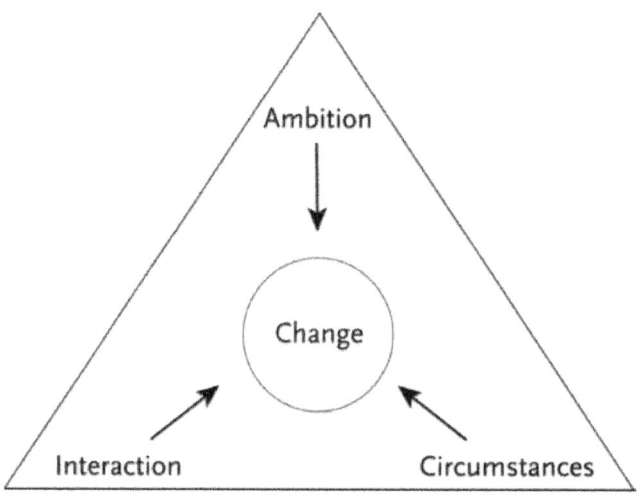

Figure 5: Planning change

Contingency planning is a well-known concept, the basis of which is concrete situations and work is done from one moment to the next. Related to this is incremental planning, dubbed 'muddling through' by Lindblom (1959). The thought behind this is that causal patterns in both social and physical reality are so complicated that centrally driven, top-down interventions have too many unintended and therefore undesired effects. Such encompassing interventions are also undesired from a normative perspective because they assume that there is one regulating point of view and preclude all other rival views. Incremental planning is based on the presumption that we may make mistakes and miss things (Frissen, 2007). With processual planning, we do have a goal (strategic intent), but the way to get there is not determined. We bet on probable situations, act and reflect on the outcomes. Furthermore, creativity, empathy, a sense of timing and even humour are important preconditions for a constructive process.

These views of change, communication and planning make clear that there are no recipes or methodologies for strategic planning, nor are there guarantees of success. However, if we take dynamics and relative unpredictability as our basis, we are better able to act consciously and respond to specific contexts more adequately. In other words, we can become better planners if we take into account our limited ability to plan. A good strategist is like a coach who follows a game closely, looks at what the players are doing and, based on that, gives instructions for moments at which action can be taken. After all, ambitions are realized in interactions with the players, and optimal use is made of the circumstances as they occur at particular moments.

CONCLUSION

Governments will never be 100% successful in steering their citizens; self-organized dynamics and other unexpected developments will continuously emerge. Therefore governments need to recognize the added value of an integral perspective and develop the capacity to apply this perspective. Although a lack of control may be uncomfortable for planners and governments, unpredictability is also an opportunity for the emergence of unexpected perspectives and ideas that may support the solution of problems in the domain of environmental planning. Several developments in the world of planning are already oriented to a broader perspective on society. Our contribution is meant to support this new thinking in a meaningful way

REFERENCES

1.   Aarts, M.N.C. & C.M.J. van Woerkum (1994). .Wat heet natuur? De communicatie tussen overheid en boeren over natuur en natuurbeleid (What is nature? Communication between .overnment and farmers over nature and nature policies)..Wageningen: Wageningen .University, Communication.Strategies Group.

2.   Aarts, N. & C. van Woerkum (2002). Dealin.g with uncertainties in solving complex .problems. In: Leeuwis, C. & R. Pyburn (eds.) .Wheelbarrows

full of frogs. Social learning in .rural resource management. .Van Gorcum, Assen,. 421-437.

3. Aarts, N. , C. van Woerkum & B. Vermun.t (2007). Policy and Planning in the Dutch .Countryside: The Role of Re.gional Innovation Networks. .Journal of Environmental .Planning and Management. .50, 6, 727-744.

4. Aarts, N. and C. Leeuwis (2010). Participatio.n and Power: Reflections on the Role of .Government in Land Use Planning and Rural Development. .Journal of Agricultural .Education and Extension., 16, 2, 131-145.

5. Abrahamse, W., Steg, L., Vlek, C., & Rothenga.tter, T. (2005). A review of intervention .studies aimed at household energy conservation. .Journal of Environmental Psychology., 25, .273-291.

6. Arts, B. & Van Tatenhove, J. (2005) Policy .and Power: A Conceptual Framework Between .the 'Old' and the 'New' Policy Idioms. .Policy Sciences., 37, 339_356.

7. De Young, R. (1993). Changing behavior and .making it stick. The conceptualization and .management of conservation behavior. .Environment and Behavior., 25, 485-505.

8. Dwyer, W. O., Leeming, F. C., Cobern, M. K., Po.rter, B. E., and Jackson,.J. M. (1993). Critical

9. review of behavioral interventions to pres.erve the environment: Research since 1980. .Environment and Behavior., 25, 275-321 .Elias, N. (1970)

10. Was ist Soziologie? (What is Sociology?).München: Juventa. .Elias, N. (1982).

11. Het civilisatieproces. Sociogenetische en ps.ychogenetische onderzoekingen (2) (The .civilization process. Sociogenetic and psychogenetic studies). .Utrecht, Het Spectrum

12. Frey, B.S. (1997). On the relationship betw.een intrinsic and extrin.sic work motivation. .International Journal of Industrial Organization., 15, 427-439. .Frissen, P. (2007).

13. De staat van verschil. Een kritiek van de .gelijkheid (The existence of difference. A .critique of eguality). .Amsterdam, Van Gennep .Glasbergen, P. (Ed.) (1995)

14. Managing Environmental Disputes. Network Management as an .Alternative..Dordrecht: Kluwer Academic Publishers. .Hajer, M. & Wagenaar, H. (2003)

15. Deliberate Policy Analysis. Understanding Governance in the .Network Society.. Cambridge: Cambridg.e University Press. .Hüsken, F. & De Jonge, H. (Eds) (2005)

16. Schemerzones & Schaduwzijden. Opstellen over .ambiguïteit in samenlevingen (Twilight Zones & .Flipsides. Essays on Ambiguity in Societies).. .Nijmegen: Roelants.

17. Idrissou, L., N. Aarts, A. van Paassen & C. .Leeuwis (2011). The Discursive Construction of .Conflict in Participatory Forest Management: The Case of the Agoua Forest Restoration .in Benin. .Conservation and Society, .9, 2: 119-131.

18. Katzev, R. D., & Wang, T. (1994). Can commitment change behavior? A case study of .environmental actions. .Journal of Social Behavior and Personality., 9, 13-26. Kingdon, J.W. (2002)

19. Agendas, alternatives, and public policies.. Longman, New York. .Lindblom, Ch.E. (1959). The Science of Muddling Through. .Public Administration Review., 19, .79-88 .

20. Lokhorst, A. M., Staats, H, Van Dijk, J., Van Di.jk, E., & De Snoo, G. (2011). What's in it for.me? Motivational differences between farm. ers' voluntary and subsidized nature.conservation practices. Applied psychology – an International Review , 60, 337-353.

21. Lokhorst, A.M., Van Dijk, J., Staats, H., Van.Dijk, E., & De Snoo, G. (2010). Usingtailored.information and public commitment to improve the environmental quality of farm.lands: An example fr.om the Netherlands.. Human Ecology., 38, 113-122.

22. Lokhorst, A. M., Staats, H., Van Dijk, J., & de Snoo, G. R. (in press a). Biodiversity on.farmlands: the crucial role of.the farmer. In: Kruse, L. (Ed.)..Naturschutz und.Freiseitgezellschaft

23. .Lokhorst, A. M., Werner, C.M., Staats, H., Van Dijk, E., & Gale, J. (in press b). Commitment.and Behavior Change: A Meta-analysis and Critical Review of Commitment Making.Strategies in Environmental Research..Environment & Behavior

24. .Petty, R. E., & Cacioppo, J. T. (1986)..Communication and persuasio.n: Central and peripheral.routes to attitude change..New York: Springer-Verlag..Scharpf, F.W. (1978) Interorganizational Policy.Studies: Issues, Concepts and Perspectives..In: Hanf, K. and Scharpf, F.W. (Eds).. Interorganizational Policy Making: Limits to.Coordination and Central Control.. London: Sage, 345 - 370..Stacey, R.D. (2001).

25. Complex responsive processes in organizations. Learning and knowledge.creation..London, Routledge.Stacey, R.D. and D. Griffin (eds) (2005).A complexity perspective on.researching organizations.. London, Routledge..Turnhout, E., S. Van Bommel and N. Aarts.(2010). How Participation Creates Citizens:.Participatory Governance as

Performative Practice..Ecology and Society.15, 4: 26. [online].URL:. http://www.ecologyandsociety.org/vol15/iss4/art26/.Van Bommel, S., Turnhout, E., Aarts, M.N.C. &.Boonstra, F.G. (2008) Policymakers are from.Saturn . Citizens are from Uranus. Involvin.g Citizens in Environmental Governance in.the Drentsche Aa Area. Wageningen: Statutor.y Research Tasks Unit for Nature and the.Environment..Van Bommel, S., N. Röling, N. Aarts & E. Turnhout (2009). Social learning for solving.complex problems: a promising solution or.wishful thinking? A case-study of multi-.actor negotiation for the integrated management and the sustainable use of the.Drentsche Aa area in.the Netherlands.. Environmental Policy and Governance., 19, 6, 400-.412. Van den Arend, S. (2007).

26.   Pleitbezorgers, procesmanagers en partic.ipanten. Interactief beleid en de.rolverdeling tussen overheid en burgers in de Nederlandse democratie (Advocates, Process.Managers and Participants. Interactive Policy and the Role Division between Government and.Citizens in Dutch Democracy).. Delft: Eburon..Van Dijk, T., Aarts, N. & De Wit, A..(2011). Frames to the Planning Game..International.Journal of Urban and Regional Research., 35, 5, 969-987.

27.   Van Lieshout, M. & N. Aarts (2008). "Outside is.where it is at!" Youth' and immigrants'.perspectives on public space..Space and Culture., 11, 4, 497-513

28.   Van Woerkum, C.M.J., M.N.C. Aarts & M.M. va.n der Poel (1999). Communication in policy .processes: reflections from the Netherlands. .European Spatial Research and Policy, .6, 75-.87.

29.   Wagemans, M. (2002) Institutional conditions fo.r transformations. A plea for policy making .from the perspective of constructivism. In C. Leeuwis and R. Pyburn (eds) .Wheelbarrows .full of frogs. Social learning in rural resource management. .Koninklijke Van Gorcum BV, .Assen, 245 - 257 .Whittington, R. (2001). .What is strategy and does it matter? .London, Routledge

Chapter 2

PHYTOREMEDIATION DYNAMIC MODEL AS AN ASSESSMENT TOOL IN THE ENVIRONMENTAL MANAGEMENT

Rafael R. Canales-Pastrana[1,2], Marlio Paredes[1]

[1]School of Science and Technology, Universidad del Turabo, Gurabo, Puerto Rico

[2]Department of Natural Science and Mathematics, Inter American University, Bayamon, Puerto Rico

ABSTRACT

Phytoremediation is considered a viable and cost effective emerging technology to clean-up trace elements. This approach has not been fully commercialized due the existence of various concerns about it. Those can be summarized as the uncertainty of the system behaviors at different scenarios, such as: contaminant, contaminant concentration and the behaviors of the physiology in the plant. Previous approaches have implemented diverse mathematical algorithms to characterize phytoremediation systems, such as: differential equation solution sets, statistical correlation and system dynamics approach. Phytoremediation Dynamic Model (PDM) employed the classical plant structure to simulate plantsoil-pollutant interaction. This model has proved its capability to mimic phytovolatilization processes of mercury chloride, obtaining more than 95% of correlation between the experimental data, and also provides the capability to know the contaminant flow rate and its concentration in plant tissue. The differential equations system which describes the model includes a comprehensive parameter which encapsulates plant bioavailability dependence in the contaminantmedia interaction as a novel approach because this has not been found on the literature previously. PDM has proved the ability to mimic plant response as a function of contaminant concentration and the applicability as an assessment tool for phytoremediation system performance.

INTRODUCTION

Since the Industrial Revolution the pollution has been exacerbated, increasing their intrusion probability in the food web [1]. Heavy metal should be a priority to environmental scientist, they are not easily degraded; rather they are bioaccumulated [2-4]. Frequently found in contaminated sites: Cd, Cr, Cu, Pb, Hg, Ni and Zn [5-7], they can be transformed by microorganism interactions into a more bioavailable forms like methyl and dimethyl compounds [8,9].

Mercury was taken as key example of heavy metal contamination, and exposure to different mercury species can inflict a variety of threats to human health, including an irreversible damage to nervous system [6,10]. The global mercury budget has increased 3.3 times in postindustrial times which can be ascribed to the exploitation of precious metals (gold and silver) and coal burning [11-13].

The environmental scientific community has the responsibility to analyze contamination issues to develop standardized protocols. Those analyses mainly consist in site contaminant characterization and construction of mathematical or graphical models, in which multivariate sequential probabilities can be exhibited and map the contaminant dispersion based on background information [12]. These components are crucial to understand their possible contaminant interactions, the establishment of the final stage goal, and the evaluation procedure on the remediation process [7,14]. These kinds of approaches have been implemented to determine the environmental hazard index or a heavy metal risk parameter linked to a specific site location map [15]. This paper discusses previous knowledge on heavy metal cleanup techniques and mathematical model for their evaluation, then presents a novel model approach to characterize phytoremediation dynamic.

PREVIOUS KNOWLEDGE ON HEAVY METAL CLEANUP TECHNIQUE AND MATHEMATICAL MODEL

Cleanup of contaminated soil is an important issue to: the environment, economy and public health. Particularly, chemical degradation affects around 12% of the total degraded soil worldwide [16]. Around the world, countries have been applying environmental strategies to prevent further soil degradation and restoring deteriorated soils, in which the cost implications are considered [17]. Besides the risk of water body contamination by soil washout runoff, there is also the risk of plants growing on contaminated soils, which then extract and translocate pollutants [18,19]. For example, vegetables have the capability to accumulate heavy metals, promoting the intrusion in the food web [19,20]. Environmental scientists have been developing traditional and non-traditional techniques.

Traditional Cleanup Technique

The traditional cleanup techniques include: flushing, chemical reduction/oxidation, excavation and capping, and stabilization and solidification [21-23]. Excavation and capping is the most commonly used and has an estimate price of $2.5 million/hectare treated [16]. Soil remediation methods for heavy metals contamination, are environmentally invasive, expensive and inefficient, especially when applied to large areas [21,24].

Non-Traditional Cleanup Technique

Bioremediation is the non-traditional cleanup technique, in which living organisms are implemented (e.g. bacteria, algae, fungi and plant) to extract or confine contaminants from the contaminated media [23,25]. The viable emerging technology to cleanup heavy metal is phytoremediation [6,26,27]. Phytoremediation employs plants for this task and has been promoted as an aesthetically pleasing and solar driven passive technique [28].

Phytoremediation

This technique can be sub-divided into: phytodegradation or phytotransformation, phytovolatilization, phytoextraction, rhizofiltration and phytostabilization [29,30]. Phytodegradation breaks down contaminants as a consequence of having a catalytic enzyme production by the root. Phytovolatilization is the uptake of a contaminant, later release through transpiration. In phytoextraction (or phytoaccumulation) plants mine and translocate the contaminant from the root to above ground tissues and fix it. Rhizofiltration refers to the absorption or adsorption of the contaminant by the plant root, while phytostabilization involves immobilization of contaminants in the root zone [31,32]. These techniques have been tested to clean up metals, pesticides and hydrocarbons on engineered wetlands [28].

Some plants, like hydrophytes, have intrinsic cleanup capabilities, but their efficiency varies significantly between species [30,33-35]. To achieve a higher efficiency, plants can be genetically modified [23,30,36]. Examples of this approach include the modification of Arabidopsis thaliana, Nicotiana tabacum and Liriodendron tulipifera with the insertion of merA and merB, two bacterial genes employed to increase the mercury remediation potential [36-40].

The implementation of this methods can cost less than one tenth of the price of traditional techniques [16,27,39] and are environmentally friendly. Therefore, their performance has been considered a highly site-specific technology [6,41], seeing the wide variation of contaminants and soil properties which affect the plant interaction [30, 34,42,43]. The most important concerns about phytoremediation are: 1) metal bioavailability within the rhizosphere; 2) uptake rate of metal by roots; 3) proportion of metal "fixed" within the roots; 4) rate of xylem loading/ translocation to shoots; and 5) cellular tolerance to toxic metals [29,30,44]. For those concerns phytoremediation has not been fully commercially implemented.

Mathematical Models

The implementation of the mathematical model on environmental science helps to evaluate different scenarios to make an objective decision without affecting the environment. Also, can bypass the human rationality, which in some cases promotes an error and/or biases [45]; particularly in complex systems such as: plant-soil interaction.

Several mathematical approaches have been used to understand the soil-plant interaction during the last forty years [46], those can be applied for modeling the phytoremediation cleanup route. Diverse mathematical algorithms have been implemented to reinforce phytoremediation process understanding. A variety of diffusion laws implementation and statistical correlations, aiming to understand the phenomena in a comprehensive way, have been found [47-52]. These models are mathematically intensive and very specialized. System Dynamic Approach (SDA) has been applied, providing a differential equations solution set, defined by models for compartmentalization of the plant physiology [53-55].

All implementations have been constructed using STELLA (system thinking software of isee systems), considering the internal interactions of the contaminant according to the plants' metabolism. However, these add an excessive complexity to the model, given the number of parameters considered, ranging from 30 to 43 variables per model [53-55]. Those variables are categorized: calibrated, estimated and assumed. These amounts of variables and their differences in the categorization enhance the model's complexity.

PHYTOREMEDIATION DYNAMIC MODEL

The construction of Phytoremediation Dynamic Model was made considering the previous model approaches; it is an implementation of SDA and a plant physiological structure. However, simpler plant structure interaction has been used. Figure 1 shows the plant schematic representation of the phytoremediation process; which is composed of four structural blocks and three processes. Each block has the intent to mimic the contaminant concentration as a function of plant physiological section (root, shoot, leaf) and soil interaction. The arrows steps are to indicate the net contaminant flow between blocks. Extraction section represents the root capability to remove the contaminant from soil. Translocation is the term typically used for the contaminant movement from the root to plant upper tissue [56]. In order to have a clear distinction, this process has been divided in two steps. Translocation 1 represents the contaminant flow from root to shoot (stem), and translocation 2 characterizes the contaminant flow from shoot to leaf.

Methodology

The development of the Phytoremediation Dynamic Model (PDM) was performed using STELLATM a dynamic software that implements the pictographic modeling representation, based upon four basic components: stocks (level variables), flows (rates), connectors (relationship) and converters (auxiliary variables) [53,55].

The plant was represented by three functional parts (root, shoot, leaf) as stocks (level variables) interconnected, mimicking its anatomy and physiology; two stocks represent abiotic factors (soil, atmosphere) of the environment (Figure 2).

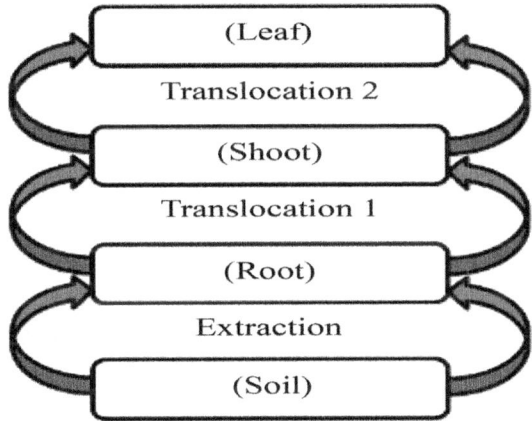

Figure 1: Basic schematic representation of plant physiology, which represents the phytoremediation process.

A similar structural representation can be found in a different phytoremediation modeling approach [11,50,55].

PDM combines, the dynamic structural diagram between biotic and abiotic environmental component with the schematic representation of the plant physiology. The model behavior will be governed by the fundamental assumption stated as follows:

1)   Fluxes (rates) depend on the contaminant concentration of the previous stocks (level variables), which relate with section rates and threshold concentration. Sections rates is a calibration variable. Threshold concentration is an estimated variable, which value establishes the minimum concentration that previous stock has to achieve to allow the contaminant flow to the next stocks. Once thresholds concentrations are achieved, the value should be maintained during the time frame modeled (Root threshold concentration, Shoot threshold concentration, Leaf threshold concentration). This works as osmotic concentration levels, which is a phenomenon observed as a function of plant species and contamina tion, as reported for plant tissues [27,30,59].

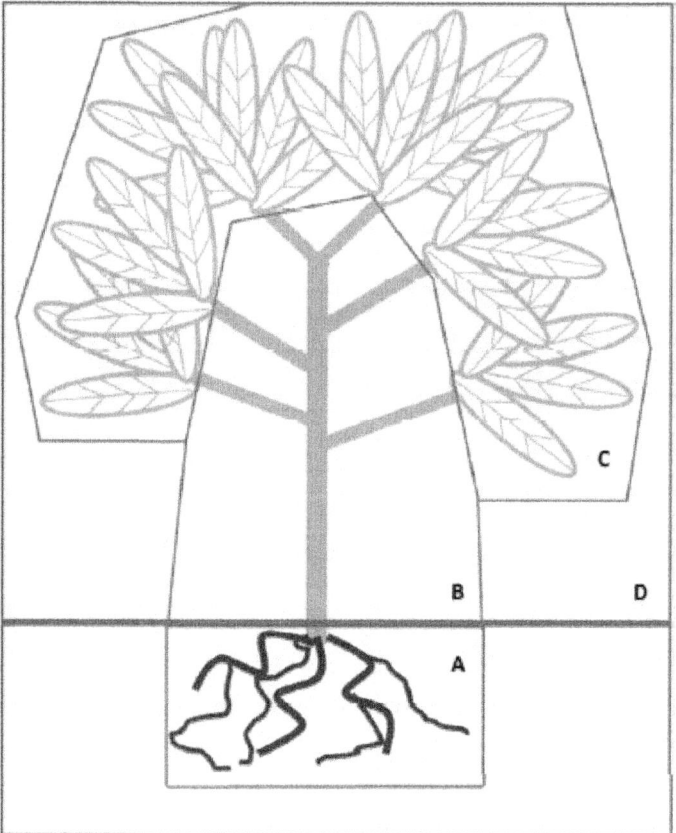

Figure 2: Dynamic structure diagram for the Phytoremediation Dynamic Model (PDM), in which the system has been divided in the compartments to be considered. The compartments can be classified as above or below the ground. The (A) Compartment represents the soil-plant interaction at the root zone, which is the below the

ground section involving two stocks: soil and root. The above ground segment; are composed by three stocks: (B) Shoots; (C) Leaf; and (D) Atmosphere.

2) Once the threshold concentration is achieved the section flow rates is constant during the time frame modeled (Extraction rate, Translocation rate, Incorporation rate, Volatilization rate), around plant transport capacity. In plant physiology it is well known that ions in solution are moved through transporters. These are characterized mainly by their transport capacity (Vmax) and affinity for the ion (Km) [56].

3) Initial level concentrations in different stocks are zero, except for the stock which represents contaminated soil.

4) Contaminant bioavailability depends on the exponential ratio between the current and initial contaminant concentration in soil. This dependence was represented in the flow equation in PMD soil section and is called Fraction. This soil-plant includes factors such as plant transporters and soil physical-chemical properties. The Km measures the transporter affinity for a specific ion, where high values represent low affinity. The contaminant bioavailability has complex interactions with soil pH, organic matter, carbonates, electrical conductivity and grain distribution [46]. The pH is one of the most important chemical properties of the soil because affects the bioavailability of the contaminant, through the modification of the cation exchange capacity [56]. The heavy metal concentration as a function of pH, has a strong correlation coefficient on a logarithmic lineal regression [28, 57,58].

Once the assumptions have been established the schematic representation of PDM was developed, using STELLATM. It was composed by five stocks, four flows and eight auxiliary variables as depicted in Figure 3. The stocks (levels variables) represent structural reservoirs of the plant physiology and environment, while flows (rates) characterize the upward net contaminant exchange between its compartments. The literature, do not make a distinction between the flows that supplies substance to shoot or leaf, both of them were called translocation as shown in Figure 1 [56]. To be explicit on PMD, translocation-2 was renamed as incorporation, which is the flow that supplies the substance to the leaf. Also, Figure 3 shows the system of differential equations, which governs the model behavior.

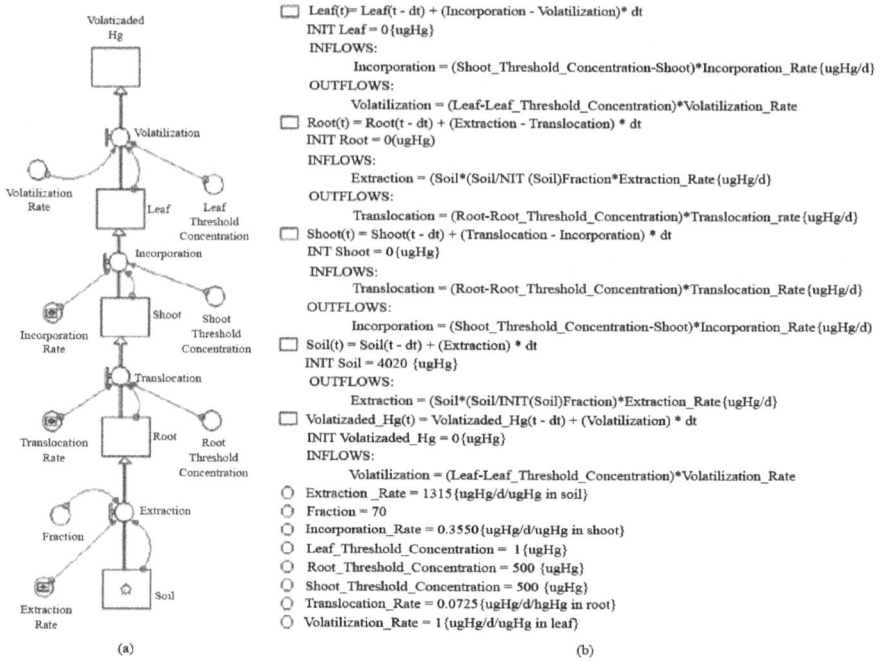

(a)

The following equations appear in part (b):

Leaf(t)= Leaf(t - dt) + (Incorporation - Volatilization)* dt
INIT Leaf = 0{ugHg}
INFLOWS:
    Incorporation = (Shoot_Threshold_Concentration-Shoot)*Incorporation_Rate{ugHg/d}
OUTFLOWS:
    Volatilization = (Leaf-Leaf_Threshold_Concentration)*Volatilization_Rate
Root(t) = Root(t - dt) + (Extraction - Translocation) * dt
INIT Root = 0(ugHg)
INFLOWS:
    Extraction = (Soil*(Soil/NIT (Soil)Fraction*Extraction_Rate{ugHg/d)
OUTFLOWS:
    Translocation = (Root-Root_Threshold_Concentration)*Translocation_rate{ugHg/d}
Shoot(t) = Shoot(t - dt) + (Translocation - Incorporation) * dt
INT Shoot = 0{ugHg}
INFLOWS:
    Translocation = (Root-Root_Threshold_Concentration)*Translocation_Rate{ugHg/d}
OUTFLOWS:
    Incorporation = (Shoot_Threshold_Concentration-Shoot)*Incorporation_Rate{ugHg/d)
Soil(t) = Soil(t - dt) + (Extraction) * dt
INIT Soil = 4020 {ugHg}
OUTFLOWS:
    Extraction = (Soil*(Soil/INIT(Soil)Fraction)*Extraction_Rate{ugHg/d}
Volatizaded_Hg(t) = Volatizaded_Hg(t - dt) + (Volatilization) * dt
INIT Volatizaded_Hg = 0{ugHg}
INFLOWS:
    Volatilization = (Leaf-Leaf_Threshold_Concentration)*Volatilization_Rate
◯ Extraction _Rate = 1315 {ugHg/d/ugHg in soil}
◯ Fraction = 70
◯ Incorporation_Rate = 0.3550{ugHg/d/ugHg in shoot}
◯ Leaf_Threshold_Concentration = 1{ugHg}
◯ Root_Threshold_Concentration = 500 {ugHg}
◯ Shoot_Threshold_Concentration = 500 {ugHg}
◯ Translocation_Rate = 0.0725 {ugHg/d/hgHg in root}
◯ Volatilization_Rate = 1 {ugHg/d/ugHg in leaf}

(b)

Figure 3.(a): The forrester diagram schematic representation of the phytoremediation dynamic model; (b) The differential equation system of the phytoremediation process.

The differential equations were rewritten according to the standard of mathematical notations, depicted in Table 1.S_, and The C_ functions represent stocks and threshold concentration, respectively. These functions have their respective sub-index (Soil, Root, Shoot, Leaf or Atm for atmosphere) to identify the model section which they represent. R_ represents the rates at which the contaminant moves, ones the threshold was attained. The function of the threshold, flux rates and the gradient in concentration between their neighbors' stocks was represented by F_. Each one of these functions has a sub-index which identifies the interaction in the model (Ext = Extraction, Tran = Translocation, Inc = Incorporation, Vol = Volatilization). The expression Init_SSoil, corresponds to the initial contaminant concentration in the soil, which is implemented as a constant to calculate the bioavailability as time evolves.

The schematic representation of PMD using STELLATM is simpler that previous UTCSP phytoremediation model [55] and generate similar systems of differential equations [11,50]. The differential equation for the soil model section can be solve by the separation of variables technique but the other equation constitutes a linear differential equation system, which can be tackle with a numerical integration solving method, such as Euler.

Qualitatively Validation

The PDM validation has been developed to mimic phy- tovolatilization because, is the most comprehensive process which includes all physiologic section of the plant.

Table 1: Differential equation system describing PDM.

Section	Mathematical representation
Soil	$$\frac{dS_{Soil}}{dt} = -F_{Ext}$$ $$F_{Ext} = \left(S_{Soil} * \left(\frac{S_{Soil}}{Init\_S_{Soil}} \right)^{Fraction} \right) * R_{Ext}$$
Root	$$\frac{dS_{Root}}{dt} = F_{Ext} - F_{Tran}$$ $$F_{Tran} = \left(S_{Root} - ThC_{Root} \right) * R_{Tran}$$
Shoot	$$\frac{dS_{Shoot}}{dt} = F_{Tran} - F_{Inc}$$ $$F_{Inc} = \left(S_{Shoot} - ThC_{Shoot} \right) * R_{Inc}$$
Leaf	$$\frac{dS_{Leaf}}{dt} = F_{Inc} - F_{Vol}$$ $$F_{Vol} = \left(S_{Leaf} - ThC_{Leaf} \right) * R_{Vol}$$
Atmosphere	$$\frac{dS_{Atm}}{dt} = F_{Vol}$$

Heavy metal accumulation and hyperaccumulation plants have been studied extensively [30], only a few research have been performed on heavy metal phytovolatilization [38,40,60,61]. The accumulated amount of heavy metal concentration in each physiological section of the plant [38,60,61], was used to establish a feasible threshold values.

Hussein et al. (2007) showed a comprehensive phytovolatilization experiment for mercury chloride ($HgCl_2$) and phenyl mercury acetate. They tested the remediation capability for two genetically modify tobacco plant in comparison of wild types. In the article, the contaminant can be found in the plants tissue and the volatilization as time evolves. The volatilization data for the two genetically modified lines are shown in Figure 4, for mercury chloride. The pLDR-merAB data set was employed for validation purposes, because they represent the simpler gene expression and present more behavioral changes in comparison with pLDR-merAB$_3$'UTR (Figure 4).

According to the environmental point of view, the amount of mercury extracted from soil is important, but as well the amount released to the atmosphere. In order to analyze amount of total mercury releases to the atmosphere (cumulative volatilized mercury) a sub model was constructed, implementing SDA (Figure 5). The qualitatively validation of PDM, was performed. In the Figure 6 depicted the likeness between the model and the experimental data values. This high similarity between the model and the experimental data validate: 1) the fundamental assumptions of the model; and 2) the value of the auxiliary variable in the base scenario which are reasonable and feasible (Table 2). The model has eight auxiliary variables that have been categorized; four as estimated and four as calibrated

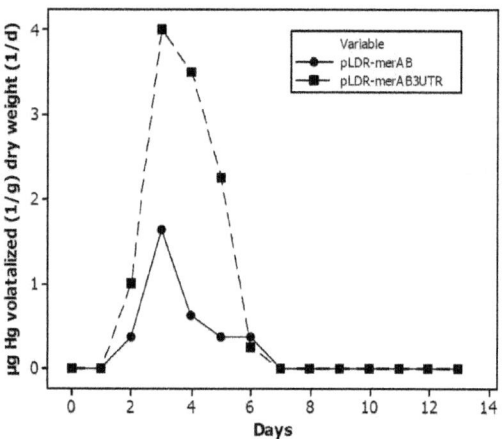

Figure 4: Volatilization data by genetically modified tobacco plant on contaminated soil with 100 µM of $HgCl_2$ (Adapted from [40]).

The categorization was performed according to the way in which their value was obtained, estimated for the value extracted from the literature and calibrated for the variables values modified to adjust model behaviors to the experimental data. Those variables are also divided in three groups: threshold, rates and bioavailability constant (Fraction).

Quantitative Validation and Statistical Analysis

The cumulative volatilized mercury concentration data was selected to perform the quantitative analysis because of the environmental relevance of those emissions that can enhance mercury concentration in the atmosphere and summarize the results of the volatilized data. Table 3 depicts the descriptive statistical analysis for the experimental and model data. The percent of difference between experimental data and model for each analysis did not exceed 0.9%. Figure 7: shows a regression fit analysis, demonstrating a strong correlation (99.4%). The slope of the regression line differed in 0.9% in comparison with the theoretical one. The analysis shows the prediction and confidence intervals as well. The prediction interval represents a range of new observation is likely to be and confidence interval represents a range that the mean will response, in both intervals that behaviors is according to the established percentage of precision. All data points achieved the 99% prediction interval; however one data point (7%) was overlapped with the line that constringes the interval, although 86% of data points are inside the confidence interval, one (7%) is touching the lines that limit the interval and another (7%) is completely outside the interval. With the results of the descriptive and regression analysis, we can be hypothesized that the difference between PDM and the experimental analysis is less than one data units. A sign test was employed as a non-parametric statistic to examine the mean difference. The Sign Test demonstrated that the null hypothesis can be rejected with a significant confidence level of 95%, having a median of 0.0200 and a p-value of 0.0001.

Table 2: Auxiliary variable categorization and base scenario values

Name (units)	Category	Value
Root threshold (μg Hg)	Estimated	500
Shoot threshold (μg Hg)	Estimated	4
Leaf threshold (μg Hg)	Estimated	1
Volatilization rate (μg Hg/(d*μg Hg in leaf))	Estimated	1
Extraction rate (μg Hg/(d*μg Hg in soil))	Calibrated	0.1315
Translocation rate (μg Hg/(d*μg Hg in root))	Calibrated	0.0725
Incorporation rate (μg Hg/(d*μg Hg in shoot))	Calibrated	0.3550
Fraction (adimensional)	Calibrated	70

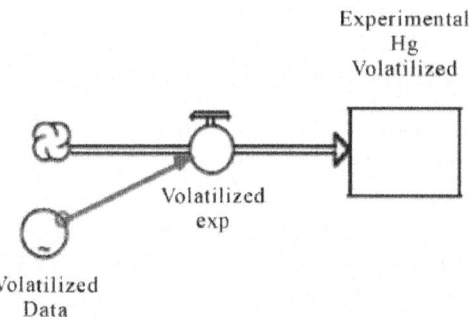

Figure 5: Schematic representation of stock (level variables) and flow model to obtain the cumulative volatilized mercury, using experimental data

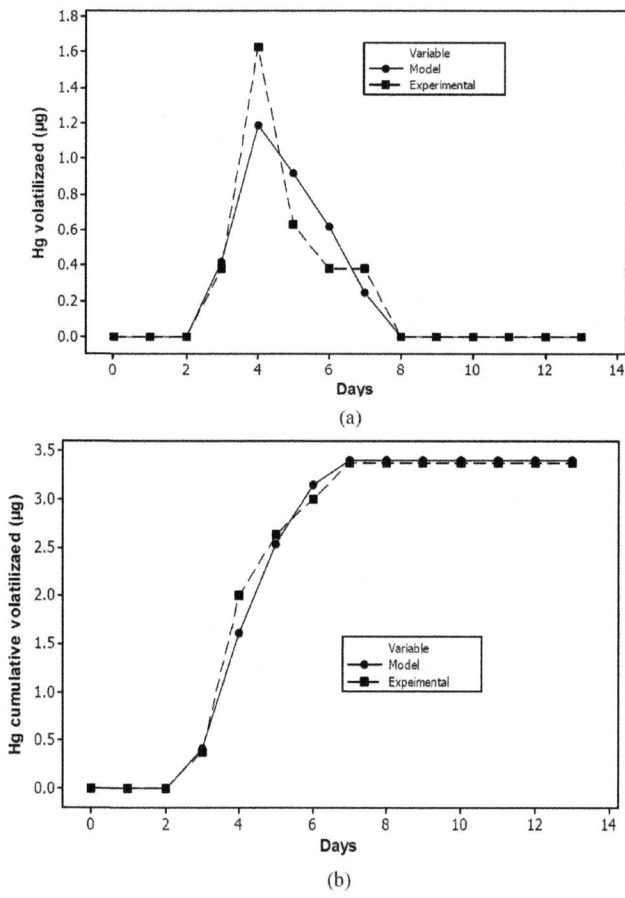

Figure 6: Comparison between experimental data and PDM. (a) Volatilized μg Hg; (b) Cumulative volatilized μg Hg

The statistical analysis demonstrates that Phytoremediation Dynamic Model (PDM) has the ability to reproduce the experimental results of phytoremediation experiment with excellent degree of accuracy and statistical significance. The differential equations system summarizes the interaction between biotic and abiotic, including bioavailability, flows rates and metal concentration. These factors are some of the most influential concerns about phytoremediation that tackles the fully commercially implementation [29,30,44]. The bioavailability factors are represented in the soil section of the model; which is governed by the Fraction calibration variable. The value of this adimensional variable represents the percentage of the contaminant, which is not available for the plant to be removed on each time steps. The calibrated value for this scenario is 70, which mean that only the 30% of the mercury chloride is accessible for the removal on each time steps. The concentration of the metal that is retained on each plants physiological section is represented in the value of the threshold variables. The contaminant flow through the phytoremediation system is characterized by the rate variables. All of those variable values were depicted in Table 2.

PDM can be also implemented as a performance tools for the technique, calculating the percentage of contaminant removed. To assess this approach, a family of runs fluctuating the mercury chloride initial concentration in the range of 10 μM to 100 μM, with an increment of 10 μM, was performed. Likewise, it have been done with two more runs ±5% of the base scenario initial contaminant concentration in soil. The Figure 8 illustrates the performance behavior for both scenarios, according to the percentage of mercury removed. The effectiveness of this phytoremediation system shows invers dependence as function of contaminant soil concentration. In the range of 10 μM to 100 μM, the amount of mercury removed varied from 31% to 13%, but close to 100 μM (± 5%) these amounts of total mercury removal varied in the hundredth. These types of analysis increased the system understanding at the time to make a decision of which kind of technique is better for a specific situation. It also provides comprehensive information for the regulators about system's functionality.

ACKNOWLEDGEMENTS

The first author would like to thank the Computational

Table 3: Descriptive statistical analysis for cumulative mercury concentration (µHg) by approach (standard deviation (s); coefficient of variation (CV)).

Approach	Mean	Range	s	CV
Experimental	2.262	3.380	1.477	65.33
Model	2.251	3.400	1.496	66.47

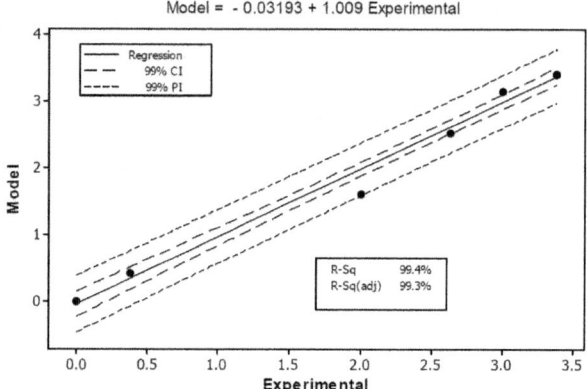

Figure 7: Regression fit analysis between experimental data and PDM, showing the prediction (PI) and confidence (CI) intervals for cumulative mercury concentration.

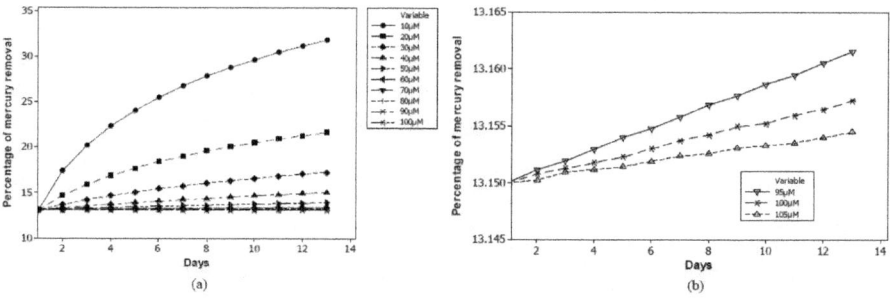

Figure 8: Percentage of mercury removal as a function of initial soil contaminant concentration. (a) From 10 µM to 100 µM, with an increment of 10 µM; (b) 100 µM ± 5%.

Scientific Laboratory at Inter American University, Bayamon Campus for providing us the computational resources and to Yamaris Pacheco-Moctezuma for reviewing this manuscript at early stages.

REFERENCES

1. Environment Canada, "Global Mercury Circulation," 2010. http://www.ec.gc.ca/mercure-mercury/default.asp?lang=En&n=A177A336-1Environment

2. N. Piorrone and K. R. Mahaffey, "Where We Stand on Mercury Pollution and Its Health Effects on Regional and Global Scales," In: N. Piorrone and K. R. Mahaffey, Eds., Dynamics of Mercury Pollution on Regional and Global Scales: Atmospheric Processes and Human Exposures around the World, Springer Science + Business Media, Inc., New York, 2006, pp. 1-21.

3. J. Sardans, F. Montes and J. Peñuelas, "Determination of As, Cd, Cu, Hg, and Pb in Biological Samples by Modern Electrothermal Atomic Absorption Spectrometry," Spectrochimica Acta Part B, Vol. 65, No. 2, 2010, pp. 97-112. doi:10.1016/j.sab.2009.11.009

4. B. Pezzarossa, F. Gorini and G. Petruzzelli, "Heavy Metal and Selenium Distribution and Bioavailability in Contaminated Sites: A Tools for Phytoremediation," In: H. Magdi, Ed., Dynamics and Bioavailability of Heavy Metals in the Roots Zone, CRC Press Taylor & Francis Group, Florida, 2011, pp. 93-127. doi:10.1201/b10796-5

5. "World Commission on Environment and DevelopmentUnited Nation Our Common Future," United Nation, 1987. http://www.un-documents.net/wced-ocf.htm

6. J. R. Henry, "An Overview of the Phytoremediation of Lead and Mercury," US Environmental Protection Agency Office of Solid Waste and Emergency Response Technology Innovation, Washington DC, 2000, p. 55.

7. I. Renberg, C. Bigler, R. Bindler, M. Norberg, J. Rydberg and U. Segreström, "Environmental History: A Piece in the Puzzle for Establishing Plans for Environmental Management," Journal of Environmental Management, Vol. 90, No. 8, 2009, pp. 2794-2800. doi:10.1016/j.jenvman.2009.03.008

8. J. M. Wood, "Biological Cycles for Toxic Elements in the Environment," Science, Vol. 183, No. 4129, 1974, pp. 1049-1052. doi:10.1126/science.183.4129.1049

9. W. P. Redley, L. J. Dizikes and J. M. Wood, "Biological Cycles for Toxic Elements in the Environment," Science, Vol. 197, No. 4301, 1977, pp. 329-332. doi:10.1126/science.877556

10. A. Shafaghat, F. Salimi, M. Valiei, J. Salehzadeh and M. Shafaghat, "Removal of Heavy Metals (Pb^{2+}, Cu^{2+} and Cr^{3+}) from Aqueous Solutions

Using Five Plants Materials," African Journal of Biotechnology, Vol. 11, No. 4, 2012, pp. 852-855.

11. S. E. Sundberg, J. J. Ellington, J. J. Evans, D. A. Keys and J. W. Fisher, "Accumulation of Perchlorate in Tobacco Plants: Developments of a Plant Kinetic Model," Journal of Environmental Monitoring, Vol. 5, No. 3, 2003, pp. 505-512. doi:10.1039/b300570d

12. N. V. Smith-Downey, E. M. Sunderland and D. J. Jacob, "Anthropogenic Impacts on Global Storage and Emissions of Mercury from Terrestrial Soils: Insights from a New Global Model," Journal of Geophysical Research, Vol. 115, No. G3, 2010, pp. 1-11.

13. E. S. Corbitt, D. J. Jacob, C. D. Holmes, D. G. Streets and E. M. Sunderland, "Global Source-Receptor Relationship for Mercury Deposition under Present-Day and 2050 Emissions Scenarios," Environmental Science & Technology, Vol. 45, No. 24, 2011, pp. 10477-10484. doi:10.1021/es202496y

14. S. Polasky, S. R. Carpenter, C. Folke and B. Keeler, "Decision-Making under Great Uncertainty: Environmental Management in an Era of Global Change," Trends in Ecology and Evolution, Vol. 26, No. 8, 2011, pp. 398-404. doi:10.1016/j.tree.2011.04.007

15. C. Franco, A. Soares and J. Delgado, "Geostatistical Modeling of Heavy Metal Contamination in the Topsoil of Guadiamar River Margins (S Spain) Using a Stochastic Simulation Technique," Geoderma, Vol. 136, No. 3-4, 2006, pp. 852-864. doi:10.1016/j.geoderma.2006.06.012

16. C. Bini, "Frorm Soil Contamination to Land Restoration," Nova Science Publisher, New York, 2010.

17. European Commission, "Soil Protection: The Story Behind the Strategy," EU: European Communities, Luxembourg, 2006.

18. M. J. McLaughlina, D. R. Parkerb and J. M. Clarkec, "Metals and Micronutrients—Food Safety Issues," Field Crops Research, Vol. 60, No. 1-2, 1999, pp. 143-163. doi:10.1016/S0378-4290(98)00137-3

19. F. Mapanda, E. N. Mangwayana, J. Nyamangara and K. E. Gillera, "The Effects of Long-Term Irrigation Using Wastewater on Heavy Metal Contents of Soils under Vegetables Harare, Zimbabwe," Agriculture, Ecosystems & Environment, Vol. 107, No. 2-3, 2005, pp. 151-165. doi:10.1016/j.agee.2004.11.005

20. Y. Cui, Y. Zhu, R. Zhai, D. Chen, Y. Huang, Y. Qui and J. Liang, "Transfer of Metals from Soil to Vegetables in an Area near a Smelter in Nanning, China," Environmental International, Vol. 30, No. 6, 2004, pp. 785-791. doi:10.1016/j.envint.2004.01.003

21. S. Kärenlampi, H. Schat, J. Vangronsveld, J. A. C. Verkleij, D. Lelie, M. Mergeay and A. I. Tervahauta, "Genetic Engineering in the Improvement of Plants for Phytoremediation of Metal Polluted Soil," Environmental Pollution, Vol. 107, No. 2, 2000, pp. 225-231. doi:10.1016/S0269-7491(99)00141-4

22. J. Hinton and M. Veiga, "Mercury Contaminated Sites: A Review of Remedial Solutions," Proceeding of National Institute for Minamata Disease, Minamata, 19-20 March 2001. http://www.nimd.go.jp/english/kenkyu/nimd_forum/nimd_forum_2001.pdf

23. G. Wu, H. Kang, X. Zhang, H. Shao, L. Chu and C. Ruan, "A Critical Review on the Bio-Removal of Hazardous Heavy Metals from Contaminated Soils: Issues, Progress, Eco-Environmental Concerns and Opportunities," Journal of Hazardous Materials, Vol. 174, No. 1-3, 2010, pp. 1-8. doi:10.1016/j.jhazmat.2009.09.113

24. E. Meers, F. M. G. Tack, S. Van Slycken, A. Ruttens, G. Du Laing, J. Vangronsveld and M. G. Verloo, "Chemically Assisted Phytoextraction: A Review of Potential Soil Amendments for Increasing Plant Uptake of Heavy Metals," International Journal of Phytoremediation, Vol. 10, No. 5, 2008, pp. 390-414. doi:10.1080/15226510802100515

25. M. H. Fulekar and J. Sharma, "Bioinformatics Applied in Bioremediation," Innovative Romanian Food Biotechnology, Vol. 2, No. 2, 2008, pp. 28-36.

26. O. V. Singh, S. Labana, G. Pandey, R. Budhiraja and R. K. Jain, "Phytoremediation: An Overview of Metallic Ion Decontamination from Soil," Applied Microbiology and Biotechnology, Vol. 61, No. 5-6, 2003, pp. 405-412.

27. C. D. Jadia and M. H. Fulekar, "Phytoremediation of Heavy Metals: Recent Techniques," African Journal of Biotechnology, Vol. 8, No. 6, 2009, pp. 921-928.

28. M. Zhang, Z. Liu and H. Wang, "Use of Single Extraction Method to Predict Bioavailability of Heavy Metals in Polluted Soils to Rice," Communications in Soil Science and Plant Analysis, Vol. 41, No. 7, 2010, pp. 820-831. doi:10.1080/00103621003592341

29. Environmental Protection Agency, "Introduction to Phytoremediation," Environmental Protection Agency, Ohio, 2000, pp. 1-72.

30. H. Sarma, "Metal Hyperaccumulation in Plants: A Review Focusing on Phytoremediation Technology," Journal of Environmental Science and Technology, Vol. 4, No. 2, 2011, pp. 118-138. doi:10.3923/jest.2011.118.138

31. A. D. Pueke and H. Rennenberg, "Phytoremediation: Molecular Biology, Requirements for Application, Environmental Protection, Public Attention and Feasibility," European Molecular Biology Organization, Vol. 6, No. 6, 2005, pp. 497-501.

32. Technology Innovation Program-Environmental Protection Agency, Environmental Protection Agency, 2008. http://clu-in.org/techfocus/ default.focus/sec/Phytotechnologies/cat/Overview/

33. X.-B. Zhang, P. Liu, Y.-S. Yang and W.-R. Chen, "Phytoremediation of Urban Wastewater by Model Wetlands with Ornamental Hydrophytes," Journal of Environmental Science, Vol. 19, No. 8, 2007, pp. 902-909. doi:10.1016/S1001-0742(07)60150-8

34. C. Lafabrie, K. M. Major, C. S. Major and J. Cebrián, "Arsenic and Mercury Bioaccumulation in the Aquatic Plant, Vallisneria neotropicallis," Chemosphere, Vol. 82, No. 10, 2011, pp. 1393-1400. doi:10.1016/j. chemosphere.2010.11.070

35. P. Zornoza, R. Millán, M. J. Sierra and E. Esteban, "Efficiency of White Lupin in the Removal of Mercury from Contaminated Soils: Soil and Hydroponic Experiments," Journals of Environmental Science, Vol. 22, No. 3, 2010, pp. 421-427. doi:10.1016/S1001-0742(09)60124-8

36. A. Harfouche, R. Meilan and A. Altman, "Tree Genetic Engineering and Applications to Sustainable Forestry and Biomass Production," Trends in Biotechnology, Vol. 29, No. 1, 2011, pp. 9-17. doi:10.1016/j. tibtech.2010.09.003

37. A. C. P. Heaton, C. L. Rugh, N. J. Wang, R. B. Meagher, "Phytoremediation of Mercury and Methylmercury-Polluted Soils Using Genetically Engineered Plants," Journal of Soil Contamination, Vol. 7, No. 4, 1988, pp. 497-509.

38. C. L. Rugh, H. D. Wilde, N. M. Stack, D. M. Thomson, A. O. Summers and R. B. Meagher, "Mercuric Ion Reduction and Resistance in Transgenic Arabidopsis thaliana Plants Expressing a Modified Bacterial merA Gene," Proceeding of National Science, Vol. 93, No. 8, 1996, pp. 3182-3187. doi:10.1073/pnas.93.8.3182

39. U. Krämer, "Phytoremediation: Novel Approaches to Cleaning up Polluted Soils," Current Opinion in Biotechnology, Vol. 16, No. 2, 2005, pp. 133-141. doi:10.1016/j.copbio.2005.02.006

40. H. S. Hussein, O. N. Ruiz, N. Terry and H. Daniell, "Phytoremediation of Mercury and Organomercurial in Choloplast Transgenic Plants: Enhanced Roots Uptake, Translocation to Shoots, and Volatilization," Environmental Science Technology, Vol. 41, No. 24, 2007, pp. 8439-

8446. doi:10.1021/es070908q

41.   N. A. Sorkhoh, N. Ali, H. Al-Awadhi, N. Dashti, D. M. Al-Mailem, M. Eliyas and S. S. Radwan, "Phytoremediation of Mercury in Pristine and Crude Oil Contaminated soil: Contributions of Rhizobacteria and Their Host Plants to Mercury Removal," Ecotoxicology and Environmental Safety, Vol. 73, No. 8, 2010, pp. 1998-2003. doi:10.1016/j. ecoenv.2010.08.033

42.   M. Israr, A. Jewell, D. Kumar and S. V. Sahi, "Interactive Effects of Lead, Copper, Nickel and Zinc on Growth, Metal Uptake and Antioxidative Metabolism of Sesbania drummondii," Journal of Hazardous Materials, Vol. 186, No. 1-2, 2011, pp. 1520-1526. doi:10.1016/j. jhazmat.2010.12.021

43.   X. Wang. L. Q. Ma, B. Rathinasabapathi, Y. Liu and G. Zeng, "Uptake and Translocation of Arsenite and Arsenate by Pteris vittata L.: Effects of Silicon, Boron and Mercury," Environmental and Experimental Botany, Vol. 68, No. 2, 2010, pp. 222-229. doi:10.1016/j.envexpbot.2009.11.006

44.   J. V. Deuren, T. Lloyd, S. Chhetry, R. Liou and J. Peck, "Remediation Technologies Screening Matrix and Reference Guide: Version 4.0. Federal Remediation Technology Roundtable," 2006.

45.   J. D. Sterman, "Misperceptions of Feedback in Dynamic Decision Making," Organizational Behavior and Human Decision Process, Vol. 43, No. 3, 1989, pp. 301-335. doi:10.1016/0749-5978(89)90041-1

46.   D. K. Benbi and R. Nieder, "Handbook of Processes and Modeling in Soil-Plant System," Food Products Press and The Haworth Reference Press, Binghamton, 2003, p. 762.

47.   S. C. McCutcheon and J. L. Schnoor, "Phytoremediation: Transformation and Control of Contaminants," Wiley-Interscience Inc., Hoboken, 2003, p. 987.

48.   B. Robinson, J. E. Ferández, P. Madejón, T. Marañón, J. M. Murillo, S. Green and B Clothier, "Phytoextraction: An Assessment of Biogeochemical and Economic Viability," Plant and Soil, Vol. 249, No. 1, 2003, pp. 117-125. doi:10.1023/A:1022586524971

49.   S. Trapp, "Plant Uptake and Transport Models for Neutral and Ionic Chemical," Environmental Science and Pollution Research, Vol. 11, No. 1, 2004, pp. 33-39. doi:10.1065/espr2003.08.169

50.   D. M. Thomas, L. Vandemuelebroeke and K. Yamaguchi, "A Mathematical Evolution Model for Phytoremediation of Metals," Discrete and Continuous Dynamical System Series B, Vol. 5, No. 2, 2005, pp. 411-422.

51. J. Japenga, G. F. Koopmans, J. Song and P. F. A. M. Römkens, "A Feasibility Test to Estimate the Duration of Phytoextraction of Heavy Metals from Polluted Soils," International Journals of Phytoremediation, Vol. 9, No. 2, 2007, pp. 115-132. doi:10.1080/15226510701232773

52. H. Qu, Q. Zhu, M. Guo and Z. Lu, "Simulation of Carbon-Based Model for Virtual Plants as Complex Adaptive System," Simulation Modeling Practice and Theory, Vol. 18, No. 6, 2010, pp. 677-695. doi:10.1016/j. simpat.2010.01.004

53. Y. Ouyang, "Phytoremediation: Modeling Plant Uptake and Contaminant Transport in the Soil-Plant-Atmosphere Continuum," Journal of Hydrology, Vol. 266, No. 1-2, 2002, pp. 66-82. doi:10.1016/S0022-1694(02)00116-6

54. Y. Ouyang, C. H. Huang, D. Y. Huang, D. Lin and L. Cui, "Simulating Uptake and Transport of TNT by Plants Using STELLA," Chemosphere, Vol. 69, No. 8, 2007, pp. 1245-1252. doi:10.1016/j. chemosphere.2007.05.081

55. Y. Ouyang, "Modeling the Mechanisms for Uptake and Translocation of Dioxane in a Soil-Plant Ecosystem with STELLA," Journal of Contaminant Hydrology, Vol. 95, No. 1-2, 2008, pp. 17-29. doi:10.1016/j. jconhyd.2007.07.010

56. M. M. Lasat, "Phytoextraction of Metal from Contaminated Soil: A Review of Plant/Soil/Metal Interaction and Assessment of Pertinent Agronomic Issues," Journal of Hazardous Substance Research, Vol. 2, No. 5, 2000, pp. 1-25.

57. M. L. Almendras, M. Carballa, L. Diels, K. Vanbroekhoven and R. Chamy, "Prediction of Heavy Metal Mobility and Bioavailability in Contaminated Soil Using Sequential Extraction and Biosensors," Journal of Environmental Engineering, Vol. 135, No. 9, 2009, pp. 839-844. doi:10.1061/(ASCE)0733-9372(2009)135:9(839)

58. J. A. Rodríguez, A. Vázquez, J. M. Grau, C. Martínez and M. López, "Factors Controlling the Spatial Variability of Mercury Distribution in Spain Topsoil," Soil & Sediment Contamination, Vol. 18, No. 1, 2009, pp. 30-42.

59. H. Yu, J. Ge, X. Zhong, M. Czakó and L. Márton, "Differential Mercury Volatilization by Tobacco Organs Expressing a Modified Bacterial merA Gene," Cell Research, Vol. 11, No. 3, 2001, pp. 231-236. doi:10.1038/ sj.cr.7290091

60. [61] S. Bizily, C. C. Rugh, A. O. Summers and R. B. Meagher, "Phytoremediation of Methylmercury Pollution: merB Expression in

Arabidopsis thaliana Plants Confer Resistance to Organomercurial," Proceeding of National Academy of Science of the United States of America, Vol. 96, No. 12, 1999, pp. 6808-6813. doi:10.1073/pnas.96.12.6808

61.   [62] O. N. Ruiz, H. S. Hussein, N. Terry and H. Daniell, "Phytoremediation of Organomercurial Compounds via Chloroplast Genetic Engineering," Plant Physiology, Vol. 132, No. 3, 2003, pp. 1344-1352. doi:10.1104/pp.103.020958

Chapter 3

ENVIRONMENTAL CHANGE AND GEOMORPHIC RESPONSE IN HUMID TROPICAL MOUNTAINS

Wolfgang Römer[1]

[1] Department of Geography, RWTH (University) Aachen, Germany

INTRODUCTION

The tropics encompass a wide variety of environmental conditions sharing high radiation and high temperatures, whilst the timing and annual amount of the rainfall and the seasonal moisture pattern enable the distinction between humid tropical, seasonal wet tropical and arid tropical zones and of the savannah and rain forest environments [40]. As a result of its great areal extent, the tropical zone encompasses a wide range of tectonic regimes, structural and lithological settings and landscapes [38].

The understanding of environmental changes in the tropics appears to be of particular importance as this zone encompasses 35 to 40 per cent of the land surface of the earth and includes about 50 per cent of the world's population [67]. Tropical countries are characterized by a rapid growth in the population and a rapid development of urban areas [6, 45]. This has resulted in increasing demands on fresh water, food, arable land and energy and mineral resources, leading to an increase in per capita consumption and severe environmental degradation.

Tropical ecosystems have been subjected to human interference for thousands of years in the form of traditional land use of many and varied kinds [87, 38]. However, rapid growth in the population and the technical advances of the last 100 years have increased the human impact on physical environments to a much higher degree than the thousands of years of human activity before that. Human interference and environmental change have been rapidly increasing since the mid-twentieth century. Gupta quotes a mean annual loss of rain forest of 174,000 km^2 during the decade 1980 to 1990 [38]. Agriculture and urbanization have modified and transformed large parts of the physical environment and have altered the operation of the geomorphic process-response systems [36]. According to [2], the annual deforestation rate of rain forests ranges from 0.38 to 0.91 per cent in Latin America, Africa and Southeast Asia with extraordinary high rates of 5.9 per cent in Sumatra and 4.9 per cent in Madagascar. More recent estimates of gross forest-cover loss in the first decade of the 21st century indicate no reversal of these trends [39]. Recent studies indicate an increase in hazards in many regions in the tropics. These appear to be linked to changes in global climate, an accelerated and disorderly process of urbanization, deforestation and the associated loss of hydrological storage capacity, particularly in mountainous domains and to the concentration of settlement activity in potential high-risk areas [44].

However, the severity of their impact varies spatially, and the intensity and course of the response to environment changes varies in the different physiographic domains depending on the nature and severity of the change and the sensitivity of the landscape. Landscapes can be viewed as systems consisting of various interconnected components or subsystems [15]. As the subsystems tend to interact on different spatial and temporal scales via different feedbacks, they may dampen or reinforce the effects of environmental changes depending on the coupling strength existing between the system components. The crossing of thresholds, on the other hand, causes a sudden change in the landscape or in the geomorphic processes, and the mutual operation of feedbacks and thresholds within the geomorphic system tends to induce a complex response to changes in environmental conditions. A consequence of these interactions is that the rate of change of landscapes as well as the severity of the geomorphic response to environmental changes is extremely variable.

Purpose and objectives

The understanding of developmental patterns in respect of the diverse and complex environmental controls and geomorphic responses in the tropics appears to be an essential prerequisite for the assessment and distinction of climatically-driven and humanly-induced environmental changes as well as for the planning of technological, social and political measures and a sustainable development. The objective of this paper is to demonstrate the role of the geomorphic response to environmental changes on a variety of temporal and spatial scales. However, a comprehensive and balanced view of the wide range of geomorphic process responses to environmental change, their causes and functional relationships is beyond the scope of this study. Instead, this study attempts to concentrate on the response of hillslope processes and their specific controls in the humid tropics, and, in particular, on the stability of hillslopes, on the role of surface wash processes in accelerating soil erosion, and on the role of weathering processes from the point of view of the availability of nutrients in the soils and the geotechnical properties of the weathered materials. A further objective of this study is to highlight some aspects of the role of the long-term development paths of the landscapes as this factor may provide some indication of susceptibility and of a potential response to environmental changes on the part of larger-scale landscape units.

The second chapter encompasses a discussion on the various factors which determine rapid mass movements in humid tropical mountains, and provides an overview of the role of extreme rainfall events in triggering landslides. As the responses of hillslopes are often predisposed by virtue of long-term evolutionary processes, the chapter also includes some case studies on the role of long-term hillslope development and of the effects of susceptibility on landslide hazards in rural and urban areas.

The third chapter is focussed on different aspects of soil erosion, land degradation and soil fertility. The fourth chapter highlights some factors which determine the intrinsic complexity of geomorphic response, interaction between human interferences, the role of changes in the frequency and magnitude of external events and the importance of interdisciplinary approaches.

LANDSLIDING AND ENVIRONMENTAL CHANGE IN TROPICAL MOUNTAINS

Landslides in humid tropical environments

Rapid mass movements are important processes in mountainous landscapes and include a wide range of types and sizes of landslides and styles of movement (Table 1). Landslides have been documented in nearly all tectonic settings within the tropical area [80, 87]. However, large single landslides and landslide events encompassing hundreds to thousands of landslides tend to occur most frequently in tectonically active mountain belts and, although with a somewhat lower frequency, in highly elevated pericratonic areas, whilst landslide events appear to be relatively rare in cratonic areas. [14, 31, 38]. The frequent occurrence of landslides in tectonically active mountains and pericratonic areas can be attributed to a specific set of conditions, which include high escarpments, long steep hillslopes in ridge and ravine landscapes, copious rainfalls (high annual rainfall totals and high short-term intensity rainfalls), and highly weathered surface materials [59]. Earthquakes, volcanism and rapidly incising streams are further factors acting as trigger mechanisms for rapid mass movements, particularly in tectonically active mountain ranges [55, 80, 87, 12, 56, 38].

Table 1: A simplified classification of rapid mass movements(Modified after [84] and [75])

Type of movement	Regolith		Rock
Type of movement	Fine-grained	Coarse grained	
Falls	Earth fall	Debris fall	Rock fall
Translational slides	Earth slide	Debris slide	Rock slide
Rotational slides	Earth slump	Debris slump	Earth slump
Flows	Earth flow	Debris flow	Rock flow
Avalanche		Debris avalanche	Rock avalanche

In many tropical mountains, landslides are part of a highly dynamic hillslope system, which is characterized by high temporal variability, cyclic changes in stability thresholds and temporal tendencies of recovery. This system is superimposed by climatic conditions, the effects exerted by the tectonic regime, lithology and structure, weathering processes and the rate of river incision. Where landsliding is the dominant formative process, changes in the environmental conditions are likely to influence the response of hillslopes by causing changes in the frequency, size and style of landsliding [58, 13, 56]. The off-site effects of large landslide events are the blocking of streams and valleys with landslide debris, and rapid sedimentation in the river channels promotes flooding in the downstream parts of the drainage basins. In a study on the impact of the hurricane Hugo in Puerto Rico it has been estimated that about 81per cent of the material transported out of the drainage basin had been supplied by landslides [47]. As events of a similar magnitude tend to occur once in 10 year the total rate of denudation due to landsliding has been suggested to range to about 164 mm/ka [47]. In Papua New Guinea, earthquakes provide an additional trigger mechanism, and estimates of denudation by landsliding indicate rates of 1000mm/ka [72, 34]. However, our understanding of the long-term contribution of landslides to the total denudation is fragmentary and the extrapolation of denudation rates to larger areas is subject to serious constraints.

Even in the case of shorter time scales, the triggering of slope failures depends on several interconnected and interacting factors. Site-specific factors such as slope, the relative relief, the degree of dissection of the landscape, the density of the vegetation cover, the geotechnical properties, the thickness of the material on the hillslopes and the intensity of land use determine the susceptibility of hillslopes to landsliding. Earthquakes and rainfall amounts are commonly the decisive triggers of landslides [76, 72, 38]. The incidence of landslides is closely associated with the timing, intensity and duration of the rainfall and the antecedent rainfall amounts [78, 1, 33]. These factors control the accumulation of moisture in the regolith and, hence, are associated with the likelihood of high pore water pressures. Hillslope steepness, on the other hand, controls the downslope directed forces and the rate of downslope subsurface water flow whilst the planform of the hillslopes determines the convergence and divergence of the surface and subsurface water flow lines and controls the size of the moisture-supplying area. The thickness of the weathering cover, its weathering degree, layering and textural characteristics, on the other hand, controls the hydrologic behaviour of the slopes and the type of movement. The accumulation of water by subsurface flow and infiltration in the regolith may also control the position of landslides on the hillslopes. In Puerto Rico most hillslopes failed at an elevation range of 600 to 800m because of the supply of water from higher elevated hillslope units [71]. Similar inferences concerning the position of landslides are indicated in studies of [48]. These indicate that slope failures resulting from extreme rainstorms are triggered on the middle or lower slope units of the hillslopes whilst landslides triggered by earthquakes tend to occur on the upper slopes.

Although landslide events are closely associated with high intensity rainfall events or periods of prolonged rainfall, there is no direct link between rainfall amount, rainfall intensity and the number and volume of landslides. Several studies have shown that rainfall events of similar order are capable of triggering different landslide volumes and of producing different landslide occurrences and landslide types [49, 37, 24]. This indicates that different thresholds are involved in the occurrence of landslide episodes. These thresholds are often interconnected by various feedbacks, resulting in complex relationships between the threshold of slope failure and the accumulation of moisture during the rainfall season and antecedent seasons, and the intensity of the rainfall event triggering the landslide [33, 62]. Instability thresholds of this type often depend on a number of site-specific factors. These may be associated with the impact of previous landslides on hillslope form, hillslope hydrology, regolith thickness, and with materials which are inherited from former landslide events.

However, even in tropical mountains with a high relief, steep hillslopes, high rainfalls and a high likelihood of high-pore water pressure, high-intensity landslide events may be rare [27]. Several factors have to work synergistically in order to trigger large-scale landslide events. Apart from bioclimatic conditions and specific structural and tectonic settings, the state of the landscape controls the response of hillslopes to environmental changes as the magnitude-frequency relationship of landslide events depends on the long-term association between overall denudation rates and the renewal of regolith by weathering. Studies on shallow landsliding in Borneo have demonstrated that under a given set of climatic, geological/structural conditions, landsliding is only possible where weathering processes are able to maintain a regolith thickness that is equal to or thicker than the threshold of critical sliding depth [27]. This indicates that on hillslopes where the regolith remains below the thickness necessary to trigger landslides, as the rate of regolith renewal is unable to keep pace with the gross denudation rate and the rate of river incision is too slow to steepen hillslopes towards a new threshold angle for landsliding, at a lower regolith thickness, large landslide events will be rare. However, this type of "regolith-supply limited" or "weathering limited" conditions appears to occur more often in tectonically active mountain belts or in terrains underlain by highly resistant rocks. As high weathering rates are characteristic features in many hot and humid tropical regions, the weathering processes and the geotechnical properties of the weathering mantles are of prime importance for an understanding of the landslide dynamics in tropical mountains.

F The important role of chemical weathering in the development of impermeable layers in the regolith, and the importance of the highly variable geotechnical properties of the saprolite, soil and colluvium on hillslopes of the Serra do Mar (Brazil) has been emphasized by several authors [32, 46]. Another set of factors is associated with the coupling strength of hillslopes and rivers, the imprints of formerly different climatic conditions including hillslope deposits with variable geotechnical properties as well as the delayed response of hillslopes to the change from dry to humid conditions in the transitional periods from the Pleistocene to Holocene and their influence on the developmental paths of hillslopes [81, 82, 61]. The interaction of these different factors may result in hillslopes which are highly prone to landsliding, though the trigger mechanisms often depend on a site-specific combination of factors as different landscape components of the mountainous terrain are affected.

Human interference in the form of deforestation and urbanization and increased rural land use coupled with infrastructural measures and construction resulting in an oversteepening or undercutting of hillslopes and changes in hillslope hydrology frequently exacerbate the susceptibility of hillslopes to landsliding. The combined sum of the effects of human modifications and alterations in the mountainous domains has increased the socioeconomic impact of landsliding and also the risks in areas with a much lower natural susceptibility to landsliding [41]. Although the contemporary landscape setting, the geotechnical properties of the material, climate and the impact of human alterations determine to a large degree the incidence of mass movements, the triggering of slope failures may be also associated with processes that occurred in the past. The landscapes in which mass movements occur are often a composite of forms and deposits that are genetically linked with actual process dynamics on the hillslopes. The long-term component in studies of mass movements has often been neglected because of the underlying assumption that the current state of a hillslope or landscape is ascertainable from an analysis of the contemporary process-response system. In many cases, this assumption appears to be justified. However, the knowledge of the long-term developmental paths of landscapes may lead to predictions of the susceptibility or sensitivity to react to environmental changes or may lead to predictions on the consequences and impacts of past events which were caused by environmental change.

Form-process relationships and geomorphic response in south-eastern Brazil

Landsliding in the Serra do Mar

The Serra do Mar forms the elevated passive margin along the Brazilian Atlantic coast and extends from Rio de Janeiro to Santa Catarina with elevations ranging from 700 to about 2000 m. Most of the area consists of folded and faulted metamorphic and plutonic rocks from the Precambrian age and landscapes range from highly elevated plateaus with steep escarpments to dissected ridge and ravine terrains, and muliconvex hilly terrains [3]. The climate is humid tropical with maximum rainfalls in the summer and without marked dry seasons in the winter. The mean annual rainfall totals range from 1500 to 2500 mm, though annual rainfall may rise locally to 4000mm [68]. About 70 percent of the annual rainfall occurs in the summer, which is also characterized by high intensity rainfalls [68]. The potential vegetation along the Atlantic coast is pluvial rain forest, which formed a highly diverse assemblage of trees, shrubs, lianas, tree ferns and epiphytes [42, 89]. Settlement and forest clearance have destroyed much of the original rain forest and estimates indicate that the remaining forests merely constitute 5 per cent of the original coverage [20]. Some local measures have attempted in recent decades to reverse these trends by the afforestation of pines and other tree species [7]. However, the destruction of forests by increasing rural land use and urbanization remains a major problem [53].

Over the last fifty years, the rapid growth of urban areas has resulted in marked changes in hillslope hydrology and the stability of hillslopes. These changes are also associated with an increasing influence of social and economic factors on risks associated with flooding and landsliding [5, 53]. In several regions, hillslopes, villages and urban areas are affected nearly every year by disastrous landslides, and particularly highly dissected terrains with steep hillslopes and highly weathered, thick regolith mantles are prone to landsliding even under undisturbed conditions [22, 19]. Many important roads cross the Serra do Mar and villages, industrial complexes lying at the foot of mountain slopes and escarpments or in basins and valleys are exposed to serious hazards caused by landsliding [53].

However, in many areas of the Serra do Mar, landslides were presumably the most important formative processes since the Late Quaternary period. Landscape evolution was probably non-uniform because of base level changes and climatic changes in the Quaternary, and the intensity of landsliding is likely to have varied as a function of climatic conditions and periods of river incision [17, 18, 61]. The various controls are often genetically linked with the sensitivity to landsliding and concern several aspects of the long-term development of hillslopes.

Some aspects of the role of long-term process-response systems

Predictions about the way hillslopes tend to respond to changes in environmental conditions may be gained from studies of the long-term development of hillslopes. Of particular importance in this respect are the roles of inherited materials and the effects of a differing hillslope-channel coupling strength. Inherited materials may provide information on the processes that have acted during past environmental changes. This enables predictions on the vulnerability of hillslopes with respect to specific slope processes or supports regional surveys on hazards with respect to the geotechnical properties of soils, weathering layers or colluvial deposits. In the Serra do Mar, several lines of evidence suggest that mass movements have occurred alongside periods of intense colluvial accumulation in the Pleistocene and early Holocene [8, 83, 54].

The accumulation of the colluvium occurred as a result of relatively dry climatic conditions in the Pleistocene and the higher frequency in the magnitude of storm events in the early Holocene. The areal extent of land surfaces currently underlain by colluvial deposits in São Paulo is estimated to be in the range of 50 per cent [30]. Today, the knowledge of the complex stratigraphy, the geotechnical properties and of the distribution pattern of the colluvial deposits is important as these deposits are often associated with debris flow hazards which often occur after vegetation clearance [46, 19].

The tendency of landscapes to react to environmental changes by landsliding may be also indicated in the hillslope development paths. Many ridge and ravine landscapes in the Serra do Mar encompass steep hillslopes, which are covered by a moderately thick weathering mantle. In southern Sao

Paulo, this terrain-type is underlain by mica schists and phyllites and often exhibits summit heights, which are dictated by the steepness of the valley-side slopes and by the spacing of the rivers [61]. These terrains are characterized by v-shaped valleys and straight valley side-slope profiles with a relative relief of 120 to 200m. The valley side slopes exhibit a narrow range of slope angles ranging from 26° to 34° for the mean slope angle and the mean maximum segment slope angle. A consequence of the geometric control of summit height by slope angle and valley spacing is that areas with similar drainage density and stream spacings are characterized by accordant summit heights [61, 60]. Such an adjustment is unlikely to result from short-term changes because the incision of the drainage net, the fixation of rivers in valleys and the development of steep valley side slopes with a mean relative relief of 120 to 200m are unlikely to have been accomplished within a period that is shorter than 105 years. Conversely, in order to maintain the geometrical expression, the hillslope processes and the hillslope-channel coupling have had to operate throughout the Holocene period.

Table 2: Geotechnical properties of the regolith on mica schistsB- textured B-HorizonT – transitional zone between B-Horizon and SaproliteS – SaproliteShear strength was determined by direct shear tests after consolidation to allow excess of pore pressure

Horizon	clay	silt	sand	cohesion	friction angle
units		weight- per cent		kPa	degree
B	54.3	26.4	19.3	10.5	31.5
B	47.2	23.8	29.0	6.6	30.2
B	67.8	15.3	16.9	13.7	29.9
T	28.8	19.1	44.9	1.9	30.4
S	20.7	28.5	50.8	0.9	38.0

Most valley side slopes in the area are covered by numerous landslide scars and landslide deposits of various ages indicating that the important formative hillslope process is shallow landsliding. The valley side slopes are covered by red-yellow podzolic soils, which show marked differences in the geotechnical properties of the soil horizons (Table 2, Figure 1, Figure 2). Particularly, at the contact of the B-Horizon to the transitional layer the decline of the cohesion tends to facilitate the development of a subsurface plane of failure. This is also indicated in the location of slip surfaces of relatively recent landslides, which occurred at a depth of 0.9 to 1.2m below the surface. This depth coincides roughly with the depth of the transitional layer.

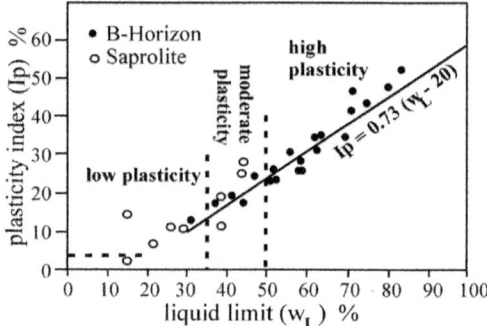

Figure 1: Range of plasticity index and liquid limits of B-Horizons and saprolitic weathering products of mica schists (modified after [61])

Figure 2: Shallow translational landslide and earthflow which resulted from a single rainstorm and the high moisture content in the regolith. (Photo Römer).

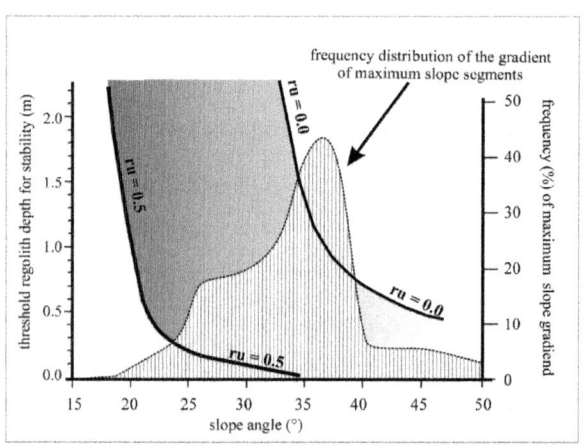

Figure 3: Limiting regolith thickness for a safety factor of slope stability (F =1.0) as a

function of pore pressure ratio (ru) and the distribution of maximum segment slope angles on hillslopes underlain by mica schists. The bulk unit weight of the regolith ($\gamma = \rho$ g) = 17.4 kPa, the cohesion (c) = 1.9 kPa, and the friction angle (φ) = 30.4°. The limiting regolith thickness has been calculated by using the infinite model for translational landslides [11]. The factor of safety (F) has been calculated by F = c + (γd cos2α - ru γd) tanφ/ (γd sinα cosα) with d = regolith depth (m), α = slope angle (°), ru = ρwg dw cos2α /γd; ρw = density of the water (kg m-3), g = gravitational acceleration (9.81 m s-2), dw = vertical height of the water table above the slide plane (modified after [61]).

A back calculation of the slope failures indicates that most valley side slopes are stable in a dry state, but tend to become instable at pore pressure ratios of 0.1 to 0.5 (Fig. 3). The close coincidences between the slope angle of the maximum segments, the threshold slope angle for failure and the threshold regolith depth indicates that the long-term formative process on the hillslopes is landsliding. This implies that as long as river incision enables the maintenance of steep slope angles, all hillslopes are likely to be affected by reoccurring landslides in the same places as long as weathering processes supply enough material to cross the threshold regolith thickness for slope failure with respect to the slope angle and the geotechnical properties again. However, the study also indicates that the form-process relationship is associated with events that are characterized by a low frequency and high magnitude reoccurring at temporal scales of several decades to centuries rather than being the result of continuously acting formative processes. It is easy to suppose that landscapes originating from such a process-response system where hillslope evolution resulted in the development of slopes close to the threshold of slope failure tend to respond violently to environmental changes and human interferences.

Extreme rainfall events and landsliding

Apart from human interferences, the high relative relief, steep hillslopes and the thick weathering layers, the most decisive factor contributing to landsliding is high rainfall. In the Serra do Mar landslide events are likely to occur independently of antecedent rainfalls and regardless of the vegetation cover and human interferences where rainfall exceeds 250 mm/24h [37]. Furthermore, the occurrence of landslides is promoted on most hillslopes which are steeper than 40% [32].

Since 1928, the Serra do Mar has been affected by about 25 to 30 extreme landslide disasters due to intense rainfall events, which have caused thousands of deaths and extensive damage to the infrastructure and various structures, though many smaller landslide events resulting in various degrees of damage tend to occur every year [23, 32, 19]. In the period from 1988 to 2000, the number of landslide fatalities in Santa Catarina, São Paulo, Rio de Janeiro,

Minas Gerais, Bahia and Pernambuco averaged between 13 to 50 and locally, in coastal areas, between 51 to 364 [5]. About 85 percent of the landslide disasters occurred during the summer season, and most of the larger events that are documented in the scientific literature concentrate on the period between December and March [46, 53, 19, 68]. However, an extraordinary rainfall event was recorded in the winter of 2004. The event was caused by a cold frontal passage which became stationary in the coastal area of south-eastern Brazil [68]. Once the initially cold post-frontal anticyclone had acquired barotropic equivalent characteristics, a persistent southerly and south-easterly flow of winds became established which was impeded along the rise of the Serra do Mar causing advection and high rainfall. The event caused serious flooding and landslides along the coastal region of São Paulo [68].

Although any generalization of the functional relationships between the incidence, type and rate of movements may be overridden by local site-specific factors, the results of studies on landsliding in south-eastern Brazil suggest that most landslides occur in the late rainy season when the accumulation of moisture in the regolith has attained a temporal maximum [1, 19]. The increase in moisture in the regolith causes a rise of the pore water pressure and hence, results in a lowering of the threshold rainfall intensity necessary to trigger landslides. However, the triggering of landslides is also a function of slope angle, slope form and of the material on the hillslopes. On steep hillslopes with a relatively thin weathering cover, shallow landslides appear to occur mostly on the middle and upper hillslope segments. This landslide type is triggered during the wet season by rainfalls of long duration and moderate intensity or at the end of the wet season during heavy storms [25]. Failure may result from the increase in pore-water pressure or from the elimination of soil-suction and the reduction of the apparent cohesion [88, 46]. Debris flows, on the other hand, are triggered in the late rain season in hillslope hollows, on the lower slope segments and on steep hillslopes when the regolith is saturated with water. The incidence of debris flows is associated with high-intensity rainfall occurring in the late rain season and appears to be strongly associated with a destruction of the vegetation cover [19].

Urbanization, environmental change and landslide hazards

The rapidly growing population in the cities in south-eastern Brazil, the unplanned growth of urban areas and the inability to house the growing number of people have resulted in human occupation of geologically and topographically hazardous terrains, which are often characterized by an inappropriate infrastructure and precarious residences [5, 53]. The combined sum of these changes has also increased the risk of landsliding even in urban areas with a much lower natural susceptibility to landsliding.

The areal extent of the alterations in urban areas has often resulted in a reinforcement of the intensity of the hillslope processes as the affected subsystems tend to work synergistically. Urbanization is associated with a sealing of the surface, a lowering of the infiltration rate, a reduction of the water storage, and an increase in surface runoff. Soil erosion resulting from vegetation-clearing measures causes the development of gullies. Large gullies tend to affect the flow pattern of rivers by decreasing the baseflow whilst the stormflow is increased [21]. This leads to more intense floods and more events where hillslopes are undercut by rivers. Deforestation of hillslopes, on the other hand, tends to increase the likelihood of debris flows as a function of the decrease in root strength [19]. Road cuts or excavations destabilize hillslopes as the material supporting the regolith or rocks on the slopes is removed. Human settlement along streams and in valleys with houses perched on steep valley side slopes next to rivers increases the risk of disasters as hillslopes are undercut by rivers. The destruction caused during a landslide event also varies with the type, size and rate of movements. Disastrous effects are often associated with large debris flows which are induced in the late rainy season by heavy rainfalls once the material on the hillslopes has become saturated with water. During the 2011 landslide disaster in the vicinity of Rio de Janeiro, cascades of mudflows and debris flows destroyed houses and buildings. As the slipped debris moved downslope, water contribution from the surrounding areas resulted in an increased fluidization of the debris, which moved rapidly into the valleys and caused an increase in sediment load and in the flooding. Flooding and landsliding resulted from unusually persistent rain and an interspersed extreme storm rainfall event which had a devastating impact along the south-eastern Brazilian coast.

SOIL EROSION AND LAND DEGRADATION

Soil erosion in humid tropical environments

Over the last four decades, deforestation and human interference with the environment have increased in nearly all tropical rain forest environments around the world [39]. The impact has caused increasing land degradation and is often accompanied by changes in the hydrologic regime, severe soil erosion and a declining productivity of cultivated areas [87, 36]. Recent developments in agricultural techniques, the increased use of agricultural machinery and the replacement of subsistence-orientated agriculture by export-orientated agriculture have resulted in a rapidly increasing and unfavourable change in environmental conditions. Most studies on the role of soil erosion in rain-forest environments indicate that soil erosion in undisturbed rain forests rarely exceeds rates of 1t ha-1 a-1 as the canopy and understorey protect the soil from the impact of raindrops [59, 38]. In rainforests, much of the rainfall is intercepted and evaporated in the canopy and understorey, and permeable litter layers support high infiltration rates. Consequently, only a small fraction of the rain water remains available for overland flow [73]. The litter cover on the surface on the other hand, tends to dampen the forces of the impact of heavy raindrops. This cover is highly permeable. The permeability results from macro-pores provided by roots, which reduce the generation of erosive runoff [59]. Under natural conditions, with a continuous cover of litter layers, the water movement occurs as over litter-layer flow and as root litter flow in pores and in shallow subsurface pipes within the root-litter carpet [19]. This water flow is mostly highly discontinuous shallow unconcentrated overland flow with a low erosive power, except in hillslope hollows, where the convergence of surface water flow lines tends to promote a concentrated overland flow.

Disturbance of vegetation in rain forest environments appear to have serious effects on erosion rates as the spatial variation in the intensity and frequency of large rainfall events tends to be higher than in savannah environments [87]. The loss of ground vegetation and litter reduces the amount of soil organic matter, which diminishes the aggregate stability and increases the vulnerability of the soil to raindrop impact and the likelihood of soil crusting [29, 36]. The destruction of the soil aggregates by raindrop impact and the formation of a fine grained crust on the soil surface tend to impede infiltration. During rain bursts, this causes a rapid increase in overland flow and favours the development of rills and gullies. Soil erosion and changes in the physical characteristics of the upper soil horizon are not the only effects of vegetation disturbance. The nutrient cycle is markedly changed as nutrients are lost by soil erosion, by leaching of the soil and by the removal of nutrients which were formerly stored in the vegetation [59]. As tropical rainforests are unable to sustain their nutrient base without sufficient vegetation, the combined effects of vegetation destruction and soil erosion tend to result in a marked depletion of the soils and in a reduction in the biodiversity [86]. The complex relationships between vegetation destruction, agricultural use, soil erosion and loss in soil fertility has been documented from several areas in the tropics, and the interaction between socioeconomic and ecologic factors appears to be of major importance.

Land degradation and soil erosion in humid tropical mountains

Land degradation encompasses various processes ranging from disturbance of the vegetation to biodegradation of the humus and litter and the deterioration of soil quality. These changes are functionally associated with the productive capacity of the soils. Measurements of soil erosion rates in tropical environments are often highly variable. Calculated erosion rates range from 0.2 to 10 t ha-1 a-1 for rain forest environments in Guyana, Brazil and the Ivory Coast [34]. In the case of erosion in the Ivory Coast, rates increased on slopes with an inclination of 6% from 0.1 to 90 t ha-1 a-1 (crop cover) and 108 to 170t ha-1 a-1 (barren) [35, p. 113].

However, a quantitative assessment of the on-site and off-site impacts of soil erosion on the landscape remains a challenge because of the wide variety of environments and the relatively small data basis. Short-term soil erosion measurements from small test-plots do not always provide representative rates for hillslopes and the extrapolation of these erosion rates to larger areas is prone to errors as physical properties of the soils, the vegetation cover and parameters such as slope length, slope steepness tend to be highly variable. A further factor is the length of the measurement period. Extreme rainfall events are highly variable in terms of space and time and hence, are often not recorded. Rainfall erosivity modelling, on the other hand, provides information on the likelihood of soil erosion, whilst the calculation of erosion rates is complicated by the high number of interacting variables [74].

Although soil erosion rates imply a continuous loss of soil, the erosive processes are triggered by separate rainfall events, and the impact of singular rainfall events on soil losses may override all preceding soil erosion rates calculated. The important role of extreme rainfall events on soil losses and on sedimentation rates on the valley floors has been documented in the drainage basin of the Tubarão river (southern Brazil) [8]. In this area, a total amount of 400mm rainfall (three days) was recorded. This event caused serious soil erosion and resulted in the accumulation of a 30 to 60 cm thick pile of sediment on the valley floor. This implies that meaningful erosion rates can be only deduced when erosion measurements are supplemented by studies of the sediment balance in the drainage basins [8]. Studies on erosion/sedimentation events in drainage basins over a longer range of time (101 to 104 years). On the other hand, are rare and often confronted with the problem of distinguishing between human induced changes and natural environmental changes. The latter applies, in particular, to cases where landforms are polygenetic and are caused by rare, high magnitude events rather than by continuous processes. However, accelerated soil erosion as a result of the increased agriculture and the destruction of the vegetation cover has been recorded in the drainage basin of the Ribeira River [9]. The drainage area of the Ribeira river covers an area of 24,200 km2. Since the 19th century, land use has increased from the equivalent of a few per cent to an area covering about 5000 km2 in the year 1979 [9].

The Ribeira drainage basin is underlain by deeply weathered metamorphic and plutonic rocks. Most of the lower valley-side slope segments and small hillslope hollows are covered with pedogenetically transformed, clay-rich colluvial sediments of the late Pleistocene and early Holocene age, which have been deposited above the "in situ" formed saprolite [8, 63]. However, in areas where the original forest has been replaced by shrubs or agricultural land use, the colluvial soils exhibit truncated soil horizons whilst gully incision into the saprolite has given rise to the development of shallow hillslope hollows and deeply dissected hillslopes (Figure 4, 5). Most of the eroded fine-grained material has been transported to the rivers and on to the flat valley floors [9]. In the drainage basin of the Ribeira River, the high influx of sediment into the valleys and onto the flat valley floors has resulted in the accumulation of 5 to 6 m thick clayey sediments, which are rich in organic matter. In the area surrounding the village of Sete Baras, 5.8m thick sediments have been deposited above the river gravel of the Ribeira River [9]. Radiocarbon age determination of the organic matter of these deposits from a layer located just above the river gravel indicates that the material above the river gravel is younger than 300 years [9]. This provides an approximate age for the start of the increased erosion episode resulting from humanly induced disturbance of the vegetation. The geomorphic analysis of erosional forms, of the degradation of the colluvial soils, and of the start of the increased vegetation clearance indicates that soil erosion has contributed to a loss of soil 170m3 ha-1 a-1 or of 235t ha-1 a-1 in the last 130 years [10, p. 65].

Figure 4: Accumulation of colluvium in a hillslope hollow in the drainage basin of the Jacupiranga River, which is a tributary of the Ribeira River. The colluvium has been

deposited on the saprolite of the mica schists/phyllites. The colluvium was formed in the Pleistocene as a result of less dense vegetation cover and drier climatic conditions. (Photo Römer)

However, the amount of soil loss on the valley-side slopes appears to have varied in different geomorphic settings depending on the relative relief, the physical properties of the regolith cover and the process domains. Studies on ultramafic rocks in the Jacupiranga Alkaline Complex, which is part of the Ribeira drainage basin provided no evidence of an increase in soil erosion even on steep hillslopes although, mining and cultivation of tea and bananas resulted in extensive destruction of the original forest cover [57]. Hillslope development in this area is primarily controlled by chemical denudation in the highly permeable weathering mantles and, to a lesser degree, by slow mass movements, whilst surface wash is limited by the lack of a significant overland flow [60]. Nevertheless, the role of the destruction of the vegetation cover cannot be underestimated as leaching processes operate at high rates in the tropics and tend to remove nutrients from the upper soil horizon, possibly reducing the fertility of the soils.

Figure 5: Piping and gully erosion in colluvium resulting from high rainfalls and vegetation disturbance in multiconvex hilly terrain in south-eastern Brazil at Jacupiranga. (Photo Römer)

Weathering and nutrient cycle

In tropical rainforests, the biomass above and below the ground contains most of the mineral nutrients. The maintenance of the nutrient level in the soil depends on the continuous cycling of the nutrients in the canopy and on the rate of decay of organic matter in the litter-layer. The latter is controlled by biological decomposition by invertebrates, and by the physico-chemical processes responsible for the release of nutrients in the upper soil horizons [59, 67]. However, the functional dependencies in the nutrient cycle appear to be stronger in soils with a low nutrient storage and low fertility and weaker in more fertile soils. Once the vegetation cover is destroyed, the supply of organic matter and the formation of the new litter on the soil surface is slowed down whilst the breakdown of the organic matter is accelerated by solar radiation [87, p. 277].

Although the physico-chemical processes controlling the productivity and fertility of the soils and the turnovers of the nutrients are not completely understood, several lines of evidence suggest that the degree of weathering and textural characteristics of the soil play an important role in the nutrient cycle. High weathering rates result in excessive base-leaching and a low pH, creating a decline in base saturation, loss of major cations and a decrease in the cation-exchange capacity [66]. This promotes the occurrence of free iron and aluminium either in the clay complexes or as amorphous iron and aluminium oxides or hydroxides in the weathering layers. As amorphous iron and aluminium oxides readily absorb, phosphorus tropical soils with a low pH are often characterized by a high phosphorus fixation capacity, resulting in a phosphorus deficiency [85, 59]. According to studies in the Amazon of Brazil the fixation of phosphorus rather than the overall nutrient decline appears in many cases to be the cause of the decline in pasture productivity [69].

Soil erosion and intense leaching in soils are responsible for several problems concerning productivity in agricultural land use. In a case study carried out in Rwanda several green farming methods applied to highly degraded soils failed to restore the fertility of the soils [77]. Improved fallow, mulching, green manure and the use of compost and cow dung were not sufficient to maintain the nutrient levels in the soil as the rapid decomposition of the organic matter at the start of the rain season resulted in a release and leaching of high amounts of nitrogen and a rapid reduction in the fertility of the soils [77]. From the point of view of the sustainability any agricultural strategies being considered, the materials used for fertilizing the soils have to be inexpensive and available from regional or local resources. The improvement of the physico-chemical properties of highly degraded soils, on the other hand, depends on several-site specific and soil-specific factors, and additional information is frequently required on the dynamics of the soil. Important improvements usually involve increasing the pH. This reduces phosphorus fixation, the disintegration of chlorite structures and reduces antagonistic effects in cation exchange.

However, any application of material has to maintain the slow dissolution of cations from dissolved minerals and has to inhibit silica dissolution which often involves an increase in pH and results in an increase in the disintegration rate of chlorite structures [77]. In relation to the requirements specified in Rwanda, several tests with calcium carbonate, travertine and volcanic tephra indicated that the combined application of cow dung and tephra represents a measure capable of improving the agricultural capacity of the degraded soils [77]. However, soil erosion, nutrient cycles and soil fertility are highly interrelated and depend often on specific local and regional factors. Although quantitative data of soil erosion rates and depletion rates are important for the implementation of effective soil conservation measures, socio-economic factors and the understanding of the traditional/cultural background appear to be of equal importance because many conservation strategies may be impractical or too expensive or are rejected as a result of limited access to the technologies required.

CONCLUSION

Over the period of the last fifty years, most tropical mountains have experienced marked changes in their environmental conditions due to the high rate of deforestation, rural land use and urban growth. These changes have often reinforced hillslope processes such as soil erosion and landsliding and have also resulted in an increase of geomorphic hazards, even in areas with a previously lower susceptibility to soil erosion or landsliding. Increased rates of soil erosion and landsliding have been documented from regions where large areas are affected by human intervention and hillslope processes are highly interdependent and tend to reinforce each other. The changes have not only affected the hillslope system but have also influenced other subsystems of the geomorphic/ecological system, which has resulted in the coupling of different responses similar to "chain reactions". Such "chain reactions" appear to occur frequently when the urban fringe expands into mountainous terrains [38]. Urbanisation and deforestation increase the runoff and hence induce soil erosion whilst the increase in storm runoff results in the undercutting of hillslopes and landsliding, thereby increasing the supply of material to the rivers, which in turn, increases the likelihood of flooding.

However, the geomorphic response displays a high degree of spatial and temporal variability. Under similar geologic and bioclimatic settings, some landscapes tend to react rapidly to ongoing environmental changes whilst others tend to absorb the effects of environmental change, as the reaction is delayed or dampened in the various interconnected geomorphic/ecological subsystems. Several factors contribute to the differences in the geomorphic response. The current state of the landscape, the degree of human modification of the landscape, the magnitude of climatically-driven events and the differing coupling strength between the long-term evolution of the hillslope system and the current hillslope processes. As geomorphic processes are triggered by separate events, the response to changes is a function of the magnitude and frequency of exogenous or endogenous events. With respect to rainfall-triggered events, the incidence of hillslope processes is often controlled by thresholds. However, these thresholds are continuously altered by human interference in the landscape, thereby increasing the risks of soil-erosion hazard and landsliding, though this interference is often necessary in that it benefits economic progress and advancement.

Disastrous landslide events are often closely associated with the expansion of the urban fringes into hilly and mountainous areas, and settlement activity in these areas has often resulted in the unsuitable modification of hillslopes, which, in turn, has increased susceptibility to mass movements [38, 53]. Although most of the recent landslide disasters are primarily controlled by geological, structural and environmental factors as well as by human interference, slope failure is often predisposed as a consequence of long-term evolutionary processes on the hillslopes. The dynamic coupling of existing controls and long-term evolutionary processes may result in the lowering of crucial thresholds. This includes the reduction of the shear strength by weathering processes, and the increase of shear stresses on the hillslopes caused by small subtle changes in slope angle and hydrology.

The intensity of human impact on tropical environments is documented in the large areas that have been subjected to deforestation. The impact has affected the geomorphic process-response system, the nutrient cycles, and biodiversity. Recent studies have shown that there is no reversal in the overall trend of tropical deforestation, though the rates of deforestation vary strongly from one decade to another and from one country to another, depending also on the methods used to assess deforestation [2, 28, 79, 39]. Estimates of the world-wide contribution of deforestation in the tropics to carbon emissions indicate a total emission of $810 * 10^6$ metric tonnes/year (period 2000 to 2005) excluding carbon emissions from logging, peatlands drainage and burning, and forest recovery [39]. However, the contribution of carbon emissions from tropical deforestation to global climatic change remains obscure as the turnover rates and recovery rates are related to various factors and the interaction between these factors is not completely understood [35, 38]. This applies also to the effects of global climatic change on the geomorphic process-response system as changes in magnitude and frequency of geomorphic processes depend also on all other environmental changes. Predictions on future climatic development trends in the tropics suggest an increase in summer monsoon and a decrease in summer rainfalls in Central America and Mexico and an increase in the number of cyclones, tropical storms and hurricanes [44]. However, human interference and climatic change often act simultaneously. This complicates predictions of crucial thresholds and the establishment of relationships between landsliding and large soil erosion events and the spatial distribution and the seasonal and annual variability of rainfall. The temporal clustering of landslide events in some regions, on the other hand, appears to indicate some associations. In Kenya landsliding was closely associated with the occurrence of El Niño circulation [52]. In southern America, on the other hand, the temporal pattern of landslide events appears to coincide with the ENSO climatic cycle. However, hillslope processes are characterized by an intrinsic complexity. Many factors appear to be capable of causing changes in both frequency and magnitude on different spatial and temporal scales.

Studies on deforestation rates in several countries of humid tropical Africa have shown that the rate of forest destruction is not only a result of the growth in population but depends also on macro-economic changes. Apart from dependence on international market prices, the extent of the agricultural area appears to depend directly and indirectly on factors such as public investment, monetary policy and exchange-rate policy, urban income levels, fertilizer subsidies, and rural-to-urban and urban-to-rural migration [50, 64, 65]. With respect to the issue of sustainable development, socioeconomic factors must also be considered.

A significant statistical relationship has been determined between the decline of the cocoa and coffee prices and subsiding governmental input, which has forced farmers in Cameroon to expand their food and crop cultivation into forested areas [51]. International prices and demands on agricultural resources, on the other hand, often result in an expansion of agricultural areas at the expense of rain forests. An example is the expansion of agricultural areas for soybean production and the increase in cultivated pastures in Brazil, which resulted from the growing importance of cattle ranching. The expansion of soybean cultivation resulted in extensive clearance of savannah forests and of tropical forests and is noted to be the second most important driver of deforestation after ranching [43].

Socio-economic factors also play an important role in establishing new methods in agriculture to improve environmental quality. Financial aspects, work expenditure, the availability of resources necessary for carrying out improvements and the consideration of traditional agricultural techniques may determine the success of sustainable developments. Socio-economic aspects are also important in the mitigation of hazards in urban areas. Hillslopes prone to landsliding are often occupied as a result of the unplanned growth of cities and increases in rent and the declining availability of land for building in the cities to house the growing population [26, 4].

The complex interaction of socio-economic, biological, geological and geomorphic aspects indicates that sustainable development requires an interdisciplinary approach. With respect to environmental planning and sustainable development in the tropics, geomorphic studies of hillslope processes may contribute to unravelling the intrinsic complexity of various hillslope hazards. This takes in a multitude of objectives, which range from assessment of the severity of influences impacting on the environment, determination of the dominant processes and hazards, assessment of the vulnerability of specific sites, determination of external triggers and predictions of events which have no historically recorded precedent.

REFERENCES

1.   A. N. Ab′ Saber., 1988 A Serra do Mar na Região de Cabatão: avalanches de Janeiro de1985. Ab′Saber, A.N. [Ed.]. A Ruptura do Equilibrio Ecologico na Serra de Paranapiacaba E.A. Polição Industrial, Brazil 74 116

2.   F. Achard, H. D. Eva, H. Stibig, J. , P. Mayaux, J. Gallego, T. Richards, J. Malingreau, P. , 2002 Determination of deforestation rates of the world′s humid tropical forests Science 297 999 1002

3.   F. F. M. Almeida, Y. de Hasui, Neves. B. B. de Brito, R. A. Fuck, 1981 Brazilian structural provinces: an introduction. Earth-Science Reviews 17 1 30

4.   R. Araki, L. H. Nunes, 2008 Vulnerability associated with anthropogenic factors in Guarujá City [São Paulo, Brazil] from 1965 to 2001. Terrae-Geoscience,Geography, Environment 3 1 40 45

5.   O. Augusto Filho, 2006 An approach to mitigation of landslide hazards in a slum area in São Paulo city, Brazil. IAEG Paper 258 1 7

6.   L. Beckel, 2001 Megacities. Geospace Verlag, Salzburg [Austria]

7.   H. Behling, V. Pillar de Patta, 2007 Late Quaternary vegetation, biodiversity and fire dynamics on the southern Brazilian highland and their implication for conservation and management of modern Araucaria forest and grassland ecosystems Philos. Trans. Soc. London B, Biol. Science 362 243 251

8. J. J. Bigarella, R. D. Becker, 1975 International Symposium on the Quaternary. Bol. Paran. Geoscienc. 33, Curitiba 1 307

9. H. R Bork, H. Rohdenburg, 1985 Studien zur jungquartären Geomorphodynamik in der subtropischen Höhenstufe Südbrasiliens. Zentralblatt für Geologie und Paläontologie, Teil 1 11/12 1455 1469

10. H. R. Bork, H. Hensel, 2006 Blätter in der Tiefe am Rio Ribeira [Brasilien]. Bork, H.-R.. Landschaften der Erde unter dem Einfluss des Menschen. Wissenschaftliche Buchgesellschaft [WBG] Darmstadt 63 65

11. E. N. Bromhead, 1992 The stability of slopes Blackie Academic & Professional, London

12. W. B. Bull, 2007 Tectonic Geomorphology of Mountains Blackwell, Singapore

13. K. Chatterjea, 1994 Dynamics of fluvial and slope processes in the changing geomorphic environment of Singapore Earth Surface Processes and Landforms 19 585 607

14. H. Chen, C. F. Lee, 2005 Geohazards of slope mass movement and its prevention in Hong Kong Engineering Geology 76 3 25

15. R. Chorley, S. A. Schumm, D. E. Sugden, 1984 Geomorphology. Methuen & Co Ltd., London

16. A. L. Coelho-Netto, 1987 Overlandflow production in a tropical rainforest catchment: The role of litter cover Catena 14 213 231

17. A. L. Coelho-Netto, 1999 Catastrophic landscape evolution in a humid tropical region [SE-Brazil]: inheritances from tectonic, climatic and land use induced changes. Geografia Fisica e Dinamica Quaternaria III 3 21 48

18. A. L. Coelho-Netto, A. S. Avelar, M. C. Fernandes, W. A. Lacerda, 2007 Landslide suceptibility in a mountainous geoecosystem, Tijuca Massif, Rio de Janeiro: The role of morphometric subdivision of the terrain. Geomorpholgy 87 120 131

19. A. L. Coelho-Netto, A. S. Avelar, W. A. Lacerda, 2009 Landslides and Disasters in southeastern and southern Brazil Latrubesse, E.M. Natural Hazards and Human Exacerbated Disasters in Latin America. Development in Earth Surface Processes 13 223 243

20. Consórció Mata Atlántica 1992 The Mata Atlántica Biosphere Reserve Plan of Action [Reserva da Biosfera da Mata Atlántica Plano de Acão]. São Paulo, Brazil 1992

21. F. M. Costa, L. Bacellar, A. P. de , 2007 Analysis of the influence of gully erosion in the flow pattern of catchment streams, Southeastern Brazil

Catena 69 230 238

22. O. Cruz, 2000 Studies on the geomorphic processes of overland flow and mass movements in the Brazilian geomorphology Revista Brasileira de Geosciécias 30 500 503

23. A. J. Da Costa Nunes, A. M. M. Costa Couto., F. Hunt, R. E. Hunt, 1979 Landslides of Brazil. Voight, B. [Ed.]. Rockslides and Avalanches, 2- Engineering sites. Developments in geotechnical engineering 14B, Elsevier, Amsterdam 419 446

24. F. C. Dai, C. F. Lee, 2001 Frequency-volume relation and prediction of rainfall-induced landslides Engineering Geology 59 155 166

25. J. De Ploey, O. Cruz, 1979 Landslides in the Serra do Mar, Brazil Catena 6 111 122

26. R. Dikau, J. Weichselgartner, 2005 Der unruhige Planet. Der Mensch und die Naturgewalten. Wissenschaftliche Buchgesellschaft [WBG] Darmstadt

27. A. P. Dykes, 2002 Weathering-limited rainfall-triggered shallow mass movements in undisturbed steepland tropical rainforest Geomorphology 46 73 93

28. FAO 2006 GlobalForest Resources Assessment 2005- Progress towards sustainable forest management. FAO Forestry Paper No. 147. Rome

29. P. J. Farres, 1987 The dynamics of rainsplash erosion and the role of aggregate stability. Catena 14 119 130

30. R. C. Ferreira, L. B. Monteiro, 1985 Identification and evaluation of collapsibility of colluvial soils that occur in the São Paulo State. First International Conference on Geomechanics in Tropical Lateritic and Saprolitic Soils. Brasilia Vol.I Brasilia 269 280

31. M. Fort, E. Cossart, G. Arnaud-Fassetta, 2010 Catastrophic landslides and sedimentary budgets. Alcántara-Ayaly, I.; Goudie, A. [Eds.]. Geomorphological Hazards and Disaster Prevention. Cambridge University Press, Cambridge 75 85

32. S. Furian, L. Barbiéro, R. Boulet, 1999 Organisation of the soil mantle in tropical southeastern Brazil [Serra di Mar] in relation to landslides processes. Catena 38 65 83

33. E. J. Gabet, D. W. Burbank, J. K. Putkonen, 2004 Rainfall thresholds for landsliding in the Himalayas of Nepal Geomorphology 63 131 143

34. A. Goudie, 1995 The changing earth. Blackwell, Oxford

35. A. Goudie, 2006 The human impact on the natural environment 6th Ed., Blackwell Publ., Oxford

36. A. S. Goudie, J. Boardman, 2010 Soil erosion. Alcántara-Ayaly, I.; Goudie, A. [Eds.]. Geomorphological Hazards and Disaster Prevention. Cambridge University Press, Cambridge 177 188

37. G. Guidicini, O. Y. Iwasa, 1977 Tentative correlation between rainfall and landslides in a humid tropical environment Symp. on Landslides & Other Mass Movement, Praga, IAEG Bulletin 16 13 20

38. A. Gupta, 2011 Tropical Geomorphology Cambridge Univ. Press, Cambridge

39. N. L. Harris, S. Brown, S. C. Hagen, S. S. Saatchi, S. Petrova, W. Salas, M. C.. Hansen, P. V.. Potapov, A. Lotsch, 2012 Baseline Map of Carbon Emissions from Deforestation in Tropical Regions Science 22 336 6088 1573 1576 10.1126/science.1217962

40. J. J. Hidore, J. E. Oliver, 1993 Climatology- an atmospheric science. New York

41. D. Higgit, 2010 Geomorphic hazards and sustainable development. Alcántara-Ayaly, I.; Goudie, A. [Eds.]. Geomorphological Hazards and Disaster Prevention. Cambridge University Press,Cambridge 257 268

42. K. Hueck, 1966 Die Wälder Südamerikas. Fischer Verl., Stuttgart [Germany]

43. A. I. Huerta, A. M. Marshall, 2002 Soybean production: competitive positions of the United States, Brazil, and Argentina. Purdue Agricultural Economics Report November 2002 4 10

44. IPCC 2007 Fourth Assessment Report: Climate Change 2007

45. F. Kraas, 2011 Megastädte. Gebhardt, H.; Glaser, R.,; Radtke, U.; Reuber, P 2011.,. Geographie. Physische Geographie und Humangeographie. 2nd Ed. Spektrum Akadem. Verlag, Heidelberg 879 885

46. W. A. Lacerda, 2007 Landslide initiation in saprolite and colluvium in southern Brazil: Field and laboratory observations Geomorphology 87 104 119

47. M. C. Larsen, Sánchez. A. J. Torres, 1992 Landslide triggered by Hurricane Hugo in eastern Puerto Rico September 1989 Caribbean Journal of Science 28 113 125

48. G. W. Lin, H. Chen, N. Hovius, M. J. Horng, S. Dadson, P. Meunier, M. Lines, 2008 Effects of earthquake and cyclone sequencing on landsliding and fluvial sediment transfer in a mountain catchment Earth Surface Processes and Landforms 33 1354 1373

49. P. Lumb, 1975 Slope failures in Hong Kong Quarterly Journal of Engineering Geology 8 31 65

50. N. Mamingi, 1997 The impact of prices and macroeconomic policies on agricultural supply: A synthesis of available results Agricultural Economics 16 17 34

51. B. Mertens, W. D. . Sunderlin, O. Ndoye, E. F. Lambin, 2000 Impact of Macroeconomic Change on Deforestation in South Cameroon: Integration of Household Survey and Remotely-Sensed Data World Development 28 6 983 999

52. W. M. Negecu, E. M. Mathu, 1999 The El-Nino triggered landslides and their socioeconomic impact in Kenya. Environmental Geology 38 277 284

53. L. H. Nunes, 2011 Landslides in São Paulo, Brazil: An integrated historical perspective. PAGES news 19

54. M. A. T. Oliveira, H. Behling, L. C. R. Pessenda, G. L. Lima, 2008 Stratigraphy of near valley head deposits and evidence of climate-driven slope-channel processes in southern Brazilian highlands. Catena 75 77 92

55. E. Parra, H. Cepeda, 1990 Volcanic hazard maps of the Nevado del Ruiz volcano, Colombia Journal Volcanol. Geotherm. Res 42 117 127

56. D. Petley, 2010 Landslide hazards. Alcántara-Ayaly, I.; Goudie, A. [Eds.]. Geomorphological Hazards and Disaster Prevention. Cambridge University Press, Cambridge 63 73

57. U. Pfisterer, 1991 Genese, Ökologie und Soziologie einer Bodengesellschaft aus Ultranasit und Bodenformen assoziierter Gesteine des süd-ost-brasilianischen Regenwaldes. Schriftenreihe Institut für Pflanzenernährung und Bodenkunde. Universität Kiel, Kiel

58. A. Rapp, 1976 Studies of mass wasting in the Arctic and in the Tropics. Yatsu, E.; Ward, A.J.; Adams, F. [Eds.]. Mass Wasting. 4th Guelp Symposium on Geomorphology. Geo Abstracts LTD, Norwich, England 1975 79 103

59. A. J. Reading, R. D. Thompson, A. C. Millington, 1995 Humid tropical environments. Blackwell, Oxford

60. W. Römer, M. Kanig, U. Pfisterer, 2002 The influence of lithology on hillslope development in the area of the Jacupiranga Alkaline Complex and its surrounding [São Paulo, Brazil]. Catena 47 151 171

61. W. Römer, 2008 Accordant summit heights, summit levels and the origin of the "upper denudation level" in the Serra do Mar [SE-Brazil, São Paulo]: A study of hillslope forms and processes. Geomorphology 100 312 327

62. W. Römer, 2012 Hillslope processes in tropical environments. Shroder,

J.Jr.; Marston, R.; Stoffel, M., [eds.]. Treatise on Geomorphology. Academic Press, San Diego, CA 7 [in press]

63. H. Rohdenburg, 1982 Geomorphologisch-bodenstratigraphischer Vergleich zwischen dem nordostbrasilianischen Trockengebiet und immerfeucht-tropischen Gebieten Südbrasiliens. Catena Suppl. Bd. 2, Braunschweig 72 122

64. M. Schiff, C. E. Montenegro, 1997 Aggregate Agricultural Supply Response in Developing Countries: A Survey of Selected issues Economic Deelopment and Cultural Change 45 2 393 410

65. U. Scholz, 2011 Strukturen und Probleme der ländlichen Räume in den Tropen. Gebhardt, H.; Glaser, R.,; Radtke, U.; Reuber, P.,. Geographie. Physische Geographie und Humangeographie. 2nd Ed. Spektrum Akadem. Verlag, Heidelberg 837 852

66. J. Schultz, 2000 Handbuch der Ökozonen. Ulmer, Stuttgart

67. J. Schultz, 2005 The Ecozones of the World. Springer, Berlin Heidelberg

68. M. E. Seluchi, S. C. Chou, M. Gramani, 2011 A case study of a winter heavy rainfall event over the Serra do Mar in Brazil Geofisica Interacional 50 1 41 56

69. E. A. S. Serrão, I. C. Falesi, Veiga. J. Bastos de, Neto. J. F. Teixeira, 1979 Productivity of cultivation on low fertility soils in the Amazon of Brazil. Sanchez, P.A.; Tergas, L.E. [Eds.]. Pasture production and acid soils of the tropics. Crentro Internacional de Agricultura Tropica, Cali Colombia 195 225

70. A. M. Silva, 2004 Rainfall erosivity map for Brazil Catena 57 251 259

71. A. Simon, M. C. Larson, C. R. Hupp, 1990 The role of soil processes in determining mechanism of slope failure and hillslope development in a humid-tropical forest, eastern Puerto Rico Geomorphology 3 263 286

72. D. S. Simonett, 1967 Landslide distribution and earthquakes in the Betwani and Trocelli Mountains, New Guinea. Jennings, J.N.; Mabbutt, J.A. [Eds.]. Landform Studies from Australia and New Guinea. Canberra 64 86

73. W. Sinun, W. M. Wong, I. Douglas, 1992 Throughfall, stemflow, overland flow and throughflow in the Ulu Segama rain forest, Sabah, Malaysia Phil. Transact. Royal Soc. London B 335 389 395

74. A. M. Silva, 2004 Rainfall erosivity map for Brazil Catena 57 251 259

75. B. J. Skinner, S. C. Porter, 1987 Physical Geology John Wiley and Sons, New York

76. C. L. So, 1971 Mass movements associated with the rainstorm of

June1966 in Hong Kong. Transactions Institute British Geographers 53 55 66

77. N. Stache, A. Wirthmann, 1998 Die Misere in der Landwirtschaft im tropischen Afrika am Beispiel einer geoökologischen Studie in Rwanda. Petermanns Geographische Mitteilungen 142 5/6 339 354

78. L. Starkel, 1972 The role of catastrophic rainfall in shaping the relief of the Lower Himalayas [Darjeeling Hills]. Geographica Polonia 21 103 147

79. R. Tavani, M. Saket, M. Piazza, A. Branthomme, D. Altrell, 2009 Measuring and monitoring forest degradation through national forest monitoring assessment Case studies on measuring and assessing forest degradation. Forest Resources Assessment Working Paper 172, Rome, Italy

80. M. F. Thomas, 1994 Geomorphology in the Tropics. Wiley, Chichester

81. M. F. Thomas, 2004 Landscape sensitivity to rapid environmental change-Quaternary perspective with examples from humid tropical areas. Catena 55 107 1

82. M. F. Thomas, 2006 Lessons from the tropics for a global geomorphology Singapore Journal of Tropical Geography 27 111 127

83. M. F. Thomas, M. B. Thorp, 1995 Geomorphic response to rapid climatic and hydrologic change during the Late Pleistocene and Early Holocene in the humid and sub-humid tropics Quaternary Science Review 14 193 207

84. D. J. Varnes, 1978 Slope movement types and processes. Schuster, R.I.; Krizek, R.J. [Eds.]. Landslides, Analysis and Control. Transportation Research Board Sp. Rep176 Nat. Acad. Scie 11 33

85. L. D. White, D. N. Mottershead, S. J. Harrison, 1984 Environmental Systems George Allen & Unwin, London

86. T. C. Whitmore, 1990 An introduction to Tropical Rainforests. Clarendon Press, Oxford

87. A. Wirthmann, 2000 Geomorphology of the Tropics Springer, Berlin

88. C. M. Wolle, W. Hachich, 1989 Rain-induced landslides in Southeastern Brazil. Proc. 12th Int. Conf. on Soil Mechanics and Foundation Engineering, Rio de Janeiro, Brazil 3 1639 1642

89. J. P. Ybert, W. M. . Bissa, E. L. M. . Catharino, M. Kutner, 2003 Environmental and sea-level variations on the south-eastern Brazilian coast during the Late Holocene with comments on prehistoric human occupation. Palaeogeography, Palaeoclimatology, Palaeoecology 189 11 24

Chapter 4

HARNESSING EARTH OBSERVATION AND SATELLITE INFORMATION FOR MONITORING DESERTIFICATION, DROUGHT AND AGRICULTURAL ACTIVITIES IN DEVELOPING COUNTRIES

Humberto Barbosa[1], Carolien Tote[2], Lakshmi Kumar[3] and Yazidhi Bamutaze[4]

[1] Universidade Federal de Alagoas (UFAL), LAPIS, Brazil

[2] Flemish Institute for Technological Research (VITO), Centre for Remote Sensing and Earth Observation, Boeretang, Mol, Belgium

[3] Atmospheric Science Research Laboratory, SRM University, India

[4] Department of Geography, Geo-Informatics and Climatic Sciences, Makerere University, Kampala, Uganda

INTRODUCTION

With the drastic advances in technology over the past decades, the availability, as well as the quantity, of large data sets for research in almost every scientific field has increased dramatically. More specifically, the availability of earth observation-based imagery data and satellite information for research purposes and practical applications has grown with many organizations such as the European Organisation for the Exploitation of Meteorological Satellites (EUMETSAT), the National Aeronautics Space Administration (NASA), the National Oceanic and Atmospheric administration (NOAA), the Flemish Institute for Technological Research (VITO), etc. for example, through GEONETCast, which is part of the core Global Earth Observation System of Systems (GEOSS), the users do not need to repeatedly build ground receiving stations for different satellites [1]. However, despite a wealth of remotely sensed data provided by GEONETCast, investments in science technology and innovations is often a low priority for decision and policy makers in most developing countries in Africa, Asia and global emerging economies like India and Brazil. Yet, most of these countries face serious environmental risks and development challenges which require reliable and timely access to accurate

Earth Observation (EO) data and derived environmental information for their sustainable development. In particular, there is a clear need for research on the integration and utility of remote sensing data and products into the risk assessment cycle, scenario development and impact forecasting, in view of global (climate) change [2].

Climate alterations, although global in nature, may have different impacts in different regions of the world. Reports of Intergovernmental Panel on Climate Change (IPCC) showed that the global average surface temperature has increased over the 20th century by about 0.6°C [3]. Global Circulation Models have projected that this rise in temperature may increase to a range of 1.5 – 5.8°C by the end of the century. Results of crop growth modeling under climate change scenarios suggest that agriculture, and thus human well-being, will be negatively affected by climate change, especially in developing countries [4]. The vulnerability to drought and land degradation has increased in the past decades and this is especially true due to increased population pressure and limited livelihood options in drought-prone areas [5]. Significant gaps in observing systems exist, especially in developing countries, and timely access to both surface-based and space observations is still a challenge in many locations.

Global-scale population growth and economic development will have a large impact on water supply and demand, and it is necessary to understand the interactions between climate change and variability, hydrology and human systems, in order to have a view on future water vulnerabilities [6]. Since developing countries will become more susceptible to climate variability and drought, it is essential to develop climate (impact) monitoring services. A climate service involves broad partnerships of producer and user organizations, climate scientists, climate service providers, economists and social scientists. It provides an opportunity to interlink global, national and regional information systems; to provide essential information to policy makers, decision takers and to the public in general at regional and local scales, and a provide for a distributed decision-relevant research and development capability [7]. The climate service for the developing countries might focus on collaborative problem solving. Also capacity building and the improvement of infrastructure, related to the acquisition of advanced remote sensing technologies and the installation of satellite receiving stations by, is needed.

The gist of this chapter is therefore to strengthen the capacities of the regional scientific community to provide stakeholders from drought to agricultural activities with satellite information that is directly useful to improve decision-making in the context of the developing countries. The chapter is based on four case studies from Africa, South America, India and Europe to demonstrate the utility of satellite imagery data obtained from free or low cost platforms in providing information to address the above environmental issues, which are critical particularly in developing countries. Based on the expertise, experience and interest of a regional network of scientists from Uganda, Brazil, India, and Europe, the chapter focuses on harnessing satellite remote sensing resources and products for drought monitoring of areas subject to or in risk of land degradation processes, agricultural productivity and drought assessment. The following paragraphs include a series of example cases where time series of satellite imagery is used to monitor the impact of climate (change):

- Degradation monitoring over South America
- Relation of NDVI with the moisture Index over the regions of different climatic types
- Sugarcane yield estimation and modelling based on NDVI data over Southeastern Brazil and
- Production and yield estimates of five major crops over Uganda

The importance of each of the above aspects in the context of assessing, monitoring and managing the natural hazards such as desertification, drought and risk management are discussed in detail in the individual sections.

All case studies make use of time series of Normalized Difference Vegetation index (NDVI), an index of vegetation activity that can be derived from broad band measurements in the visible and infrared channels onboard satellite instruments and which is directly related to the photosynthetic capacity of plants [8]., Satellite sensors such as NOAA Advanced Very High Resolution Radiometer (AVHRR), Moderate Imaging Spectroradiometer (MODIS), Système Pour l'Observation de la Terre (SPOT) Vegetation etc. provide NDVI data on different intervals (8-day, 10-day, monthly and seasonal). NDVI value varies minus one (-1) to plus one (+1), whereby low NDVI values (< 0.2) reflect sparse vegetation, and higher NDVI values (> 0.4) reflect high vegetation densities. Several studies looked into the interannual variability and trends of NDVI in relation to meteorological parameters such as rainfall, temperature etc [9-11].The long time series of NOAA AVHRR has been widely used to relate the synoptic meteorology/climatology to understand the vegetation dynamics, vegetation response to climate and climate vegetation feedback mechanism [5, 12-14]. The studies of [9, 15-17] concluded that AVHRR NDVI is a valuable tool to monitor and asses large scale agricultural droughts. Other studies use NDVI time series derived from MODIS [18-20] or SPOT-Vegetation [21, 22].

In this work, the authors made use of NDVI imagery datasets derived from different satellites to relate the respective crop and climatic parameters to understand the sensitivity of NDVI with these parameters and further to use them for vegetation monitoring over large areas which contributes to risk management.

TIME SERIES ANALYSIS FOR DESERTIFICATION MONITORING

In the last decades, remote sensing technologies started to contribute enormously in documenting changes in land cover and monitoring desertification, drought and agricultural activities on regional and global spatial scales. Although desertification is highlighted as one of the most important global environmental issues, both desertification and greening processes have been reported on global scale [23-26]. These processes are related in many ways with other environmental issues, such as climate change and the carbon cycle, loss of biodiversity and sustainability of agriculture [27]. Also in South America environmental change is an important concern. In the last decades, South American ecosystems underwent important functional modifications due to climate alterations and direct human interventions on land use and land cover [28]. In South America, the main forest conversion process in the humid tropics in the period 1990-1997 was the clear-cutting of closed, open, or fragmented forest to make room for agriculture at a rate of approximately 1.7 million ha per year [29]. Apart from deforestation, also forest degradation occurs, a process leading to a temporary or permanent deterioration in the density or structure of vegetation cover or its species composition. Land degradation in arid, semi-arid and dry sub-humid areas is called desertification, and is the result of various factors, including climatic variations and human activities [30].

To determine desertification conditions, this case study focuses on vegetation dynamics in South America over a long time period based on a time series of low spatial resolution, high temporal resolution NDVI derived from SPOT-Vegetation, and to recognize to which extent this variability can be attributed to variability in rainfall, since rainfall is one of the most determinant factors of vegetation growth. In general, and especially in semi-arid regions, strong correlations between precipitation and the NDVI can be found. Therefore, the NDVI can be used as an indicator for vegetation status and vegetation response to precipitation variability. This study expands the analysis, as was performed on the Andes region [31] to South American continental level. In the first phase, trends of vegetation and precipitation indices are analyzed. In a next step, through correlation of NDVI and precipitation dynamics, these areas where the evolution of vegetation is not related to climate only and human induced impacts play an imperative role can be identified. The time series of SPOT-Vegetation 10-daily composite NDVI data (April 1998 – March 2012) at

1 km resolution (http://www.vgt.vito.be/) was smoothed [32] and consequently synthesized to monthly images using the maximum value composite technique. Also 10-daily rainfall estimates at 0.25° resolution, available from the European Centre for Medium-Range Weather Forecasts (ECMWF) through MeteoConsult and the Monitoring Agricultural ResourceS (MARS) unit, were combined to retrieve monthly composites. The spatial resolution of the NDVI time series was degraded in order to fit the rainfall estimates using a weighted average approach. Many authors remove seasonality by integrating the data into annual values (e.g. [25, 28, 33]). In this study, in order to remove seasonal vegetation changes and thus facilitate the interpretation through the historical record, deviations from the 'average' situation were calculated for the NDVI time series using the Standardized Difference Vegetation Index (SDVI) [34] and for the precipitation time series using the Standardized Precipitation Index (SPI) [35]. In a next step, for each pixel a correlation analysis is performed on the monthly NDVI and SPI datasets, in order to identify the temporal scale at which the environment is most sensitive to precipitation anomalies (the so-called 'best lag').

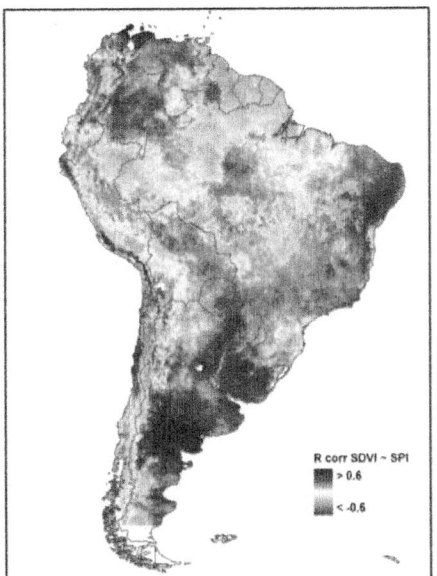 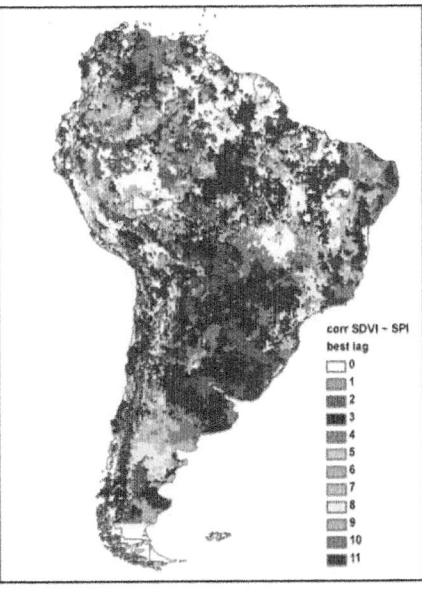

Figure 1: Correlation analysis between NDVI and SPI. Best lag expressed in months.

The results of the correlation analysis between NDVI and SPI are shown in Figure 1. Positively correlated (blue) areas suggest precipitation-vegetation coupling. In general, high positive correlation with best lags between 3 – 6 months are found in the semi-arid regions of South America, while weak (or even negative) correlation at all time scales was found in both the hot and humid zones and the deserts and high mountainous areas. These positive relationships between vegetation greenness and rainfall in drylands, where biomass production is determined by the amount of rainfall, and the opposite in humid and cold regions, where rainfall is not the limiting factor for vegetation growth, and deserts, where there is no rainfall at all, is consistent with findings of other authors, such as [33, 36]. In order to identify if a pixel is greening or degrading, and after identifying the best lag for each pixel, linear least squares trend analyses were performed on the SDVI time series and the SPI time series, taking into account the accumulated rainfall over the respective best lag. Only trends with Pearson correlation coefficients significantly different from zero (at significance level $p < 0.05$) are considered significant trends. Figure 2 shows the slope of the significant linear trends of vegetation greenness and precipitation anomalies, SDVI and SPI respectively. SDVI show slight but significant ($P<0.05$) positive trends in large areas in the northern part of South America, but significant negative trends in Argentina and the Peruvian coast. The results are comparable, but far more pronounced than results from annual series trend analysis, such as from [25, 28]. Also the SPI shows positive trends in the north-west of South America and the centre of Brazil. Negative precipitation trends are found in Argentina, the Peruvian coast and north-east of Brazil.

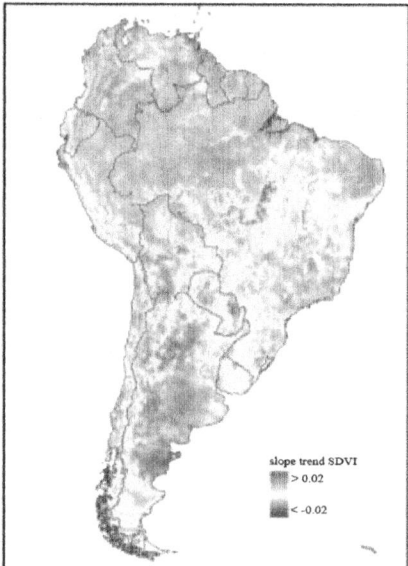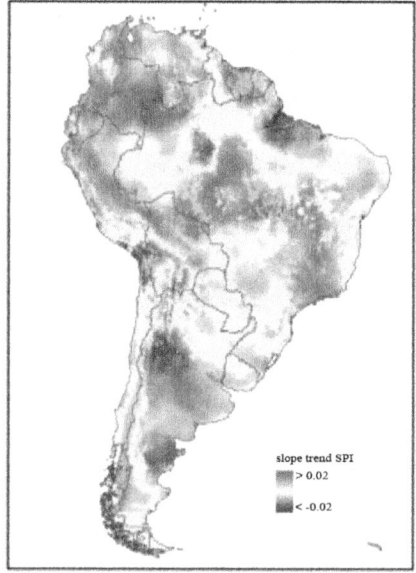

Figure 2: Slope of the trend analysis of SDVI (left) and SPI (right). Non significant trends are masked in grey.

Following the trend analyses, a decision tree approach is adopted in order to interpret the results. Five queries are stipulated: (1) Does the time series of SDVI show a significant trend? (2) Are significant trends in SDVI coupled to significant correlations between NDVI and SPI? (3) Are significant trends in SDVI linked to significant trends in SPI? (4) Does the SDVI show a positive trend? (5) Do trends in SDVI correspond to trends in SPI?

The first question is meant to identify the focus area, i.e. pixels where the de-seasoned vegetation index (i.e. SDVI) shows a significant linear trend over time. The second and third queries are used to identify areas that differ in their relationship between vegetation and precipitation trends, respectively, while queries 4 and 5 are meant to distinguish positive and negative trends that are linked to trends in precipitation. On the other hand, it is possible to identify regions where positive or negative trends in vegetation are not linked to changes in precipitation and other climate variables or human impact play an imperative role. In step 4, pixels are divided in greening or degrading pixels. In the last step, both the trends in SDVI and SPI are evaluated.

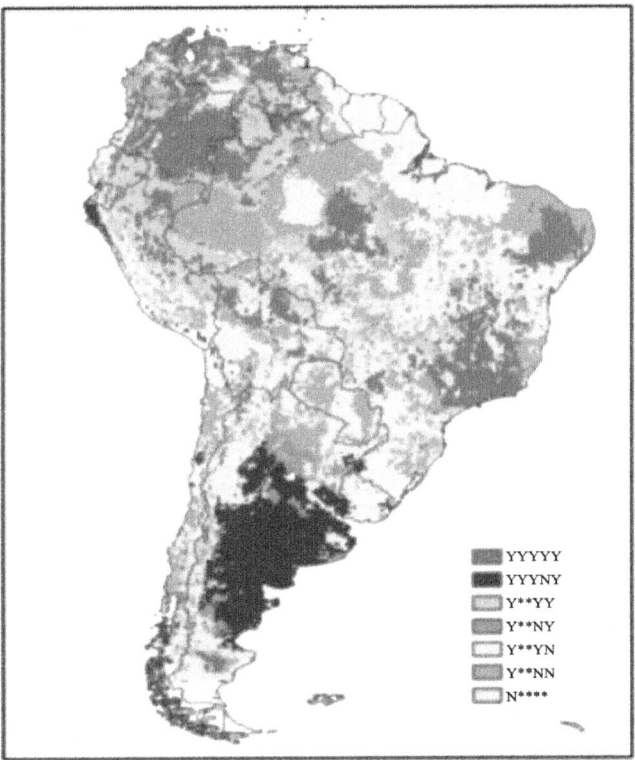

Figure 3: Result of the decision tree approach. Pixels are classified according to the 5

queries described in the text (where Y and N refer to a 'yes' or 'no' answer to the query, and * to either Y or N): green areas show an increase in SDVI, coupled to a significant positive trend in SPI (dark green) or not (light green); red areas show a decrease in SDVI, coupled to a significant negative trend in SPI (dark red) or not (light red); yellow areas show a positive trend in SDVI, but a negative trend in SPI; orange areas show a negative trend in SDVI, but a positive trend in SPI. Pixels without a significant trend in SDVI are masked in grey.

The results of the decision tree analysis are shown in Figure 3. Green classes show a significant positive trend in SDVI coupled to an increase in SPI. Red classes show a coupling between a decrease in SDVI and SPI. The yellow and orange classes are classes where SDVI and SPI show an opposite trend. Argentina is clearly suffering from vegetation degradation linked to a decline in precipitation, which confirms the findings of [35]. The opposite is going on in Colombia and some parts of Brazil, where a process of greening seems to be linked to an increase in precipitation. Nevertheless, Brazil shows a patchy result, with some areas showing an increase in SDVI, although the precipitation decreases (yellow areas in Figure 3), and other areas showing a decline in vegetation, although precipitation increases (orange area in Figure 3), probably related to deforestation or forest degradation.

The resulting map can be used to estimate the coupling between vegetation (SDVI) and precipitation (SPI), shown in Figure 4. Three estimations are made: (A) a large estimation where significant trends in SDVI are coupled with trends in precipitation when SPI shows the same trend (significant or not); (B) an average estimation where significant trends in SDVI are linked to trends in precipitation when SPI shows a significant trend; and (C) a conservative estimation where significant trends in SDVI are linked to trends in precipitation when the SPI shows a significant trend and the correlation between NDVI and SPI is significant.

In estimation A, also weak trends in SPI are taken into account. These non-significant trends in precipitation are most probably not by itself responsible for significant trends in SDVI. Together with other variables like temperature change or human impact, these weak changes in precipitation might however give an extra impulse to greening or degradation processes. It is also possible that an increase in precipitation does not result in higher vegetation cover because the area is covered with climax vegetation or, the other way round, a decrease in precipitation does not result in further degradation because the area is already covered with minimal vegetation growth. Estimations A and B show little difference. The difference between estimation B and C is based on the significance of the correlation between NDVI and SPI, and shows larger differences, mainly in the greening pixels. In many pixels that show a positive trend in SDVI, vegetation greenness is not significantly correlated to SPI. It is therefore not certain that in these areas the increase in SDVI is coupled to an increase in precipitation.

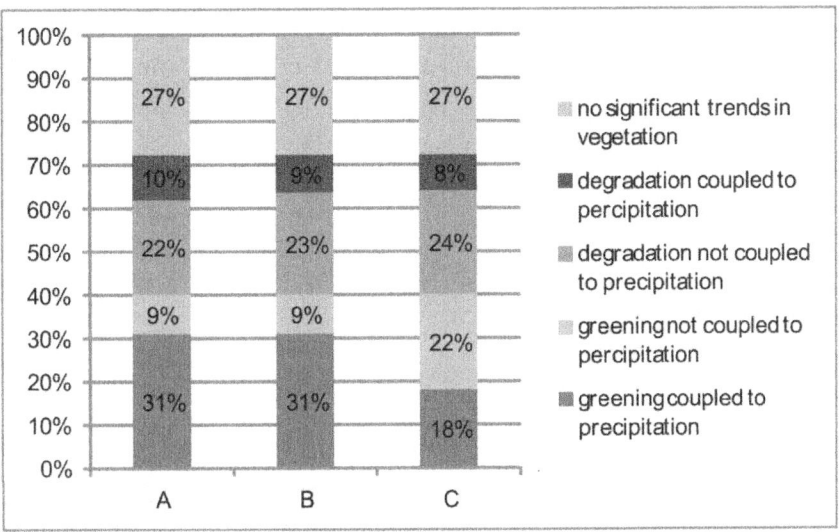

Figure 4: Large (A), average (B) and conservative (C) estimation of the linkage of greening and degradation to changes in precipitation over time. In case of estimation B and C, the yellow areas in Figure 3 are included in the light green fraction, while the orange areas in Figure 3 are included in the light red fraction.

From the conservative estimation, we can conclude that in 8% of South America, vegetation degradation is coupled to a significant decrease in the amount of precipitation in the last 14 years. Our results corroborate with the findings of [37]. In contrast, in 18% of the subcontinent, vegetation greenness has significantly increased over the last 14 years, coupled to an increase in precipitation. For 46% of the study area, significant degradation or greening processes could not be linked to changes in precipitation over time, indicating human impact or the influence of other climatic factors, such as temperature. Finally, and without taking into account the link with trends in precipitation, 40% of the subcontinent is showing an increase in photosynthetic activity over time, while desertification is taking place in 32% of the area.

TIME SERIES ANALYSIS FOR DROUGHT MONITORING

Drought is a recurrent feature of the Indian climate and usually begins at any season and can prolong for many years. The study of drought characteristics is to ascertain the spatial and temporal distribution of droughts, synoptic meteorological conditions associated. The definition of drought mainly depends on the precipitation deficiency. Studies on droughts in India have been reported by many scientists based on the rainfall anomalies over a particular region [38-40]. According to India Meteorological Department (IMD) guidelines, drought is defined as the consequent rainfall deficiency (below 19% of normal) for a period of 2 consecutive weeks. However, this criterion varies from country to country, based on the meteorological/climatological conditions such as percentage of moisture present, land topography etc.

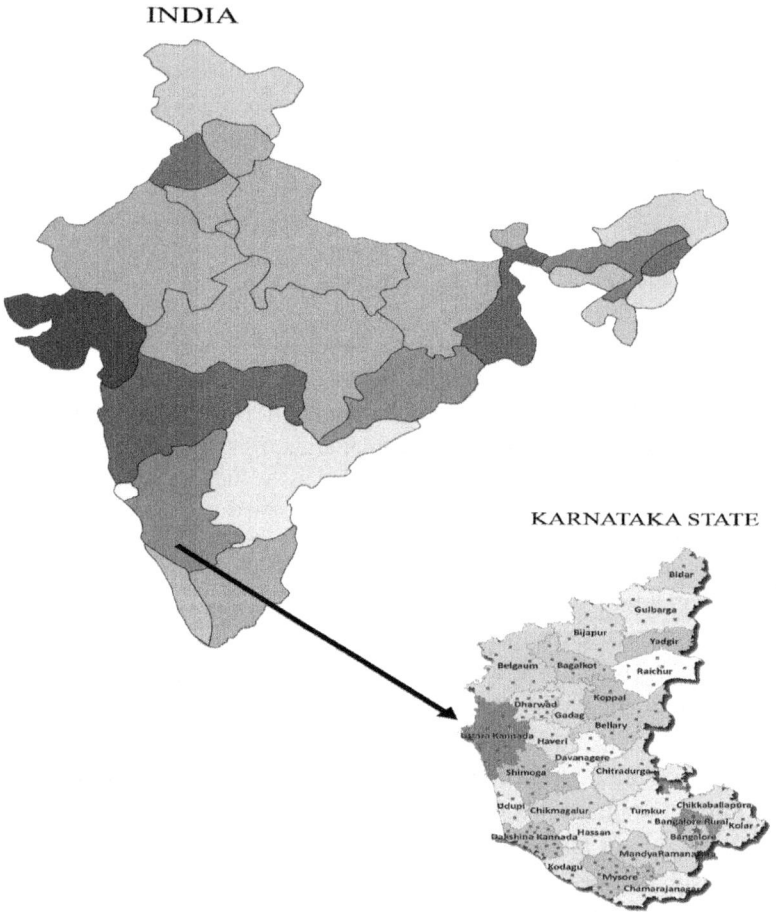

Figure 5: Study area of the present investigation

Drought is of different types including meteorological droughts, agricultural droughts, hydrological drought and socioeconomic droughts. These are based on the variations of rainfall, crop water, surface water and economic conditions respectively. These droughts show variation with respect to the climatology which prevails of that region. The climatology of a region is a replica of the severity of drought. The climatology can be derived from the Thoronthwaite Climate System by taking the inputs of rainfall and water need. The model gives the amount of moisture (annual/seasonal) from which the classification of climate can be studied.

In view of the above, the present study focuses in obtaining the climates in different parts of Karnataka state which is located at the western half of the Deccan Palateau of India. The Moisture Index values which are the basis for delineating climatic type were compared with the AVHRR NDVI to understand the drought climatology in different test regions of Karnataka, India (Figure 5).

The rainfall (P) and potential evapotranspiration (PE) data for the period 1982 to 2000 was downloaded from [41]. This data is based on the global rainfall and temperature (PE can be calculated from temperature) data sets of Climate Research Unit, University of East Anglia, United Kingdom. This was averaged for all the districts of Karnataka state till the year 2000 and uploaded website.

Taking the inputs of P and PE, we run awater balance model and derived the monthly Aridity Index (IA) and Humidity Index (IH). Moisture Index (IM) which is the basis to tell the climatology of a region can be obtained by subtracting IA from IH.. Table 1 below shows the infered climate types based on IM as per Thoronthwaite Climate Approach [42, 43].

Table 1: Classification of climatic types based on Thoronthwaite Approach

Moisture Index(IM)	Climatic Type	Notation
100 & above	Perhumid	A
80 to 100	Humid	B4
60 to 80	Humid	B3
40 to 60	Humid	B2
20 to 40	Humid	B1
0 to 20	Moist Subhumid	C2
-20 to 0	Dry Subhumid	C1
-40 to -20	Semiarid	D
-60 to -40	Arid	E

The selected districts for the study are Chikkamagaluru, Belgaum, Chamrajnagar and Gulbarga of which climates are Humid, Dry subhumid, Semiarid and Arid. A comparative study was made with the seasonal values of IM and NDVI using time series and correlation analysis.

Table 2 shows the climatology of the four selected test sites during the period of 1982 to 2000. The overall climate of the test regions for the study period represented the humid, dry subhumid, semi arid and arid for Chikkamagaluru, Belgaum, Chamrajanagar and Gulbarga districts respectively.

Table 2: Climatic types of test regions from 1982 to 2000

Year	Climate Type			
	Chikkamagaluru	Belgaum	Chamarajanagar	Gulbarga
1982	C1	C1	D	E
1983	B1	C2	D	E
1984	C2	D	D	E
1985	C1	D	D	E
1986	C2	D	D	E
1987	C1	D	D	E
1988	B2	C1	D	E
1989	C2	D	D	E
1990	C1	D	D	E
1991	B1	C1	D	E
1992	B2	C2	D	E
1993	B1	B1	C1	E
1994	B2	C2	D	E
1995	C2	C1	D	E
1996	B1	D	D	E
1997	B1	C1	C1	E
1998	B2	C2	C2	E
1999	B1	B1	C1	E
2000	B1	C2	D	E

Chikkamagaluru (Humid region)

The climatic types of Chikkamagaluru were dominated by humid type which is followed by the dry sub humid type. Chikmagalore (Humid region) recorded a maximum IM of 450 during the year 1994 (Figure 6). The variation of IM shows that it has increased from June to August in all years and recorded a comparative less value in the month of September. In a similar way, the NDVI progressed from June to September in all years with a maximum value of 0.55 for September, 1992. It is conspicuous that the variability in NDVI is more than IM and the trends of IM and NDVI were increasing. The correlation coefficient of these two indices is +0.08 which is very poor. The studies of [38, 39] also suggested that understanding the relation of NDVI with rainfall and its by-products is a very difficult task in humid regions. Since the plenty of moisture is already available in the soil, the vegetation can utilize the moisture for its growth and in such case it may not directly/immediately dependant on rainfall derived indices in humid regions.

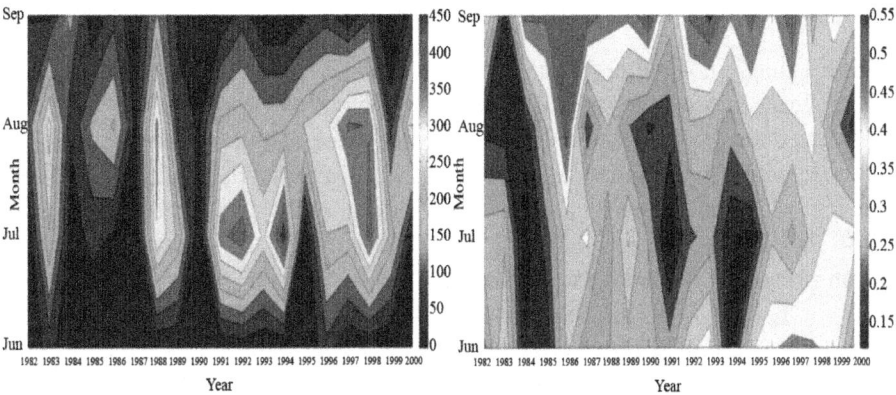

Figure 6: Variation of IM and NDVI – Chikkamagaluru

Belgaum (Dry subhumid)

The climate of Belgaum varied from semi arid to 1st humid type during the study period. 10 years of the study period show the subhumid climates followed by semi arid in 7 years. The variation of IM with NDVI over Belgaum unraveled the low variability of IM associated with high variations of NDVI (Figure 7). The maximum NDVI is 0.55 that is during September of 1984 and 1992 linked with the very low values of IM. The IM values were around zero during all the years of study period in June and July months countered by the low values of NDVI. There is no sea – saw relation found between IM and NDVI over this region. The correlation in this case was also found to be very poor and insignificant.

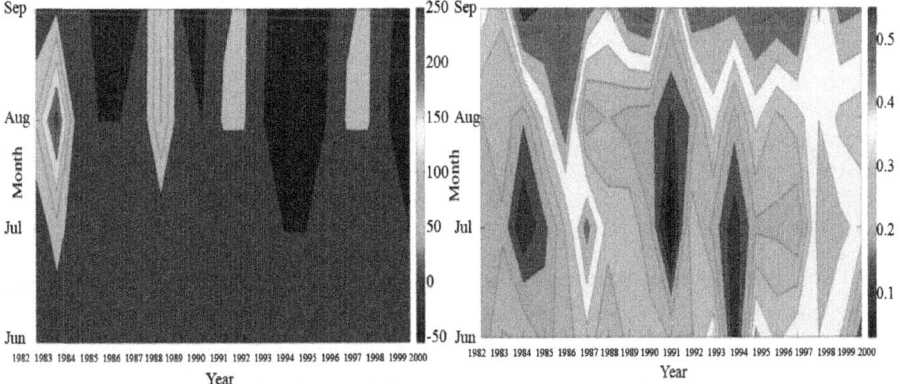

Figure 7: Variation of IM and NDVI – Belgaum

Chamrajnagar (Semiarid)

From Table 2, it is noticed that Chamrajanagar shows the arid climate category in 15 years of the study period. The years such as 1993, 1997, 1998 and 1999 displayed subhumid climates. The climatic types of this region are the good representative of climatic features over the study period. In the case of semi arid region of Chamrajnagar, the variability of IM and NDVI is better as compared with the humid and dry subhumid regions (Figure 8). The overall IM is varied from -10% in the year 1984 to a maximum of 44% during the year 1998 for the south west monsoon season. Accordingly NDVI, also varied from 0.26 to a high value of 0.42 in the year 1996. The values of NDVI are less in this region than previously mentioned areas but the trends of IM are NDVI were positive with slopes of 0.824 and 0.002 for IM and NDVI respectively along with the standard deviations of 11.5 and 0.05. The correlation of these two data sets is 0.42 at 0.05 level of significance which infers the good agreement of IM and NDVI. From this analysis, it can be noticed that the deficiency of moisture which is represented by IM in this study was well reflected by low NDVI values and adequate moisture conditions are supported by the moderate vegetation conditions.

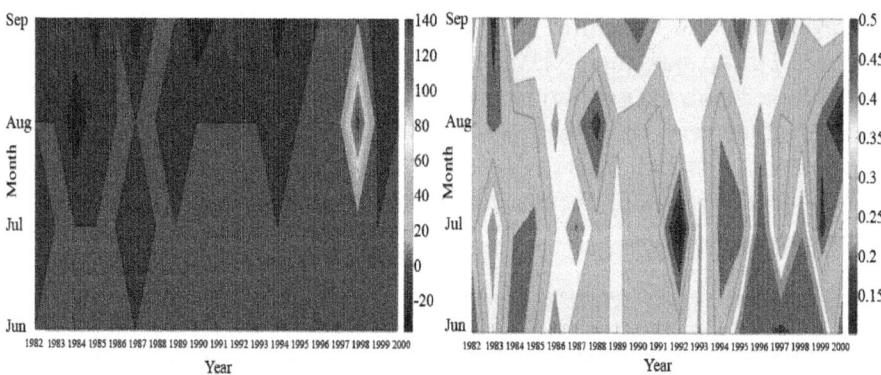

Figure 8: Variation of IM and NDVI – Chamrajanagar

Gulbarga (Arid)

All the years of study period are dominated by the arid category for this region. The comparison of IM and NDVI yielded good results in the arid region of selected test sites (Figure 9). The interannual variability was found to be very high both in IM and NDVI where as IM varied from -77% during August of 1982 to zero value during September month in the years 1983 and 1992 with the corresponding NDVI values of 0.055 and 0.127 respectively. The time series plot for IM and NDVI for the total south west monsoon season display the one to one linear agreement where the trends of both were highly increasing than other test regions (Figure 8). The slopes of the trends were 1.68 and 0.007 respectively. The standard deviations of 14 and 0.05 infer that the interannual variability of IM is more than NDVI from which it can be noticed that the vegetation over a region may not respond immediately to the rainfall/available moisture despite there is a dependence of vegetation on rainfall/available moisture. The time series plot of IM and NDVI (Figure 10) shows that the maximum of amount of moisture of-12% have seen in the year of 1998 with the NDVI value of maximum NDVI of 0.269 that is recorded during the entire study period. The correlation in this case is +0.64 which is at 0.01 level of significance which shows the strong agreement between IM and NDVI.

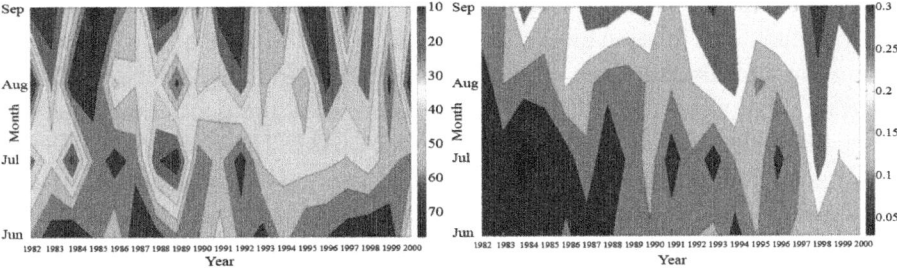

Figure 9: Variation of IM and NDVI – Gulbarga

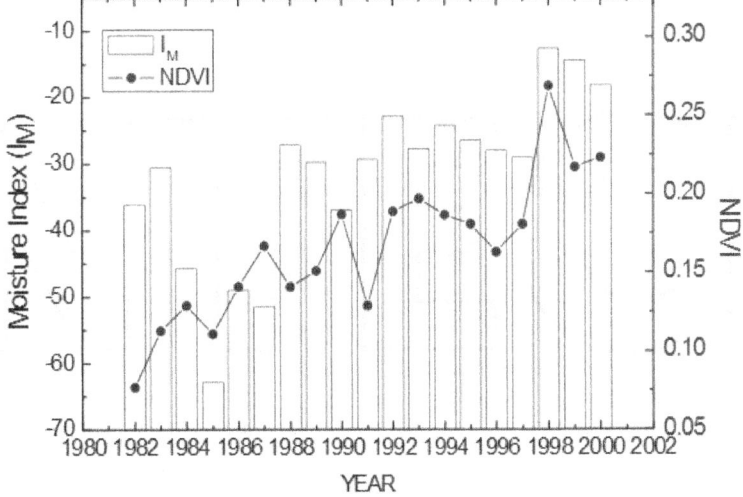

Figure 10: Time series of IM and NDVI for southwest monsoon

The study commenced with the retrieval of IM values from the water balance model on monthly basis over different selected test regions those represent the various climatic types such as humid, dry sub humid, semi arid and arid. The knowledge of different climatic types enables us to understand the climatology of the test regions during the study period. Since these climates were derived from the IM values, they replicate the status of the moisture content available which is very essential input to decide the crops fate. The comparison study of IM with the satellite derived NDVI shown very interesting features of sensitivity of NDVI with IM over different climatic types. The study inferred the poor correlation such that no linear and significant relation of IM with NDVI over humid and dry subhumid regions. The reason for this could be the plenty of available moisture over these regions and even temporary perturbations of land surface conditions may not affect the crops/agriculture. The relation grown up to strong when the comparison closes from semi arid to arid regions. Especially, Gulbarga, arid region displayed very strong relation of IM with NDVI which unraveled the poor/good vegetative conditions associated with low/high values of IM. The correlation of +0.65 is a good supporting factor to say that the relation is substantial. The overall analysis of the present study suggested that the relation of IM with NDVI is very strong and it is of immense use for the studies of drought monitoring in the arid areas as compared with the other climatic types.

QUANTIFICATION OF SUGARCANE CROP PRODUCTIVITY: A STUDY CASE IN SOUTHEASTERN BRAZIL

Agriculture represents an important segment of the economy of Brazil. Over the past 30 years, Brazilian agricultural growth and development has been guided by policies and technologies based on research for development. Remote sensed imagery plays an important role in agricultural crop production over large area, quantitatively and non-destructively, because agricultural crops are often difficult to access, and the cost of ground estimating productivity can be high. The recent development of GEONETCast–EUMETCast data has allowed us to obtain frequent and accurate measurements of a number of basic agrometeorological parameters (e.g. evapotranspiration, surface albedo, surface temperature, solar radiation, rainfall etc.). The GEONETCast–EUMETCast real-time and on-line data dissemination systems represent global network of satellite-based data dissemination systems designed to distribute space-based, air-borne and in situ data, metadata and products to diverse communities.

To determine agriculture productivity, this case study aimed to develop a GEONETCast-EUMETCast product-based method of estimating the productivity of sugarcane using an agrometeorological spectral model. The study was carried out in the Municipalities of Barretos and Morro Agudo, located in the state of São Paulo, Southeastern Brazil (Figure 11). The analysis was performed for 2009/2010 and 2010/2011 year's crop.

The values of sugarcane parameters used such as Respiration Factor (RF) (0.5 for temp. $\geq 20°C$ and 0.6 for temp $<20°C$); Agricultural Productivity Factor (APF) (2.9), Yield Response Factor (Ky) and Crop Co-efficient (Kc) were taken from [44-46]. The EUMETCast service is installed at Laboratory of Analysis and Processing of Satellite Images (LAPIS) at Federal University of Alagoas (UFAL). The remote sensing data of NDVI S10, Production of Dry Matter (DMP)

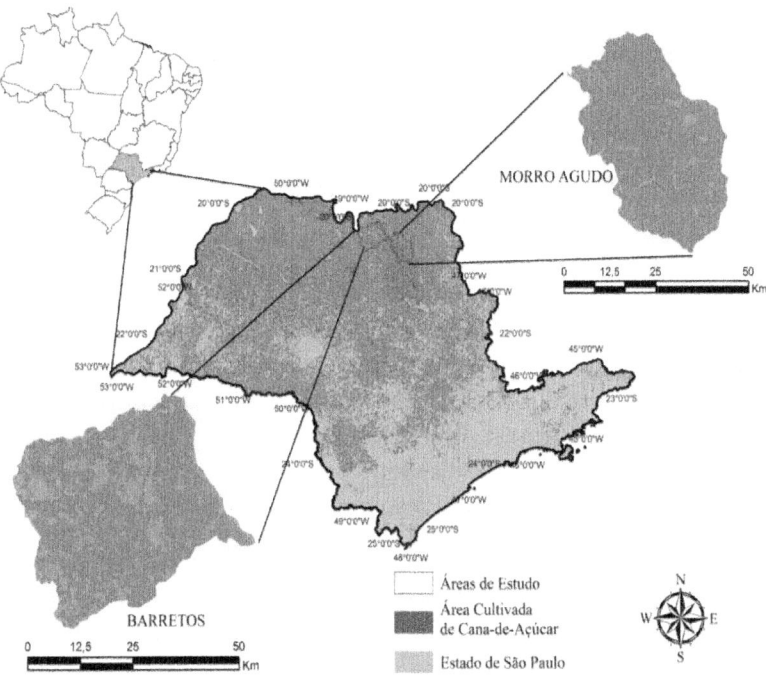

Figure 11: Spatial variability of crop yields (2010/2011) of Barretos and Morro Agudo in Sao Paulo, Brazil.

This application proposes to test a remote sensing approach to quantify estimates of sugarcane productivity over the Coruripe municipality with the Integrated Land and Water Information System, (ILWIS, 3.7.1) GIS software. ILWIS was used to compute sugarcane crop estimates for each pixel in NDVI DMP and ETp images by applying radiative, aerodynamic and energy balance physics in 7 computational steps. These images are currently provided over both daily and 10 day composites at about a 3km and 1km spatial resolution, by EUMETSAT and VITO respectively.

Step 1: Input NDVI and DMP databases using algorithm adapted from GEONETCast Toolbox

To implement a methodology for the ingestion of both the NDVI and DMP databases (raster) into ILWIS, specific routines of GEONETCast–toobox are adapted to import the datsets. For the ingestion procedure, based on a GIS approach using open source components, it requires additional work on corrections using overlays (Status map and LOG–file) to mask all appropriate areas of sugarcane crops over the Coruripe municipality.

Step 2: Computation of Fractional Vegetation Cover (FVC) from NDVI

For each pixel, the NDVI is converted to Fractional Vegetation Cover (FVC) by means of the the formula of [47]. The FVC is the one biophysical parameter that determines the contribution partitioning between bare soil and vegetation for surface evapotranspiration, photosynthesis, albedo, and other fluxes crucial to land–atmosphere interactions.

$FVC = 1.1101*NDVI - 0.0857.$

Step 3: Computation of Leaf Area Index (LAI) from FVC

For each pixel, the FVC is converted to Leaf Area Index (LAI) by means of the formula of [48]. The LAI, defined, as the total one-sided leaf area per unit ground area, is one of the most important parameters characterizing a canopy. Because LAI most directly quantifies the plant canopy structure, it is highly related to a variety of canopy processes, such as evapotranspiration, interception, photosynthesis and respiration.

$LAI = -2Ln (1 - FVC).$

Step 4: Computation of growth factor from LAI

[49] developed a simple approach for deriving growth rate equation from LAI. Experimental evidence indicated that the growth rate of several agricultural crop species increases linearly with increasing amounts of LAI, when soil water nutrients are not limiting [46]. The following equation is used:

$CGF = 0.515 - e[-0.667 - (0.515*LAI)]$

where CGF = Corrected Growth Factor. Experimental evidence indicated that the growth rate of several agricultural crop species increases linearly with increasing amounts of LAI, when soil water nutrients are not limiting.

Step 5: Computation of maximum yield potential (Yp)

The final equation that was used to derive maximum yield potential (Yp) includes evaporative fraction corrected growth factor (CGF), respiration factor

(BF), agricultural productivity factor (APF) and production of dry matter (DMP) product.

$Y_p = CGF*BF*APF*DMP$

where Y_p is the maximum yield potential (kg ha-1).

Step 6: Retrieval of evapotranspiration (ET_p) via Land Surface Analysis –Satellite Application Facility (LSA SAF) ET_p product

The crop coefficient is defined as the ratio of crop evapotranspiration, ET_r, to reference evapotranspiration, ET_p. Kc is crop specific and ranges from zero to over unity, depending on the crop growth stage. Crop evapotranspiration at any time during the growing season is the product of reference evapotranspiration and the crop coefficient.

$ET_r = ET_p * K_c$

Crop coefficients was developed for nearly all crops by measuring crop water use with lysimeters and dividing the crop water use by reference evapotranspiration for each day during the growing season of 2009/2010 [50].

The recent development of LSA–SAF products has allowed us to obtain frequent and accurate measurements of a number of basic agrometeorological parameters (e.g. surface albedo, surface temperature, evapotranspiration). The satellite estimated agrometeorological parameters have several advantages compared to conventional measurements of agrometeorological data in ground meteorological network.

Step 7: Estimation of sugarcane productivity

The sugarcane yield estimation model over the growing season, on a biweekly basis, is accomplished by using an agrometeorological model integrated to ILWIS according to [46]:

$Ye = Yp[1 - ky(1 - ETr/ETp)]$

where Ye is the estimated yield (kg ha-1), Yp the maximum yield (kg ha-1), ky the yield response factor; ETr the actual evapotranspiration (mm) and ETp the maximum evapotranspiration (mm). Maximum yield (Yp) is established by the genetic characteristics of the crop and by the degree of crop adaptation to the environment. The resulting map of the estimated yield (Ye) is clipped to mask the Coruripe municipality boundaries in the State of Alagoas, Brazil. To establish correct coordinates, Map calculation within the ILWIS is used to implement this procedure. Flow diagram of methodology of quantifying sugarcane productivity via satellite products is shown in Figure 12.

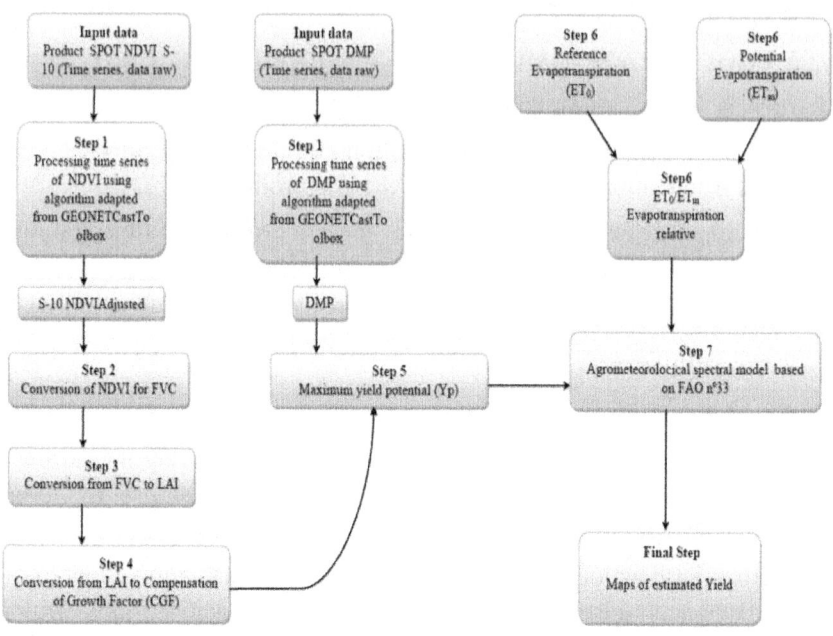

Figure 12: Flow diagram of methodology of quantifying sugarcane productivity via satellite products.

Figure 12 shows the spatial variations in sugarcane production over the Barretos and Morro Agudo municipalities for 2009/2010 and 2010/2011. The figure clearly indicates high spatial patterns in yield variability. This could be due to the mixing of significant fraction of observed pixels for the "arable pixel" and "non-arable pixel" within the municipalities. The quantified results give sugarcane yield mean range of 50 to 135 Ton ha-1.The results obtained here represents a first step towards an operational use of ILWIS tools in Brazil using NDVI S-10, DMP SPOT and ETo for operational estimating of sugarcane productivity. Overall, the model was able to identify (Figure 12) and quantify (Table 3) the spatial variability of agricultural production over the municipalities analysed. Therefore, the methodology is useful for developing estimates of operational support for the sugarcane productivity [51].

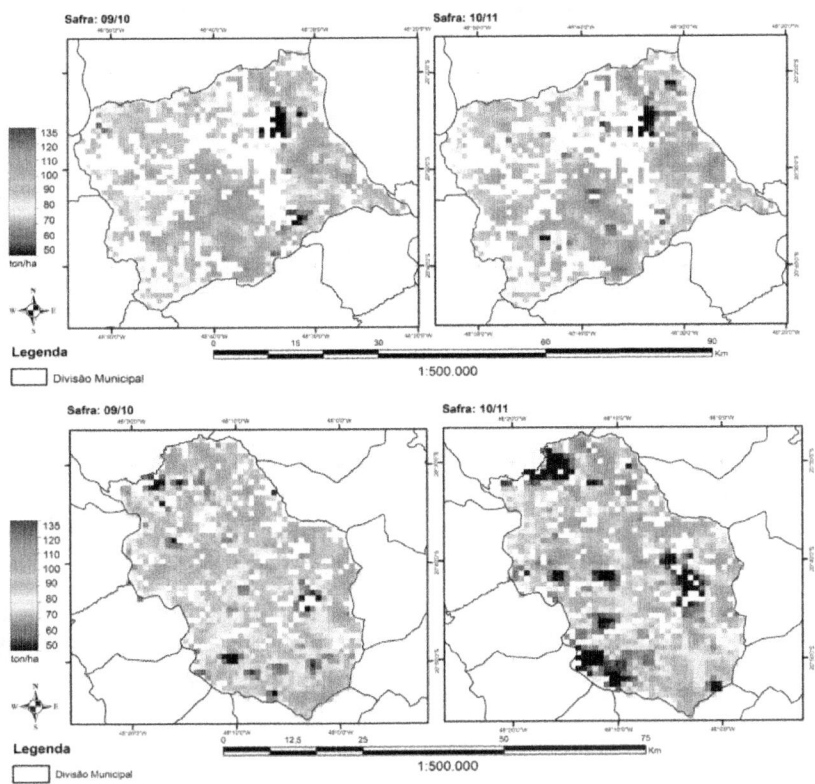

Figure 13: Spatial variability of crop yields over the Barretos and Morro Agudo municipalities for 2009/2010 and 2010/2011.

Table 3: Comparison between the productivity of sugarcane using an agrometeorological spectral model and harvested crop yield from National Food Supply Company (CONAB).

Mean Productivity (t/ha)		
Barretos	Crop	
	2009/2010	2010/2011
Agrometeorological model	93,15	93,31
CONAB	96,21	93,29
Mean Productivity (t/ha)		
Morro Agudo	Crop	
	2009/2010	2010/2011
Agrometeorological model	87,96	84,10
CONAB	93,60	87,25

TIME SERIES ANALYSIS FOR AGRICULTURE MONITORING: UGANDA

The economy of Uganda and its development goals are heavily premised on agriculture. Over 79% of the households are engaged in agriculture while 73% are directly or indirectly employed in the agricultural sector. Uganda's agriculture is however almost entirely rain-fed and very susceptible to climate risks. Studies indicate that Uganda's agricultural sector will be adversely affected by climate variability and projected climatic changes making real time monitoring of crop growth and crop productivity very important for better adaptation to climate variability and climate change. Quantitative analyzes reveal that the agricultural sector in Uganda needs to grow at an annual rate of 7% to effectively contribute to national development. Currently, the rate of agricultural growth in Uganda is below the population growth. With the threats of climate change and variability, Uganda needs to among other things harness geo-information technologies (remote sensing and geographical information systems) to improve agricultural productivity.

Remotely sensed images are powerful tools in monitoring crop productivity and yields. In most developed countries and emerging developing countries like Brazil and India, remote sensing has been greatly harnessed to plan for agriculture production, monitor crop growth and estimate yields. This is paramount in the sense that timely interventions can be taken and obviates possibilities of famine and food insecurity. Although there have been strides taken to improve the utility of remote sensing in the agriculture some developing countries, a lot remains to be done to make it more efficient, relevant and more productive. An investigation of the causative factors of the low utility and uptake of remote sensing in the agricultural sector in Uganda implicates a number of factors ranging from low capacities to expensive images. Recent developments have however extended numerous opportunities in utilizing remote sensing in the agricultural sector.

The onset of utilization of remotely sensed techniques in Uganda was in the early 1990s spearheaded by the National Biomass Project and focused largely on land use and land cover mapping. The activities of the National Biomass Project were later taken on by the National Forestry Authority (NFA) but the domains and scope remained largely the same with more focus on land use, land cover and related aspects being given priority. Apart from the NFA, academic institutions of higher learning and to some extent some research institutions like the National Agriculture Research Laboratory Kawanda (NARL) and the National Environment Management Authority (NEMA) have some remote sensing application either for teaching or research. In general, the remote sensing applications in Uganda in the agricultural field are scattered and more project based. This is partly due to the fact that there is lack of a government agency with a clear mandate to spearhead and propel the utility of remote sensing application in the country. Nevertheless, some efforts through the government cooperation with UN agencies such as the FAO regularly provide some information analyzed at the regional level for early warning in the agricultural sector. Some of the historical constraints to efficiently harnessing remote sensing in natural resource management in Uganda are generally those also experienced in other developing countries in Africa including the high costs of imagery data, processing software, coarse resolution of images, inadequate physical and human capacities and weak institutions. To-date, most of the issues to do with data costs and software have been significantly resolved with many freely available images, open source versatile software or special low for developing countries on commercial software. The contemporary challenge now is more of institutional/agency capacities, human capacities and policy environment for enhancing the utilization of remote sensing in the country.

A range of great opportunities, hitherto unavailable exist now for effectively using remote sensing in agriculture and natural resource management, notably through; (a) datasets disseminated through the Geonetcast Platform (b) freely available and downloadable datasets (c) open source softwares and low cost commercial softwares. Details of the Geonetcast is fully described in various sources (e.g. [48-50]). In brief, its a low cost facility which enables dissemination of near real time satellite imagery data. It is part of the emerging Global Earth Observation system of Systems (GEOSS), led by the Group on Earth Observation (GEO), for environmental analysis [54]. The Geonetcast does not require internet connectivity which is always a major constrain in developing countries and the data is disseminated at a very high temporal resolution through a ground recieving station, making monitoring easy. The facility streams diverse datasets which can broadly be used in environmental monitoring covering agriculture, water, soils, fire forestry etc. The data can be processed using the ILWIS software, where a specific toolbox has been developed.

In this case study, we demonstrate the utility of relatively low spatial but high temporal resolution satellite images from earth observation systems in monitoring and assessment of agricultural productivity in Uganda. There are two main inputs i.e. production data and remotely sensed data. Production data was obtained online from the FAOSTAT [55]. We extracted the annual yield, production and harvested area of the top five crops produced in Uganda according to FAOSTAT; (1) plantain/banana (2) cassava (3) sweet potatoes (4) sugarcane and (5) maize for 10 years spanning from 2001 to 2010. Bananas/ plantain (Musa spp) are largely grown in central, western and eastern (highland areas) parts of Uganda. As perennial crops, banana are year round crops. Cassava (Manihot esculenta) is an important food security crop in Uganda with the largest production coming from eastern and northern Uganda. Cassava accounts for approximately 13% of the daily caloric intake in Uganda.

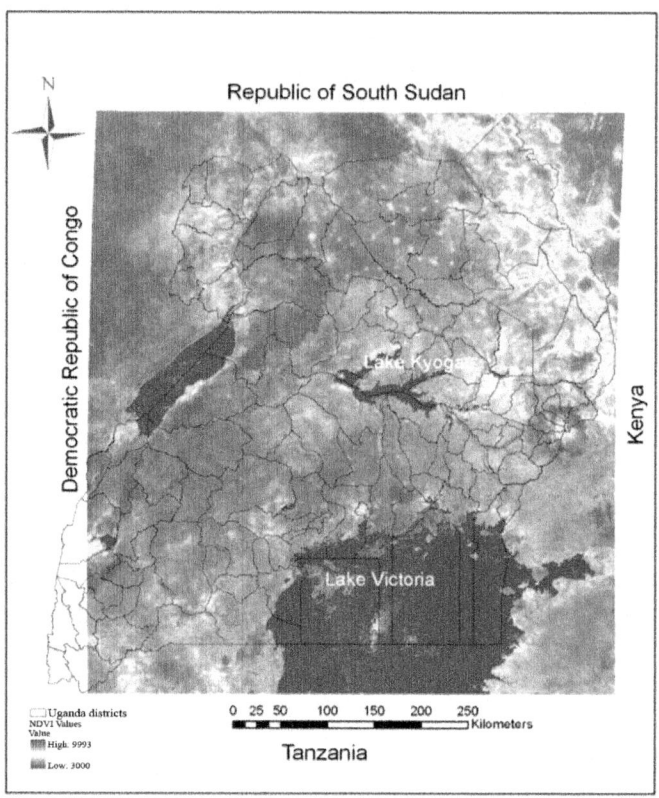

Figure 14: Scope of study

Cassava is commonly planted in the first season which is around February-march in most parts of the country and its also a perennial crop. Sweet potatoes (Ipomoea batatas) are also a food security crop in Uganda grown largely in the mid to high altitude regions of the country (1000-3000 meters above sea level). They are annual crops grown twice a year with the first season between February and June, while the second season stretches from September to November. They thrive well in the deep volcanic soils of Southwesten Uganda and Eastern Uganda. Sugarcane (Saccharum officinarum) in Uganda is largely grown on large plantations mainly in the near east and western Uganda. There are also a couple of out growers who are supported by sugar companies. It is mainly an income generating perennial crop. Maize (Zea mays L.) is grown in almost every part of the country and is a major staple food crop. It is an annual crop grown twice a year (March to June and September to November) in areas of the country where biophysical conditions are supportive. To ease the analysis, the production and yields for the five crops were compounded into one annual value.

The results on harvested area, production and yields are shown in Figures 15, 16 and 17 respectively.

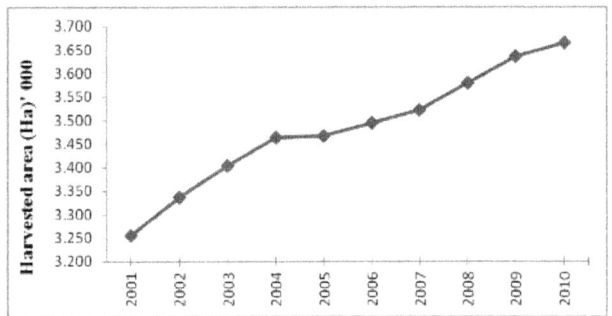

Figure 15: Harvested area of five crops between 2001 and 2010(Data obtained from FAOSTAT)

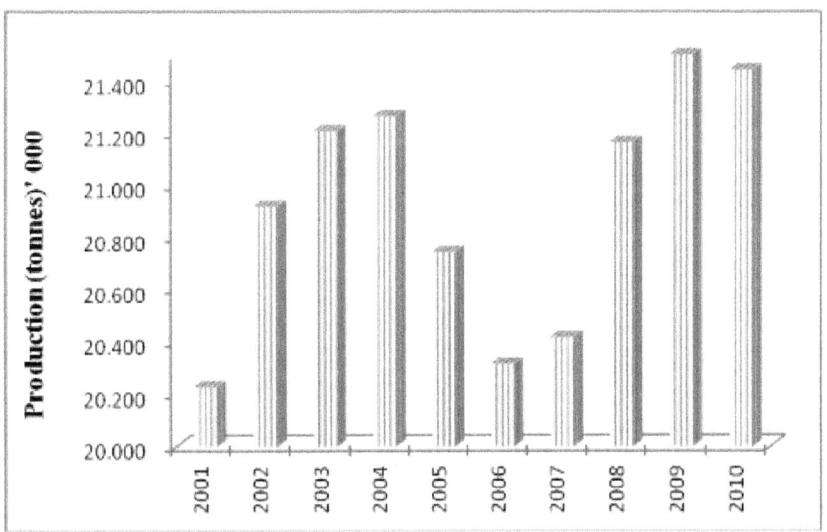

Figure 16: Production (compunded) trends of five selected crops between 2001 and 2010 compounded(Data obtained from FAOSTAT)

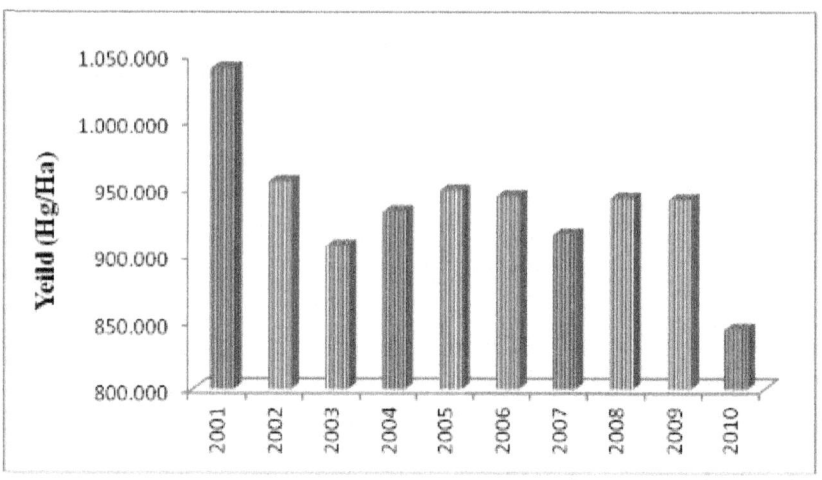

Figure 17: Yields (compounded) of five major crops between 2001 and 2010(Data obtained from FAOSTAT)

The results based on the three factors i.e. area harvested, production and yields do not depict a definitive trend. In terms of harvested area, there is an increasing trend, implying that more areas are being converted for cultivation of the specified crops (Figure 15). Production between 2001 and 2010 has modestly increased. Yields per hectare are however more variable and actually show and generally declining trend. Bearing in mind that production is increasing, it becomes explicit that the increments in production are related to extensification rather than intensification. In most cases, extensification entails conversion of ecologically sensitive and fragile areas such as wetlands or reclamation of forest area which has its environmental implications. Subjected to a statistical analysis, the results revealed a strong and positive correlation between the yields and production area ($r2=0.52$, $p<0.05$).

Remote sensing analysis was on the MODIS NDVI data, which has a spatial and temporal resolution of 250 m and 16 days respectively. For each year 23 images are available in a decal arrangement. For the 10 year period, we downloaded a total of 230 images in HDI format and processed them in ERDAS Imagine where file format conversions were undertaken and later ILWIS for arithmetic analysis. Individual images (23 decades) for each year were stacked to generate a single profile for each year. Relevant statistics such as the mean, standard deviation, coefficient of variation were late extracted. The spatial distributions of average NDVI for selected years are shown in Figure 18, while Figure 19 gives the temporal average NDVI dynamics for the 10 years. Mean average NDVI value is 0.56. In spatial terms, the southern part of the country registers higher NDVI values than the northern part. This is not surprising in light of the coverage in terms of natural cover and the crops grown which significantly entail banana and a range of annual crops grown in two seasons. The North Eastern part is particularly poorest in terms of annual average NDVI values. Understandably it is a semi arid region and generally more tailored to livestock enterprises than cropping enterprises. Annual NDVI values for the whole country were subjected to a correlation analysis with production and yield data, resulting into poor and insignificant correlations (r2=0.19 to 0.2.1). The low coefficient are partly explained by the fact that some crops like sugarcane are eother irrigated or are grown in areas almost permanently under water (wetlands). However when the data was collapsed into the growing seasons and the water bodies excluded from the analysis, better and significant correlations were obtained (r2 0.46 to 0.61, P<0.05) demonstrating the efficacy of using NDVI for crop monitoring and yield prediction. In light of the expected variability and changes in climate, coupled with the availability of data in real time, the NDVI analysis represents a great potential in sustainable adaptation where from both a policy perspective and direct intervention. This has positive implications in timely provisioning of information to farmers, adaptation to climate change and variability as well as enabling science based policy options for appropriate interventions.

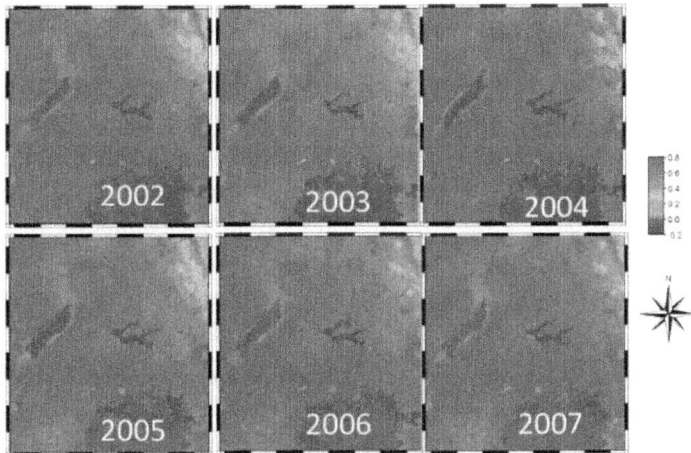

Figure 18: Average annual NDVI for selected years

Specific prediction coefficients for different crops and regionalized to the climatic conditions can be helpful to local governments where timely interventions can obviate social instability related to crop failures. On the other hand, predictions of higher yields can also enable relevant agencies to solicit for markets for the produce, improving the welfare and livelihood of the farmers, who in the context of Uganda are largely small holder farmers. All these can only be realized if there is a good policy framework that ties all the relevant pieces in the chain i.e. science, production, markets and institutions.

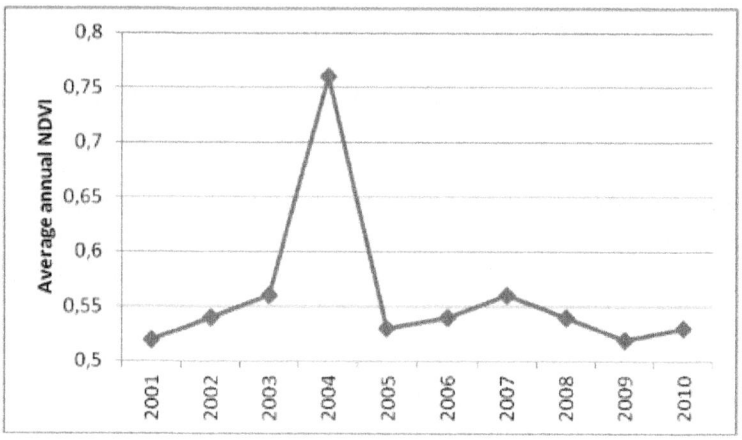

Figure 19: Variation of annual NDVI values between 2001 and 2010

OUTLOOK AND CONCLUSIONS

We have provided four multi-disciplinary case studies on the power of using remote sensing technologies, and more specifically time series analysis of low resolution satellite derived indicators, for monitoring and analysing land cover changes, desertification, drought and agricultural activities on different spatio-temporal scales. Generating this information at finer temporal resolutions is crucial for reducing risks to disaster, preparedness and formulation of strategies for better adaptation to climate change particularly the increasing dramatic hydro-meteorological events in developing and emerging countries.

This chapter provides a variety of methodologies of processing chains over satellite data, allowing the monitoring of areas subject to or in risk of desertification and land degradation processes. This chapter provides new insights related on the use of remote sensing data for climate (change) impact monitoring, which will contribute to the advance of warning systems and adaptation measures in developing and emerging countries. The focus of future activities should however focus on institutional support and capacity building for impact assessments for Africa, South America and India. The importance of training and joint cooperation with local providers and users cannot be overestimated.

One of the most robust, multi-purpose and yet simple remote sensing index is the NDVI. NDVI imagery data is widely available for immediate use at almost no cost. This has been given emphasis in this chapter through demonstration of its utility in various environmental and production domains. The section 2 of the chapter mainly emphasizes the relation between trends of vegetation greenness and rainfall over a long term period, taking into account the time lag between rainfall and vegetation response. As a result, areas of greening or degradation can be identified, and the process can be linked or not to changes in precipitation.

The section 3 of the chapter tells the mode of relation between NDVI and moisture index over different climatic regions. The relation was found to be poor over humid and dry subhumid regions where as it is improving in semi arid and arid regions. The relation of above cannot be taken as granted in the humid regions though it is implicitly understood that NDVI maintains positive relation with IM. The study infers that the NDVI and IM relations cannot be used to characterize the drought over humid regions but can be taken as an indicator in arid and semi arid regions. This is particularly relevant for adaptation purposes in semi arid regions which cover big chunks in Africa, India and some parts of Southern America.

Section 4 of the chapter mainly focuses on the estimation of sugar cane yields in Southeastern Brazil by using spatial tools which have been integrated in ILWIS 3.7.1, open source software. This study underpins that the NDVI data along with the other meteorological data is of immense use for the estimation of crop yields. This gives a business orientation on the utility of spatial tools, but also has a livelihood implication where small scale farmers or out growers are involved in sugarcane production. Interestingly, sugarcane is a major crop in all the case study countries in this chapter.

The last section of the chapter also gives more emphasis on yield estimates of five major crops in Uganda. The results of the study showed that the production between 2001 and 2010 has modestly increased with the variability in yields. Also, this analysis showed that the extensification of crops is dominated by intensification and it is implied that the increments in production are related to extensification.

In a nutshell, the chapter demonstrates how remotely sensed data available in the public domain freely or at very low cost can be harnessed to address critical challenges in developing countries pertaining to environment, agricultural productivity, drought, desertification and ultimately climate change adaptation. The chapter shows that relating the satellite derived vegetation indices with existing models and parameters can be useful proxies to understand the various phenomena of the crops. However, despite the availability of the technology, full benefits from available remotely sensed imagery resources for developing countries can only be realized when enabling policies are formulated and implemented and concerted capacity development is undertaken to establish a critical human resource base. This will enable the policy makers to go for the risk managing practices such as agricultural crop reinsurance schemes, drought defining criteria etc.

In light of the resource constraints in developing countries, cooperation and collaboration is important to develop a nucleus of future demand and contributing to new scientific insights related to projected changes in drought drawing information from satellite data, which will contribute to the improvement of warning systems and adaptation measures in developing and emerging countries.

REFERENCES

1. L Wolf, and M Williams, GEONETCast- Delivering environmental data to users worldwide (September 2007IEEE Systems Journal 2008;23401405

2. UNEPClimate change strategy, United Nations Environment Programme 2008

3. IPCCClimate Change 2007Synthesis Report- Contribution of Working Groups I, II and III to the Fourth Assessment Report of the Intergovernmental Panel on Climate Change, Geneva, Switzerland 2007.

4. G Nelson, M Rosegrant, J Koo, R Robertson, T Sulser, T Zhu, C Ringler, S Msangi, A Palazzo, M Batka, M Magalhaes, R Valmonte-santos, M Ewing, and D Lee, Climate Change: Impact on Agriculture and Cost of Adaptation," IFPRI, 200919

5. Barbosa H.A; Lakshmi Kumar2011Strengthening regional capacities for providing remote sensing decision support in drylands in the context of climate variability & change. In: Young, S. S.; Silvern, S.E. Environmental change. USA, 979-9-53307-109-0

6. C Vörösmarty, P Green, J Salisbury, and R Lammers, Global water resources: vulnerability from climate change and population growth. Science 2000289284

7. WMO2011Climate Knowledge for Action: A Global Framework for Climate Service- Empowering the Most Vulnerable. World Meteorological Organization, WMO-1065

8. P Sellers, Canopy reflectance, photosynthesis and transpiration. International Journal of Remote Sensing 19856813351372

9. Sarma A.A.L.N & Lakshmi Kumar, T.2006*Studies on crop growing*

period and NDVI in relation to water balance components, Indian Journal of Radio & Space Physics, 35, 424-434.

10. C Bhuiyan, F. N Kogan, 2008*Monsoon dynamics and vegetative drought patterns in the Luni basin under rain-shadow zone*, Internation Journal of Remote Sensing, 3132233242

11. Lakshmi Kumar T.V., Humberto A. Barbosa, K. Koteswara Rao & Emily Prabha Jothi(2012*Studies on the frequency of extreme weather events over India*, Journal of Agriculture Science & Technology, 1413431356

12. J Cihlar, St. Laurent, L., & Dyer, J.A., (1991*The relation between normalized difference vegetation index and ecological variables*, Remote sensing of Environment, 35279298

13. M. L Davenport, S. E Nicholson, 1993*On the relation between rainfall and the Normalized Difference Vegetation Index for diverse vegetation types in East Africa*, International Journal of Remote Sensing, 14, 2369.

14. H. A Barbosa, M. d. S. & T. V Mesquita, Lakshmi Kumar (2011*What do vegetation indices tell us about the dynamics of the Amazon evergreen forests?*, Geophysical Research Abstracts, 13, EGU 2011, 12894.

15. F. N Kogan, 1997*Global Drought Watch From Space.* Bulletin of the American Meteorological Society, 78, 621636

16. L. S Unganai, and F. N Kogan, 1998*Southern Africa's Recent Droughts From Space.*Advance in Space Research, 21, 3507511

17. P. S Ramesh, R Sudipa, and F Kogan, 2003*Vegetation and temperature condition indices from NOAA AVHRR data for drought monitoring over India.* International Journal of Remote Sensing, 24, 43934402

18. Z Wan, P Wang, and X Li, 2004*Using MODIS land surface temperature & normalized vegetation index for monitoring drought in the sourthern Great Plains, USA*, Int. J. Remote Sens., 256172

19. J. K Knight, R. L Lunneta, J Ediriwickerma, S Khorram, 2006*Regional scale land cover characterization using MODIS- NDVI. 250 m multi-temporal imagery: A phenology based approach.* GIScience and Remote Sensing (Special issue on Multi- Temporal Imagery Analysis), 43(1), 1-23

20. Chris Funk & Michale E. Budde (2009*Phenologically- tuned MODIS NDVI based production anomaly estimates for Zimbabwe.* Remote Sensing Environment. 113115125

21. F Lupo, I Reginster, and E Lambin, Monitoring land-cover changes in West Africa with SPOT Vegetation: impact of natural disasters in 1998-1999. International Journal of Remote Sensing 2001221326332639

22. W Immerzeel, R Quiroz, and S De Jong, Understanding precipitation patterns and land use interaction in Tibet using harmonic analysis of SPOT VGT-S10 NDVI time series. International Journal of Remote Sensing 2005261122812296

23. J Xiao, and A Moody, Geographical distribution of global greening trends and their climatic correlates: 1982-1998. International Journal of Remote Sensing 2005261123712390

24. A Anyamba, and C Tucker, Analysis of Sahelian vegetation dynamics using NOAA-AVHRR NDVI data from 1981-2003. Journal of Arid Environments 2005633596614

25. U Helldén, and C Tottrup, Regional desertification: a global synthesis. Global and Planetary Change 200864169

26. Z Bai, D Dent, L Olsson, and M Schaepman, Global Assessment of Land Degradation and Improvement- 1. Identification by Remote Sensing, Report 2008/01, ISRIC- World Soil Information, Wageningen 2008

27. E Lepers, E Lambin, A Janetos, R De Fries, F Achard, N Ramankutty, and R Scholes, A synthesis of information on rapid land-cover change for the period 1981-2000. BioScience 2005552115124

28. G Baldi, M Nosetto, R Aragón, F Aversa, J Paruelo, and E Jobbágy, Long-term satellite NDVI data sets: evaluating their ability to detect ecosystem functional changes in South America. Sensors 200885397

29. F Achard, H Eva, H-J Stibig, P Mayaux, J Gallego, T Richards, and J-P Malingreau, Determination of deforestation rates of the world's humid tropical forests. Science 2002297999

30. UNCCDDesertification: a visual synthesis, United Nations Convention to Combat Desertification (UNCCD) Secretariat, Bonn 2011

31. C Tote, K Beringhs, E Swinnen, and G Govers, Monitoring environmental change in the Andes based on SPOT-VGT and NOAA-AVHRR time series analysis. In 6th International workshop on the Analysis of Multi-Temporal Remote Sensing Images, 2011

32. A Klisch, A Royer, C Lazar, B Baruth, and G Genovese, Extraction of phenological parameters from temporally smoothed vegetation indices. 2006

33. Z Bai, and D Dent, Global assessment of land degradation and improvement: pilot study in Kenya, Report 2006/01, ISRIC- World Soil Information, Wageningen 2006

34. A Peters, E Walter-shea, L Ji, A Vina, M Hayes, and M Svoboda, Drought monitoring with NDVI-based standardized vegetation index.

Photogrammetric Engineering and Remote Sensing 20026817175

35. N Guttman, Accepting the standardized precipitation index: a calculation algorithm. Journal of the American Water Resources Association 1999352311322

36. A Lotsch, M Friedl, B Anderson, and C Tucker, Coupled vegetation-precipitation variability observed from satellite and climate records. Geophysical research Letters 2003

37. Z Bai, and D Dent, Land Degradation and Improvement in Argentina-1. Identification by remote sensing, ISRIC- World Soil Information, Wageningen 2007Thoronthwaite and Mather, 1955)

38. Lakshmi Kumar T.V., Uma, R, Aruna, K & Emily Prabha Jothi(2011On the relation of NDVI with rainfall in Western Ghats, India, RAISE Conference Proceedings, 4245Dec 2011.

39. Lakshmi Kumar T.V., Uma, R, Koteswara Rao K & Humberto Barbosa(2012Variability of NDVI in relation to southwest monsoon over Western Ghats, India, International SWAT Conference, July 2012.

40. Liu Xiaomang Dan Zhang., YuzhouLuo&Changming Liu(2012Spatial and temporal changes in aridity index in northwest China: 1960 to 2010, Theoretical and Applied Climatology, DOIs00704-012-0734-7, Aug 2012.

41. www.indiawaterportal.org

42. Thoronthwate C W & Mather J.R(1955The water balance model, Publ in Clim, Drexel Institute, 8,1, 104

43. C W Thoronthwaite, 1948An approach toward a rational classification of climate, Geographical Review, 38(1), 55-94.

44. J. R. F Gouvea, 2008Mudancas Climaticas e a expextiva de seus impactos da cultura de cana- de acucar na regiao Piracicaba- *SP*, Dissertacao (Mestrado em Fisica do Ambiente Agricola)- Escola Superior de Agricultura " Luiz de Querioz", Universidade de Sao Paulo, Piracciaba, 99.

45. B. F. T Rudorff, 1985Dados Landsat na estimativa da produtividade agrícola da cana-de- açúcar.114, Dissertação (Mestrado em Sensoriamento Remoto)- Instituto Nacional de Pesquisas Espaciais- INPE, São José dos Campos.

46. J Doorenbos, A. H Kassam, 1979Yield response to water. Roma: Food and Agriculture Organization of the United Nations, 193.

47. J. C. J Munoz, J. A Sobrino, L Guanter, J Moreno, A Plaza, P Matinez, 2005Fractional Vegetation Cover Estimation from Proba/CHRIS

Data:Methods, Analysis of Angular Effects and Application to the Land Surface Emissivity Retrieval. 3rd ESA CHRIS/PROBA Workshop, ESRIN, ESA SP-593, Frascati, Italy.

48. J. M Norman, M. C Anderson, W. P Kustas, A. N French, J Meicikalski, R Torn, G. R Daik, T. J Achmugge, Remote Sensing of Evapotranspiration for Precision-Farming Applications. In: IEEE International Geoscience and Remote Sensing Symposium, 2003Tolouse. Proceedings: IGARSS 2003OrganizingCommittee, 21-25.

49. L. M. S Berka, B. F. T Rudorff, Y. E Shimabukuro, Soybean yield estimation by an agrometeorological model in a GIS. Scientia Agricola, 60n. 3; 4334402003

50. Toledo Filho M. R.1988Probabilidade de suprimento da demanda hídrica ideal da cultura de cana- de-açúcar (Saccharum spp.) através da precipitação pluvial na zona canavieira do estado de Alagoas. 1988.72f. Dissertação (Mestrado)- Progama de Pós-graduação em Agronomia, Escola Superior de Agricultura "Luiz de Queiroz", Universidade de São Paulo, Piracicaba.

51. Barbosa H. A; Rocha, D; Lakshmi Kumar, T.V., Bamutaze, Y.Quantification of sugarcane crop productivity using agrometeorological-spectral model: A study case in Northeastern Brazil. Revista AGROLLANIA, 1690-80669Enero-Diciembre, 62-75, 2012

52. http://www.earthobservations.org/geonetcast.shtml,

53. http://www.eumetsat.int/Home/Main/DataAccess/EUMETCast/ SP_201005191146224675?l=en,

54. http://www.itc.nl/Pub/WRS/WRS-GEONETCast/GEONETCast-toolbox.html

55. http://faostat.fao.org/

Chapter 5

CLIMATE CHANGE AND FOOD SECURITY

Christopher Kipkoech Saina[1], Daniel Kipkosgei Murgor[1] and Florence A.C Murgor[1]

[1] Chepkoilel University College, School of Environmental Studies, Department of Applied Environmental Social Sciences, Eldoret, Kenya

INTRODUCTION

Climate change is possibly the most significant environmental challenge of our time and poses serious threats to sustainable development in the world and more so in most developing nations. Impact of climate change affects ecosystems, water resources, food and health. As such inter-related government policies must be designed to avoid conflicts in policy design and implementation. There is a direct link between climatic changes and global food insecurity more so in developing countries where climate change compounded with poverty has exacerbated the impacts. In order to address the challenges posed by climate change, it is necessary to examine the factors contributing to climate change and how such influence food production globally. Climatic factors like precipitation, evaporation, humidity and sunshine duration form the basis for improvement of food security. There is need for policy makers, communities and aid providers to incorporate evidence based technologies in food systems and knowledge. Evidence based technologies are those that have empirically been tested and used. They include zero tillage, integrated soil fertility management; irrigation technologies for example drip irrigation, seed improvement, water harvesting, organic agriculture and incorporation of indigenous knowledge. The impact of some of the technologies can be seen in the light of global improved grain yield through use of integrated soil fertility management, rain-fed and irrigated environment technologies. Drought-tolerant grain crops are also likely to help increase yields.

The results of this study are pertinent to policy makers in the field of food security and livelihood sustainability. Mitigation and adaptation measures must be effective, affordable and appropriate for environmental sustainability and development. This review advocates for the integration of conventional agro-science based systems with traditional agricultural knowledge in order to mitigate the severity of climate change and its impact on food security and livelihoods sustainability. Integration of agro-science and traditional agricultural systems is important if food security is to be sustained.

GLOBAL CLIMATE CHANGE

The expression of the term "climate change" according to many people means the alteration of the world's climate as a result of human activities through fossil fuel burning, clearing forests and other practices that increase the concentration of greenhouse gases (GHG) in the atmosphere. This is in line with the official definition by the United Nations Framework Convention on Climate Change that states that climate change is the change that can be attributed "directly or indirectly to human activity that alters the composition of the global atmosphere and which is in addition to natural climate variability observed over comparable time periods" [1]. The Intergovernmental Panel on Climate Change (IPCC) defines "climate change" as "a change in the state of the climate that can be identified by changes in the mean and or the variability of its properties, and that persists for an extended period, typically decades or longer" [2]. Climate change can be defined as a systematic change in the key dimensions of climate including average temperature and wind and rainfall patterns over a longer period of time. In recent usage, especially in the context of environmental policy, climate change usually refers to changes in modern climate. It may be qualified as anthropogenic climate change, more generally known as "global warming" or "anthropogenic global warming" (AGW).

IMPACT OF CLIMATE CHANGE

Due to prevailing nature of enhanced greenhouse effect in the atmosphere, the following effects have occurred at the global, regional and national levels.

Based on [3]; [4], there has been evidence of increase in global temperatures that has led to climate change at global, regional and national levels over the past 100 years. Increase in global temperatures experienced over the past century is as a result of accumulation of greenhouse gases in the atmosphere leading to global warming. Using complex climate models, the "Intergovernmental Panel on Climate Change" in their third assessment report has forecast that global mean surface temperature will rise by 1.40C to 5.80C by the end of 2100. Multiple datasets show essentially the same global warming trend over the past 100 years, with the steepest increase in warming in recent decades. The evidence of human-induced climate change goes beyond observed increases in average surface temperatures; it includes melting ice in the Arctic, melting glaciers around the world, increasing ocean temperatures, rising sea levels, acidification of the oceans due to excess carbon dioxide, changing precipitation patterns, and changing patterns of ecosystem and wildlife functions. Reduced agricultural productivity with the resultant food shortages has been experienced. Studies have shown that with higher concentrations of CO_2, plants can grow bigger and faster. However, the effect of global warming may affect the atmospheric general circulation and thus altering the global precipitation pattern as well as changing the soil moisture contents over various continents.

There has been an increase in sea level observed in some parts of the world due to excess heating of air - which has caused large scale melting of ice covers. Large scale flooding of California in 1999 and parts of western coast in India in the last 5-8 years are testimonies to effects of sea level rise. If the sea level rises by 80-90 cm, perhaps many of coastal cities of the world will be washed away besides great changes in harbours and their facilities, in sea routes and in fishery industry [3]. Loss of fertile agricultural land occasioned by flooding impacts on food security and livelihoods at household and national level. There has been an increase in drought and floods globally. Ironically, changes in the climate due to excess greenhouse gases are causing both increased drought and increased flooding. Violent storm activity increase as temperatures rise and more water evaporates from the oceans. This includes occurrence of more powerful hurricanes, pacific typhoons, and an increased frequency of severe localized storms and tornadoes. These storms often result in flooding and farmland damage hence causing food insecurity. Warming also causes faster evaporation on land leading to drought induced famine.

1. Change/shifts in seasons and seasonal characters have been experienced through out the globe due to change in air temperature and rainfall patterns. Some seasons have either been shortened or prolonged. Winters have extended in many places, while summer is more severe in other places. The degree of dependability has reduced and an element of uncertainty has increased. This disorientates the farmers in the rural community who have hitherto depended on indigenous knowledge in predicting weather patterns in food production.

2. Major changes have occurred in water resources of the world due to disturbances in hydrological cycles. Heavy rainfall tracts are gradually converted into low rainfall tracts with many humid areas being transformed into arid areas. Similarly, ground water depletion is high and recharging is very low.

3. There has been a shift in disease/pest cycles of plants and animals. Many insignificant pests / diseases are attaining major proportions because composition of microbial population is affected by shift in temperature

and hydrological cycles. These have impacted on food production output and post harvest loss occasioning food shortages and loss of livelihoods.

4.   Ecosystems change. Changes in climate will cause some species to shift from one region to another and in combination with other stresses such as development, habitat fragmentation, invasive species, could have negative consequences on biodiversity and the benefits that healthy ecosystems provide to humans and the environment. Water hyacinth, an invasive species in Lake Victoria has tremendously reduced fishing activities impacting on livelihoods.

Figure 1: Depicts typical ravages of climate change in Kerio Valley, Kenya May 2012

HOW CLIMATE CHANGE WILL AFFECT LIVELIHOODS

The economy and environment could be affected as a result of climate change especially in the absence of countermeasures. The following are some of the livelihood sectors that are likely to face some impacts according to [5] and other literature:-

The health sector impacts will affect the populations by altering the health status of millions of people, including through increased deaths, disease and injury due to heat waves, floods, storms, fires and droughts. Increased malnutrition, environmentally related diseases such as cholera, dysentery, meningitis, lymphatic filariasis, yellow fever, malaria, TB among others in some areas will exert great pressure on the public health resources and development goals will be threatened by longer-term damage to health systems from disasters.

In the water sector, it is predicted that climate change will lead to an intensification of the global hydrological cycle hence having an impact on water resources. A change in both volume and distribution of water will affect both ground and surface water supply for industrial and domestic uses, irrigation, hydropower generation, navigation, in stream ecosystem and water based recreation. Drought-affected areas will likely become more widely distributed.

In the event of drought in Arid and Semi arid environments sensitive to slight changes in climate, there is bound to be human – human and human - wildlife conflict with respect to use of the scarce resources. Heavier precipitation events are very likely to increase in frequency leading to higher flood risks. By mid-century, water availability will likely decrease in mid-latitudes, in the dry tropics and in other regions supplied by melt water from mountain ranges. More than one sixth of the world's population is currently dependent on melt water from mountain ranges.

Figure 2: Fragile ecosystems are destroyed during heavy rainfall; a consequence of climate variability. Kerio Valley, Kenya, May 2012.

Biodiversity and biomass energy will be affected by climate change by affecting both plants and animals distribution, population sizes, physical structure, metabolism and behaviour. Climate induced changes will largely influence the distribution of tree parasites and pathogens which will ultimately play an important role in determining future tree distribution. Climate change alters conditions in ecosystem making species unable to cope with sudden changes. There has been direct and indirect impact on the forest ecosystem. The direct impacts result in water stress in plants due to prolonged dry spell. This in turn increases fire hazards in forest areas. The biomass energy sub sector is sensitive to climate variability as productivity is a function of rainfall and temperature. Increased temperatures leads to drying of biomass hence declining biomass productivity like decreased fuel wood supply. Globally, most households rely on wood fuel for the cooking and heating especially in sub-Saharan Africa thus endangering livelihoods and environmental sustainability.

Agriculture/Food sector will see those mid-latitude and high-latitude areas initially benefit from higher agricultural production and many others at lower latitudes, especially in seasonally dry and tropical regions. The increases in temperature and the frequency of droughts and floods are likely to affect crop production negatively, which could increase the number of people at risk from hunger and increased levels of displacement and migration in search of livelihoods.

Human settlement and business are the most vulnerable especially those located in coastal areas and river flood plains, and those whose economies are closely linked with climate-sensitive resources. This applies particularly to locations already prone to extreme weather events, and especially areas undergoing rapid urbanization. Where extreme weather events become more intense or more frequent, the economic and social costs of those events will increase. Cities loom as giant potential flood or other disaster traps.

Livestock sector is affected by climate changes by interfering with the distribution, production size and frequency of disease and pests. Livestock industry depends on the balance of pasturage and water resources supply and any form of change in climate will impact negatively the livestock sector. The impacts have largely been felt by pastoralist communities who have been forced to change their livelihood from pastoralists to sedentary agro pastoralist to survive.

Climate causes instability in the tourism sector for those countries whose economies are largely dependent on tourism. Addition or reduction in precipitation leads to wildlife devastation and reduces the aesthetic value of sceneries hence impacting on livelihoods dependent on the industry.

Gender is similarly impacted by climate change. The gender poverty link show that 70% of the poor in the world are women and their vulnerability is accentuated by race, ethnicity and age. Most climate change policies, issues and programs are not quite gender neutral despite the fact that women and men are affected differently due to tradition, socially constructed roles and responsibilities.

CLIMATE CHANGE MITIGATION AND ADAPTATION MEASURES

Mitigation is "an anthropogenic intervention to reduce the anthropogenic forcing of the climate system; it includes strategies to reduce greenhouse gas sources and emissions and enhancing greenhouse gas sinks [1]. To ensure that environmental sustainability is maintained some urgent measures need to be put in place to help in sustainability of the environment by way of mitigation. Some examples of mitigation actions include developing new low-energy technologies for industry and transport, reducing consumption of energy-intensive products and switching to renewable forms of energy such as solar and wind power. Natural carbon sinks, such as forests, vegetation and soils, can be managed to absorb carbon dioxide, and technologies are being developed to capture carbon dioxide at industrial sources and to inject it into permanent storage deep underground. There is need also to manage the impact as have occurred on the environment. It is true that future impacts on the environment and society are now inevitable, owing to the amount of greenhouse gases already in the atmosphere from past decades of industrial and other human activity and to the added amounts from continued emissions over the next few decades until such time as mitigation policies become effective. Taking steps to cope with the changed climate conditions is called "adaptation".

Adaptation is "the adjustment in natural or human systems in response to actual or expected climatic stimuli or their effects, which moderates harm or exploits beneficial opportunities" [6]. The measures to be undertaken are those that will try to address the adverse impacts that may occur as a result of non action by humans with the aim of attaining a sustainable environment. Some of these measures may include some of the following; conducting risk assessments, protecting ecosystems, improving agricultural methods, managing water resources, building settlements in safe zones, developing early warning systems, instituting better building designs, improving insurance coverage, developing social safety nets and enhancing public awareness and education. All these measures are intrinsically linked to sustainable development as they reduce the risk to lives and livelihoods and increase the resilience of communities to all hazards. Both, adaptation and mitigation should be considered jointly, as some adaptation measures can contribute to reducing greenhouse gas emissions, while conversely mitigation measures can be planned to help reduce the impacts.

GLOBAL FOOD SECURITY

Food security exists when all people at all times have physical or economic access to sufficient safe and nutritious food to meet their dietary needs [7]. Global food security refers to the situation where each person, member of any household is having physical and economic access to sufficient, safe and nutritious food to meet their dietary needs and food preferences for an active and healthy life – (Food and Agriculture Organization definition). That food has to be affordable, safe and healthy, culturally acceptable, meeting specific dietary needs of the people and is obtained in a dignified manner, produced in ways that are environmentally sound and socially just. Thus there should be no perceived inadequacy of the household food supply expressed through concerns about running out of food and not having enough food to make a meal or malnourished as a result of physical unavailability of food, lack of socio-economic access to food or inadequate food utilization.

The World's efforts to meet the Millennium Development Goal of cutting hunger in half by 2015 are far from reach. With the world's population set to reach 9 billion by 2050, agricultural production will need to increase by 70% in order to meet demand. Climate change adds a new dimension of this challenge as it is one of the key drivers of change affecting the food system and contributing to rising food prices. It leads to changes in growing seasons and rainfall patterns and the increased frequency of extreme events such as droughts and floods. It has been estimated by the United Nations Environment Programme that up to 25 per cent of world food production could be lost by 2050 as a result of climate change, water scarcity and land degradation [8].

In the developing world, nearly 1 billion people are unable to meet their dietary needs. Another 5-10% is at risk of 'acute' food insecurity in times of crisis. Despite improvements, the millennium development goal on hunger is likely to be missed by a wide margin in areas like Sub-Saharan Africa, where persistent food insecurity is compounded by widespread political instability, conflict and the HIV/AIDS pandemic. The UN Food and Agriculture Organization [9] estimate that 820 million people in developing countries are suffering from malnutrition. In a world in which nearly half the population survives on $2/day or less, more than 800 million people go to sleep hungry any given night, and a child dies every five seconds due to hunger-related complications, the need to respond to the needs of the poor for food is ever-present and widespread. Despite the universal recognition of every person's right to food, vulnerability to hunger remains a daily reality for many [10].

The United States of America and European Union together provide about two thirds of global food aid deliveries. The United States of America is by far the most important donor of food aid both for bilateral programme aid and as the main contributor to the World Food Program. The World Food Program (WFP) is the primary agency responsible for administering multilateral food aid. The WFP and various NGOs administer project food aid to support a wide range of developmental projects targeting the poor in developing countries. Others include the UN Food and Agriculture Organization (FAO), the Food Aid Convention (FAC) and the World Trade Organization (WTO). All these organizations have different mandates and are concerned with different aspects of the provision of food aid. Many organizations around the world are working to find ways to produce the food needed in a sustainable way, within the limits of what our ecosystems can support for current and future generations, and to safeguard this production from the impacts of climate change. Food security is the outcome of food system processes all along the food chain. Climate change will affect food security through its impacts on all components of global, national and local food systems. It is important to note that climate change variables influence biophysical factors, such as plant and animal growth, water cycles, biodiversity and nutrient cycling, and the ways in which these are managed. Agriculture is the primary source of livelihood among many rural populations [11].

CAUSES OF GLOBAL FOOD INSECURITY

There are many causes of global food insecurity and most of them are region specific, except climate change where impacts are felt globally. Some are human causes like destruction of fertile lands and others are non-human like natural disasters (e.g. floods). While human destruction of his mother nature through careless management practices have impacted negatively on his ability to produce enough food for the population, the contribution that humans make to climate change is potentially causing increased food insecurity globally. Industrialization, modern agricultural practices and the need to produce for the market without thinking about environmental sustainability has had disastrous effect as climate change has set in. In some areas, sea levels have risen and increased incidence of extreme events posing danger and threatening livelihoods and at the same time increasing the vulnerability to future food insecurity in the world. Coastal freshwater are being contaminated with salt water and people do not get fresh water for domestic use and plants or food crops that are not salt-tolerant cannot be grown. Storm surges become common occurrences together with flooding. This hinders people living along coast from growing food crops. Increased temperature also leads to heat stress for plants which increase evaporation and lowers productivity. Climate change has also been a cause of more frequent and more intense extreme weather events like increasing irregularities in seasonal rainfall patterns. The changing growing seasons have shifted ecological niches.

Rainfall is becoming more unpredictable and unreliable and has become a common occurrence leading to greater uncertainty among farmers and their traditional agricultural knowledge and coping strategies. The phenomenon has had immediate impact on food production, distribution, infrastructure, livelihoods and human health in all parts of the world. Where climate change has influenced rainfall and temperature patterns, the suitability of land for different types of crops and pasture is affected, including the health and productivity of forests; the distribution, productivity and community composition of marine resources and the increased incidence of pests and diseases. It also affects the functioning of biodiversity and ecosystem of natural habitats; and the availability of good-quality water for crops, livestock and inland fish production. It may also increase aridity of arable lands, induce drought and deforestation, can increase fire danger with consequent loss of the vegetative cover needed for grazing and fuel wood [12]. It also leads to depletion groundwater and induce the internal and international migration thereby triggering resource-based conflicts and civil unrest in either areas of origin or destination. Conflicts over water resources will have implications for both food production and people's access to food in conflict zones [13]. Production from both rain-fed and irrigated agriculture in dry land ecosystems accounts for approximately 25 percent, and rice produced in coastal ecosystems for about 12 percent [14].

Natural fall-back mechanisms during food crisis may also be affected. These include disappearance of traditional fruits, herbs, vegetables, mushrooms, wild foods and other coping mechanisms. It is expected that as the world climate continues to change, 5 000 plant species in sub-Saharan Africa will decrease in size or shift to other agro-ecological zones due to climate change (Levin and Pershing, 2005). All these impacts have negative influence on food security of the people living in the affected areas. At the same time people may resort to unfamiliar ways of surviving and countering food insecurity. It was observed that people may decide not to migrate but find new, unfamiliar ways of earning a living [15]. FAO has been instrumental in assisting various communities all over the World develop the right food production technologies aimed at boosting food security. They possesses technical expertise relevant to climate change adaptation in a variety of ecosystems, including agro-ecosystems (crops, livestock, grasslands), forests and woodlands, inland waters, and coastal and marine ecosystems. It works to build national, local and community-level capacities to raise awareness of and prepare for climate change impacts, assists member countries in identifying potential adaptation options and helps local people understand which are the most applicable to their particular circumstances.

Since 2002, FAO has been promoting National and Regional Programmes for Food Security (NPFS and RPFS) as instruments that help countries enhance productivity and diversify the livelihoods of rural people on a scale sufficient to achieve the 2015 targets set by WFS and the Millennium Development Goals (MDGs). This is an organization committed to eliminating food insecurity in the World. From their advice, people have created artificial microclimates, breed plants and animals with desired characteristics, enhance soil quality, and control the flow of water. Advances in storage, preservation and transport technologies have made food processing and packaging a new area of economic activity. This has allowed food distributors and retailers to develop long-distance marketing chains that move produce and packaged food. The countries that are not able to adopt new technologies need to be assisted. Developing countries urgently need more assistance to help them prepare for the impacts of climate change that are unavoidable. The transfer of the industrialised countries' best energy efficiency and renewable energy technology and assistance with disaster preparedness, agricultural productivity improvements, water management, conflict prevention, reforestation, preventing deforestation and critical infrastructure would be most appropriate.

Other important practices for addressing food insecurity include changing consumption patterns and food preparation practices, efficient water use, improving soil quality, capacity to withstand extreme events and carbon sequestration. Others include promotion of agro biodiversity for local adaptation and resilience, reducing uncertainty by improving the information base and devising innovative schemes for insuring against climate change hazards.

People need to adopt practices that enable the vulnerable to protect their existing livelihood systems, diversify their sources of income, change their livelihood strategies or migrate. Sustainable livestock management practices for adaptation and associated mitigation should also be given high priority.

INTEGRATING INDIGENOUS KNOWLEDGE AND SCIENCE BASED TECHNOLOGIES

Integration of both indigenous knowledge and science based technologies is important in combating the effects of climate change on food security. Farmers need to combine the best of their traditional approaches with modern agroscience based technology. Towards this end documenting traditional knowledge of the world is paramount if an integrated approach is to be effective as most traditional knowledge remains as tacit knowledge.

INDIGENOUS FOOD SYSTEMS

Communities have created and developed unique food systems over time in their specific local ecosystems. Communities have used these technologies for a long period of time and have perfected them to improve their livelihoods. Traditional knowledge in the management of agriculture and food production is important. Indigenous people in Sub-Sahara Africa have developed methods of surviving droughts and emergency. However, aid providers have ignored this vital knowledge inherent in the local people during their programme intervention. According to [16], one of the most important methods in combating desertification is the recognition of the value of traditional knowledge in drought management.

Severe changes in global climate are projected to affect livelihoods systems and consequently food security. Droughts and floods have become frequent, occasioning loss of livelihoods migration and insecurity. Compounding the vagaries of climate change is the upward trends of population increase. These are often marginalized, poverty stricken persons. Poverty poses a serious environmental threat as people exploit natural resources with inappropriate technology in order to survive. Population increase occasions agricultural practice in marginal lands. This causes resource degradation and environmental degradation leading to poor harvest and food insecurity. Indigenous food system knowledge is required to reduce the effects of soil erosion leading to poor food production. Family and tribal structures and their autonomous traditional practices of resource management and land tenure have broken down. Therefore, traditional land tenure systems and agricultural practice of improved shifting cultivation should be incorporated into policy. This is because community elders control parcels of land to be used for food production and grazing. Grass stripes are used as a form of land management in Swaziland and grazing rotation by Lake Victoria communities thus, soil erosion and biodiversity loss is checked [17].

Drought influences availability of water for crop production. This is projected to be a major constrain to food security, and economic development in the future. In Africa climate change is expected to intensify the continents critical food situation. Reduction of water quality and availability will increase food insecurity [18]. Governments must therefore fund research on crops that require less rain, are faster in maturity and pest resistant. These are the almost extinct traditional indigenous food crops. This is corroborated by the UN/ISDR 2007. Governments' agricultural policies must allow for diversification of food crops to cushion populations against loss of livelihoods in the face of climate variability. Increased production in traditional root crops and food legumes and lentils for sustainable agriculture and food security must be incorporated into agricultural policies. Production of food grain or root crops must be increased to decrease dependency on food export in Africa. Similarly, value chain actors that include suppliers, farmers, traders and processors must be strengthened in the sector of food production. This calls for the cultivation of more indigenous food crops to reduce aid dependency. Incorporating modern food production methods with indigenous food systems is ideal as people are better able to adopt new ideas when these can be seen in the context of existing practices.

Indigenous systems of crop protection against pests must be integrated into agricultural policy. This is cheap and more available system to the rural poor farmer. The wide spread use of indigenous material, such as agrochemical plants to combat pests that normally attack food crops has been reported. It is likely that climate change will alter the ecology of disease vectors and as such indigenous practices of pest management would be useful for adaptation strategies. Other indigenous systems that are adopted by local farmers include controlled bush clearing. Smoking of seeds to deter stock borers, use of green manure and green mulch spray as herbicide, all improve food production and reduces the vagaries of food insecurity. Pastoralists in Arid and Semi Arid Lands (ASAL) in Africa use a multi-species composition of herds to survive climate extremes. Traditionally they forego large grazers for small browsers like goats and sheep since the feeding pattern of the latter is lower [19].

Men and women play different roles at the household and community levels, climate change affects women disproportionately than men. Women interact with nature more but have limited mitigating and adaptive capacities. Climate change and variability are therefore likely to amplify existing patterns of disadvantage. Women in Africa are custodians of culture and customs. This vast knowledge must be used to mitigate climate change and food insecurity. Women workload in rural Africa is always based on the maintenance of household food security particularly in hardship seasons [20]. It is through this role that while modernizing policies in food production, we must adjust male bias to avoid falling into the trap of food insecurity. Women must have access to land titles, inputs and credits and must be deliberately involved in agricultural extension [21]. Inclusive policy will design and implement programmes that lead to gender equity and food security.

Figure 3: The photos depict impact of climate change and livelihood coping strategies. The photographs were taken the arid areas of Kerio Valley, Kenya May 2012.

Poverty driven environmental degradation has been exacerbated by the erosion of tradition knowledge by westernization. Traditionally, cultural norms and practices and taboos were used to regulate and ensure sustainable exploitation. Poverty drives communities to farm in marginal fragile ecosystems, using rudimentary technology leading to environmental degradation, poor yields and hence food insecurity. It is therefore considerable to conclude that the spiral events of poverty, poor technology, lack of inputs and land tenure culminate in unsustainable underdevelopment. Governments must therefore invest in their populations to eradicate poverty, thus providing a springboard to address food security for sustainable development. A healthy citizen is in a position to adapt new technology and address environmental challenges of climate change.

THE ROLE OF TECHNOLOGY IN FOOD SECURITY

With the world population expected to reach over 9 billion by 2050, the global demand for food is projected to increase by at least 2.5 times over current levels [22]. The challenges of feeding over 9 billion people by 2050 in a sustainable and environmentally friendly way cannot be met without the continued innovation and adoption of new technologies. To achieve notable increases in agricultural productivity, technology and innovations must be applied to the entire agricultural supply chain. From notable advances in biotechnology that can make more efficient use of water and fertilizers and reduce pesticides, to bio-fortification, improved crop varieties and best practices to reduce post-harvest losses and improve irrigation methods, a science-based approach to new and existing technologies must be applied to maximize their potential benefits worldwide. These technologies, along with traditional breeding approaches, are all essential to meeting the demands being placed on agricultural productivity [22].

Throughout the last century, the potential increase in agricultural productivity has been achieved by adopting and applying innovative agricultural technologies worldwide. These increases have not only bolstered food security, but have helped minimize the environmental impacts of agriculture. This is critical, because loss of biodiversity and habitat in turn lead to desertification, loss of fresh water sources, and greater food insecurity. Major innovations in mechanization, the use of fertilizers and pesticides, and plant and animal management and breeding techniques provided the basis for the fivefold increase in US agricultural output over the twentieth century. This increase in output was achieved with less land and labour, and in recent years with less energy and chemical use per unit of output [23].

In many parts of the world – particularly South and East Asia – growth in agricultural productivity has been rapid, largely as a result of the extensive adoption of new agricultural technologies. For millions of poor people, particularly in Asia, the technological advances of the Green Revolution (complemented by a massive increase in irrigation) provided a route out of poverty through: directly increasing producer incomes and wages;

lowering the price of food; and generating new livelihood opportunities as success in agriculture provided the basis for economic diversification. Asian industrialization was in essence agriculturally led [24]. Despite decades of investment in new agricultural technology however, hunger and poverty continue to plague large areas of the developing world. The problem is particularly acute in areas of the world dependent upon rain-fed agriculture, in particular sub-Saharan Africa, where the impact of new technologies has been less apparent and agricultural productivity has at best stagnated, and may even have fallen in some areas.

INDIGENOUS TECHNOLOGY

According to [25], traditional agricultural practices in Africa provide valuable lessons to be learned from local farmers who, through their own innovations and experimentation on farms, have perfected tools such as the hoe and the plough, developed seeds and plants through preservation and selection, and designed crop mixtures and rotations leading to improved productivity. Practices like fallow, terracing, ridging mixed farming, and intercropping were practiced by local people long before the introduction of the Green Revolution. These practices provide advantages that have been identified as those that have ensured soil fertility, controlled pests as well as diversifying sources. Some examples of traditional technology method that have helped guard against food insecurity at household and national level and helped in the sustenance of the environment include but not limited to the following:-

Farmer's seed saving strategy and water harvesting

For seeds to be of quality and viable in relation to its germination rate, storage and drying are of great importance. One of the common ways of ensuring the dryness of grains even in rainy season is simply to hang on the maize cobs on the ceiling above the cooking fire. Similarly, rain water harvesting through collection pond, irrigation during the evening, water erosion control through plantation, canalling of water through the hard rock area by using wooden conduits, aquifer recharging, etc. are some examples of farmers innovations to ensure sustenance of the food production.

Solar drying

Rural communities in Africa and south East Asia have used solar to dry their foods as a form of preservation to bolster food security at the home. Fresh vegetables are immersed in salted boiling water for a few minutes and then dry them in the sun for about 3 days. Similarly, edible insects such as white ants, termites, and caterpillars, mushrooms and tomatoes are stored in the same way. According to [7], drying is also often used to preserve meat, fish, and roots. Cassava and bananas are also preserved by fermentation followed by drying. The drying helps prevent growth of the microorganisms and stops biochemical activities that cause foods to decay. This helps in nutrition and energy needs when fresh supplies are not accessible.

Storage of roots and tubers

Fresh roots and tubers are highly perishable and cannot be stored for long periods thus Cassava has to be processed within 2 days of harvesting to avoid damage. Fresh cassava, therefore, is best left unharnessed until needed. As an alternative, yams, coco yams, and cassava may be stored in underground pits after harvesting. In some instances, root crops such as cassava can be grown as a food reserve, left in the ground for up to 2 years and used as the main source of energy during lean times [26].

Hunting and gathering

This is a food procuring method used by many communities in their quest to meet their food demands. The whole exercise was done in a sustainable manner so that there is more left for the next season. Hunting was controlled to also maintain the ecosystem. Great care was taken not to kill unnecessarily.

Fermentation process

In most Africa countries, fermentation is a traditional method of preserving vegetable surpluses which, when used, enhances the overall flavor of the meal. The technique provides a suitable environment for lactic acid bacteria to grow, thus imparting an acid flavor to the vegetable. Cassava and sweet potatoes are the most commonly fermented foods. Alcoholic beverages constitute the largest category of fermented products in Africa [27]. Most of these beverages are processed from fruits. Banana beer, a popular drink in Eastern and Western Africa region is made by allowing banana juice to ferment. Palm and coconut wine are manufactured in the same way.

Home garden

Home or family garden normally run by the women play a great role in increasing small-scale production of micronutrient-rich foods. The home garden is the most direct means of supplying families with most of the non-staple foods they need year-round. These gardens have saved many families from glaring hunger because being small, they are usually planted early hence the crops ripen at the very time of need though in small quantities Nevertheless, indigenous technology can be integrated with modern technology to come up with appropriate technologies for communities. Such technology developed with indigenous peoples input is easily adopted, less costly and uses available local knowledge. The Marakwet people of Kenya have had traditional furrow systems that they have been using for irrigation and these systems have been integrated with modern systems for best results.

Modern technology

There is need to adapt science-based technological innovations that are affordable and have positive improvement on global food security and have no or insignificant impact on climate and environmental sustainability. The use of modern technologies to boost food production and thus sustain the global population requires political will and sufficient investments in modern agriculture. In the 21st century, many determinants of food security are trans-boundary and multilateral agreements towards this cause, are paramount. Appropriate technological innovations are required and be implemented at all levels.

The Green Revolution drove widespread shifts in the agricultural sector from subsistence and low external input agriculture to mono-cropping with high yielding varieties. By the 1970s, Green Revolution-style farming had replaced the traditional farming practices of millions of developing country farmers. By the 1990s, almost 75% of Asian rice areas were sown with these new varieties. Overall, it is estimated that 40% of all farmers in developing countries were using Green Revolution seeds by this time, with the greatest use found in Asia, followed by Latin America [28]. The rapid spread of Green Revolution agriculture throughout most countries of the South was accompanied by a rapid rise in pesticide use [28]. This was because the High Yielding Varieties were more susceptible to pest outbreaks. Promising increases of yield were thus offset by rising costs associated with increased use of chemical inputs. Modern agriculture practices, such as precision farming, would help convert this concept of "evergreen agriculture revolution" into a reality. There is need to provide modern agriculture implements at reasonable costs, high-tech agriculture knowledge, agronomic support and agriculture extension services and help with farm planning and crop management, allowing farmers to increase their food output and net incomes in the world. Technological advancement and adaptation is vital for food security.

Sustainable development requires that technologies developed to improve food security situation in the world have least negative impact on the environment while maximizing benefits of improved food production and welfare of humans. There is need to focus on the effect of climate change on domestic production in food-insecure countries, assess climate change impacts on foreign exchange earnings, determine the ability of food surplus countries to increase their commercial exports or food aid and analyze how the incomes of the poor will be affected by climate change.

Synthetic fertilizers, pesticides and herbicides are made from non-renewable raw materials such as mineral oil and natural gas or from minerals that are depleting such as phosphate and potassium. As the price of petroleum increases, so does the cost of external inputs and machinery, forcing small farmers who are dependent on these inputs into debt. The production of agrochemicals is also an important source of greenhouse gas (GHG) emissions. In particular, fertilizer production is energy intensive, accounting for 0.6-1.2% of the world's total GHGs [29]. Industrial, chemical-intensive agriculture has also degraded soils and destroyed resources that are critical to storing carbon, such as forests and other vegetation.

The rise in use of chemical inputs has also had adverse environmental and health impacts on farm workers and consumers. A substantial portion of pesticide residues ends up in the environment, causing pollution and biodiversity decline. The extensive use of pesticides has also resulted in pesticide resistance in pests and adverse effects to beneficial natural predators and parasites [30]. The Green Revolution also brought about a shift from diversity to monocultures. When farmers opted to plant Green Revolution crop varieties and raise new breeds of livestock, many traditional, local varieties were abandoned and became extinct. And yet, maintaining agricultural biodiversity is vital to long-term food security as it is vital insurance against crop and livestock disease outbreaks and improves the long-term resilience of rural livelihoods to adverse trends or shocks [31].

In genetic engineering technology, genetically modified organisms (GMO) are created by altering the DNA of an organism, in this case a food producing plant; this is done in order to change the characteristics of the plant. Through this process of genetic engineering (GE) a plant can be made to produce a higher yield, be more resistant to pesticides, require less water and still be fast growing. The problem of food security seemed to be solved by producing plants which produce more food and are resistant to pests, so with very little testing and no real case studies and field trials, genetically engineered seeds began to be produced to grow genetically modified crops. An American company called Monsanto took the lead and became the largest producer of GM seeds as well as their famous herbicide called 'Roundup'. Monsanto made the winning combination; a very successful weed killer and their GM seeds, which are tolerant to their herbicide [32]. A lot of food that we eat today contains genetically modified ingredients and usually without our knowledge. Supporters of this technology maintain that it ensures and sustains food security around the world as the population increases. However, a debate on the socioeconomic ramifications of the way such science is marketed and used continues. Critics believe that the problem of food shortages is a political and economic problem, food shortages and hunger are and will be experienced by the poorer nations and that GE Food is an expensive technology that the farmers of the developing nations would not be able to afford easily.

Substantial improvements are possible in rain-fed agriculture, particularly in sub-Saharan Africa and South Asia. Tapping this potential requires innovative strategies to manage the sudden excesses of water and frequent dry spells. Integrating soil and water management focused on soil fertility, improved rainfall infiltration, and water harvesting can significantly reduce water losses, and improve yields and water productivity. Water storage has the greatest potential to deliver more water for food. Apart from dams, storage can also mean holding water in natural wetlands and reservoirs, in groundwater aquifers, soils, and in small tanks and ponds. Modern irrigation technologies, such as sprinkler and micro irrigation, have potential for adaptation to smallholdings; particularly where farmers are growing high-value marketable crops and where water is scarce.

Affordable systems, such as bucket and drum drip irrigation kits, have been developed for small plots and vegetable gardens predominantly cultivated by women. The introduction of treadle pumps, originally developed in Bangladesh, has revolutionized water lifting [33].

Conservation agriculture technology on the other hand is a farming practice being piloted in Kenya by the government. The method contributes to sustainable agricultural production and environmental conservation, by maintaining a permanent or semi-permanent organic soil cover; through the use of mulches or cover crops, employment of zero or minimum tillage and crop rotation. Weed control is done using herbicides or shallow cultivation resulting to minimal soil disturbance, water and nutrients retention. Some of the benefits of conservation agriculture technology are reduced labor and farm-power requirements, improved soil fertility, crop yields increase over time compared to conventional farming, livelihood improvement, decreased carbon dioxide in the atmosphere and reduction of climate change.

Conservation agriculture technology acknowledges the importance of creating and maintaining a healthy soil and integrates various approaches to the management of weeds, pests, diseases, and plant nutrients. Adoption of conservation agriculture technology will help crops adapt to changing climatic conditions and ensure harvest despite unreliable rainfall. This is an innovation whose time has come and cannot be stopped.

ADOPTING EVIDENCE BASED INNOVATIVE TECHNOLOGY FOR ENVIRONMENTAL SUSTAINABILITY

There is need for countries and communities to adopt research based innovations that have proved effective in addressing food insecurity and environmental sustainability. The following are some of the examples;

Integrated pest management: The aim is to produce quality crop yields with techniques that minimize environmental impacts. Pest outbreaks can thus be prevented or limited, by developing and using green mechanical, biological, chemical and other controls only as needed.

Fertilizers: The use of fertilizers helps increase cultivated soil carbon reserves by increasing the photosynthetic conversion of CO_2 to biomass that is subsequently converted to soil organic matter.

Water management: Agriculture depends on water availability and water quality, thus it will be increasingly important to develop innovative strategies for sustainable water management. Innovative methods for conserving water on the farm-level will be important, such as improved irrigation techniques and indigenous furrow/pans irrigation.

Improved seeds: Where appropriate, improved seeds, including those derived through biotechnology, have the potential to make a major contribution to increasing crop yields, nutritional content, and productivity, and mitigating environmental impacts such as climate change. Drought-tolerant crop varieties, for example, have the ability to help protect yield potential when water is scarce, while other crop varieties can be produced with genetics that protect against yield losses due to flood conditions. Salt-tolerant crops can be developed to allow land that has become unproductive for crops to be used for food production. Breeding of plants with improved water efficiency will be important. Plants with an improved nitrogen-efficiency can grow and produce high yields with lower amounts of fertiliser or have much higher yields under the same fertiliser input. Such plants would also help to minimise the emission of nitrous greenhouse gases (GHG) and save energy on the production of nitrogen fertiliser an energy intensive process [34].

Reduction of GHG from livestock: Livestock waste products are a source of GHG emissions. There are a number of examples of how best practices can help reduce emissions. For instance, research to reduce GHG from livestock is looking at selective breeding and biological means of reducing emissions. Examples include biogas production from animal waste by using co-digestion.

Using information technology (IT) for agriculture development: Cell phones offer a means of providing valuable information and advice to farmers in remote places. IT applications in agriculture are limited in the developing countries thus there is significant potential for maximising gains in agriculture through various IT applications, such as drought and flood management coupled with climate and weather information, waste reduction, risk mitigation and market development. Local operators of ICT can search for answers in a central database and provide information on either crop prices, weather forecasts for irrigation, water management and plant diseases. Therefore, it is critical to build capacity among farmers and create conditions that would allow them to access and apply these IT applications.

Minimizing harvest losses: The reduction in pre and post harvest losses would in itself contribute in a major way to food security. There is an urgent need for replacing the rudimentary pre and post-harvest practices with innovative, scientific and low cost models.

Adjustment in farm practices: Farm mechanization will be essential for increasing food production in developing countries. Machinery and implements have to be tailor made to the conditions in each of the agro-climate zones. In addition, the development of prediction tool models and on-site diagnostics can optimize farm practices by minimizing the inputs (fertilizer, water, agrochemicals) and maximizing the yield.

Figure 4: The photo shows adaptive agricultural mechanization in the Kerio Valley Region of Kenya, 2012

Carbon Sequestration: The process of transferring atmospheric CO_2 into soil and biotic pools can enhance soil quality, increase agronomic productivity, improve quality of natural waters, and lower rates of anoxia (decrease in the level of oxygen) or hypoxia (dead water) in coastal ecosystems.

Conservation agriculture: Conservation agriculture techniques such as low or no-till agriculture, made possible through the use of herbicides and biotechnology-derived crops, prevents wind and water erosion and loss of ground moisture, improves soil biodiversity, increases soil fertility, and in appropriate, carefully managed cases has the potential to reduce carbon emissions. In addition, by limiting soil disturbance and promoting a permanent soil cover, conservation agriculture can contribute to limiting emissions from agriculture by increasing soil carbon content (i.e. reducing emissions) and preventing erosion.

Enzyme applications: The number of enzyme applications in food applications has been growing. Enzyme technologies can improve the quality and quantity of food products. Some examples include reducing the content of unsaturated fat in fat spreads, improving vegetable flavour, increasing cheese yield, improving phosphorous use by certain animals, enhancing fiber digestion, and slowing the staling of baked goods [34].

A low-carbon economy would be beneficial to the world. Research, development and deployment of clean technologies would be the most appropriate for the present and future generations not only in food production and distribution but also in all spheres of development. Governments need to encourage this innovation by all players with strong investment frameworks to harness the power of markets and stimulate research, development and deployment. A mechanism to accelerate technology development and transferring support of countries' action on adaptation and mitigation was established at the United Nations climate change talks in Cancun (2010). The mechanism was to be guided by countries' most urgent needs, priorities and national circumstances. They also provided direct in-country advice and support to facilitate prompt action on the deployment of technologies based on identified needs, including through a network of national, regional and international technology centres, networks and organizations. While developed countries have a responsibility to support developing countries to acquire clean technologies, it is important to recognize that new technologies come from all over the world.

CONCLUSION

This study analysed the importance of integrating indigenous knowledge systems and modern science based agricultural technologies to attain a food secure population in the face of climate change hence securing livelihoods and environmental sustainability. The study has shown that it is important to approach food security issues and climatic changes in a multi-faceted approach. The paper argues for an explicit recognition of a hybrid evidence based approach which recognizes the need for the integration of traditional food systems, modern food systems and technologies globally. To address global food insecurity, there is a need for countries to adopt protocols and treaties pertaining to climate change mitigation. There is also a need to translate the available knowledge on climate change mitigation into action through design and implementation of evidence based interventions. Countries are encouraged to implement environmental sustainability best practices that include low carbon emission energy technologies and promote the use of energy efficient processes. Capacity building and awareness of the interrelationship of intricate chemical, physical and biological systems should be enhanced to ensure that communication about climate change and food security is meaningful. This allows people to make informed and responsible decisions towards sustainable food security and environment.

Diversification of livelihoods, adaptation of agricultural technologies, enhancing early warning systems, drought monitoring and seasonal forecasts with respect to food security is important. Improved management of cultivated land and livestock management, the use of new, more energy-efficient technologies by agro-industries and protection of ecosystems are also necessary actions towards sustainable use of the environment for food security

The main challenge to the adoption of an integrated approach is the fact that traditional knowledge exists within diverse communities with diverse traditions and most of it is not documented. Access to such knowledge is limited and therefore more research is necessary to document traditional knowledge for effective utilization in a future hybrid system. Adopting modern technologies is expensive and the process of integrating technologies requires experts who will be supported through local and community based research. The process of capturing and translating traditional knowledge into action will remain a significant challenge in most developing countries.

Recommendations for further research

1. We recommend further studies on approaches to integration of indigenous knowledge systems and the science based technologies (Hybrid system) towards improved food security and environmental sustainability

2. How communities can be involved in policy design, implementation and evaluation in relation to the hybrid system

3. More research on community food systems and coping strategies in face of climate change especially in most developing countries

4. More research on intellectual property rights in relation to indigenous knowledge systems as related to food security and climate change.

REFERENCES

1. UNFCCC Article 1, Definitions: Accessed 12The August 2012

2. IPCC Fourth Assessment Report, Working Group I, Glossary of Terms: http://ipcc-

3. wg1.ucar.edu/wg1/Report/AR4WG1_Print_Annexes.pdf.

4. EPA, 2009. EPA's Endangerment Findings Accessed 2010 June (http://www.epa.gov/

5. climatechange/endangerment/downloads/EndangermentFinding_ClimateChange-

6. Facts.pdf)

7. Mauna Loa Observatory, Hawaii. Consequences of Enhanced Greenhouse Effect Ac-cessed 2010 June (http://www.legitimatemillions.com/globalwarming.html)

8. NEMA 2009. State of Environmental Report Kenya, 2006/7 "Effects of Climate change and coping mechanisms in Kenya", Nairobi, Kenya

9. IPCC (2001c) Climate change 2001: Mitigation. Contribution of working group iii to the third assessment report of the Intergovernmental Panel on Climate Change

10. FAO. 1997. Agriculture, food and nutrition for Africa. A Resource Book for Teachers of Agriculture

11. IPCC. 1995. Climate change: a glossary by the Intergovernmental Panel on Climate Change Accessed July, 2012 (www.ipcc.ch/pdf/glossary/ipcc-glossary.pdf).

12. FAO (1980) Principles of Surplus Disposal and Consultative Obligations of Member Countries, Rome.

13. Thielke, Thilo 2006. Starvation in Africa: Kenya's Deadly Dependency on Food Aid; Accessed 2010 (http://www.spiegel.de/international/spiegel/starvation-in-africa-ken- ya-deadly-dependency-on-food-aid-a-396031.html)

14. ILO. 2007. Chapter 4. Employment by sector, In Key indicators of the labour market (KILM), 5 th

15. Edition

16. Laurence, W.F. & Williamson, G.B. 2001. Positive feedbacks among forest fragmenta- tion, drought and climate change in the Amazon.

Conservation Biology

17. Gleick, P.H. 1993. Water in crisis: A guide to the world's fresh water resources. New York, Oxford University Press.

18. Millennium Ecosystem Assessment, 2005 Accessed on 15 The August 2012 (http:// milleniumassessement.org/en/Condition.aspx)

19. FAO. 2007. Adaptation to climate change in agriculture, forestry and fisheries: Per-spective, framework and priorities. Report of the FAO Interdepartmental Working Group on Climate Change, Rome

20. UNO, 2008. Desertification Accessed 2010. (http://www.goodplanet. info/eng/pollu-

21. tion/desertification/(theme)/308)

22. UNEP (2008) Nairobi. Kenya Accessed on June, 2011 (http://www. unep.org)

23. UNEP, 2009. Clearing the Waters: A focus on water quality solutions Accessed in September 2010 (http://www.unep.org/PDF/Clearing_the_ Waters.pdf)

24. Seo, Sungno and Mendelsohn, Robert 2006. The impact of climate change on live-stock management in africa: a structural ricardian analysis, Accessed on (http:// www.ceepa.co.za/docs/CDPNo23.pdf)

25. IPCC Fourth Assessment Report: Climate Change 2007. Indigenous knowledge in mitigation and adaptation, Accessed on 2010 (http://www. ipcc.ch/publica- tions_and_data/ar4/wg2/en/ch9s9-6-2-2.html)

26. Gabriele, Geier (1995) Food Security Policy in Africa between Disaster Relief and Structural Adjustment Reflection on the Conception and Effectiveness of Policies; theCase Study of Tanzania

27. Reeves, Tim 2009. A Sustainable Green Revolution for Global Food Security

28. Embracing Science-Based Technologies, Accessed August 2012 (http:// www.global-harvestinitiative.org/index.php/policy-center/embracing-science-based-technologies)

29. Timmer, C.P. (1988). The agricultural transformation, In: H. Chenery and T. Sriniva-san (eds),Handbook of Development Economics, Vol. 1

30. Oniang'o, Ruth, Allotey, Joseph And Malaba, Serah 2006. Contribution of indigenousknowledge and practices in food technology to the attainment of food security in Af-rica –InJournal of Food Science, Vol 69

31. Issue 3 Accessed on August 2012 (http://onli-nelibrary.wiley.com/ doi/10.1111/j.1365-2621.2004.tb13346.x/pdf).Katz HS, Weaver WW. 2003.

32. Encyclopedia of food culture Volume 1.RANDFORUM/UNDP. 1995. Sourcebook on African Food Technology. Production

33. and Processing Technologies for Commercialization

34. Rosset, P., Collins, J. and Lappe, F.M. 2000. Lessons from the Green Revolution: Do we need new technology to end hunger?Tikkun MagazineVol.15, No.2

35. Bellarby, J. et al. 2008. Cool farming: Climate impacts of agriculture and mitigation potential.Pimentel, D. 2005. Environmental and economic costs of the application of pesticidesprimarily in the United States, InEnvironment, Development and Sustainability7:229-252.

36. Pimbert, M. 1999. Sustaining the multiple functions of agricultural biodiversity: FAO ackground paper series of the Conference on the Multifunctional Character of Agri-culture and Land, 1999 FAO

37. Taylor, Yvette 2011. Food Security and GMO Foods Accessed on may, 2012 (http://www.earthorganization.org/articles/Library/Food_Security_ and_GMO_Foods/default.aspx)

38. Kay, Melvin 2011. Water Smart: The role of water and technology in food security,Accessed August 2012 (http://www.intracen.org/Water-Smart-The-role-of-water-and-technology-in-food-security/)

39. BIAC, 2009. Innovation to Address Food Security accessed on June 2012 (http://www.biac.org/statements/agr/FIN09- 11_Agriculture_and_ Innovation.pdf)

Chapter 6

A FLOOD CONTROL APPROACH INTEGRATED WITH A SUSTAINABLE LAND USE PLANNING IN METROPOLITAN REGIO

Paulo Roberto Ferreira Carneiro[1] and Marcelo Gomes Miguez[2]

[1] Universidade Federal do Rio de Janeiro, Instituto Alberto Luiz Coimbra de Pós-Graduação e Pesquisa de Engenharia (COPPE/UFRJ), Laboratório de Hidrologia e Estudos do Meio Ambiente, Ilha do Fundão, Rio de Janeiro/RJ, Brazil

[2] Universidade Federal do Rio de Janeiro, Escola Politécnica (POLI/UFRJ), Rio de Janeiro/RJ, Brazil

INTRODUCTION

The Brazilian National Water Resources Policy, instituted by Law no. 9.433 in 1997, is based on six fundamental principles that structure the whole National Water Resource Management System: 1) water is a commodity in the public domain; 2) water is a limited natural resource, endowed with economic value; 3) in situations of scarcity, the priority water resources use is for human consumption and watering animals; 4) the management of water resources must always provide multiple water uses; 5) the hydrographical basin is the territorial unit for the implementation of the National Water Resources Policy and the activities of the entities belonging to the of National Water Resources Management System; 6) the water resource management must be decentralized and have the participation of public authorities, water users, civil society and communities.

This Law and its regulatory texts incorporate municipalities, along with users and civil organisations, into the management system, ensuring a greater balance of power on water resource committees and boards. However, no legal text has clearly defined the relation between water management, which is a state or federal attribution, and land use planning, which is responsibility of the municipalities. In this sense, there remains a lack of definition regarding the fundamental role of municipal administrations as formulators and implementers of urban policies with impacts on water resources, whether through direct investment, or by means of actions of regulatory nature.

Besides the gap pointed out above, the occurrence of conflicts of competency is also observed in the hydrographical basins related to metropolitan areas, given that the 1988 Brazilian Constitution did not establish clear management rules for these territories. The definition of the needed and related administrative organisation for the metropolitan areas is left to the federative states. On the other hand, overlaps is observed in the attributions of the local, state, or even federal administrations, and various undefined roles are identified, which make the task of coordination and sharing of the responsibilities even more complex.

Based on these elements, and departing from Brazilian reality, the proposed chapter deals with the need of integration of land use planning with water resource management, seeking to establish relations between the types of land use, urban settlements and the problems involving urban flooding.

A case study was developed for the Iguaçu-Sarapuí River Basin, located in the western portion of the Guanabara Bay Basin, which lies at the Rio de Janeiro State Metropolitan Region, in Brazil, and is one of the most critical areas in the state in relation to urban flooding. In this region, urban expansion dynamics is, in general, marked by irregular occupation of risk areas, without the appropriated infrastructure in terms of land tenure.

The significant investments in infrastructure in progress in the region, mainly the construction of the Metropolitan Ring Road[1] - will bring substantial transformation to the region current urban configuration. The scenarios built with the aid of mathematical modelling demonstrate that the disorderly urban expansion, induced by the accessibility to the rural areas in the interior of the region, may be degrading for the medium and long term urban flooding control in this basin.

THE ROLE OF THE MUNICIPALITY IN WATER RE-SOURCE MANAGEMENT IN BRAZIL

The competence of municipalities in federated countries is concentrated on functions that, in general, are related with the allocation or rendering of local public services and with the functions of planning, incentive and inspection of the territorial order, environmental protection and also with some level of regulation of economic activities [1]. In the case of Brazil, recently, municipalities with greater capacity of investment have begun to incorporate functions related with the provision of more comprehensive social services, which, traditionally, were restricted to the state and federal spheres.

In the specific case of water resource management, however, municipal participation in basin committees has been the main form, if not the sole, of interaction with other public and private actors related with water. Many factors hinder the municipality action in the water management sphere, the main one being the legal impossibility, by Constitutional definition, of the municipalities directly managing water resources, even in the case of basins entirely contained by their territories. The exceptions may be associated to the transfer of some specific attributions through cooperation agreements with the states or the Federal Government.

Although local administrations are closer to local populations, their politico-administrative role does not allow a systemic vision of the territory in which they lie. More effective participation of local governments in water management is hindered, or even made unviable, also by the absence of clear definitions about its nature and functions, and by the fact that the majority of municipalities have limited budgetary autonomy, bearing in mind that they depend heavily on fund transfers from the other levels of government administration.

Regarding the financial restrictions [2], it is alarming that most of the multilateral financial agencies, except the Global Environment Facility – GEF, still have not included, in their agenda, projects of integrated natural resources management articulated to land use planning, particularly in urban areas. There are few planning experiments implemented articulating water conservation and/or preservation measures and land use regulation, despite the dysfunctions of urban growth.

Another aspect is that the sectoral nature of local government interests makes them act more as users than as "impartial" managers of water resources [3]. The debility and lack of institutional hierarchy of local governments confronted by actors wielding greater power would lead to greater vulnerability and to the possibility of capture and politicisation in water management [3]. These aspects are aggravated in metropolitan areas, where municipal administrations often express antagonistic interests and priorities among themselves, creating atmospheres of dissension with little space for cooperation.

Although there are restrictions on the participation of municipalities as direct managers of water resources, there is no doubt related to the importance of local governments in territorial planning, as well as in its consequences to water resources conservation. It is the attribution of municipalities to devise, approve and inspect instruments related with territorial order, such as master plans, zonings, development of housing programs, delimitation of industrial, urban and environmental preservation areas, among other activities with impacts on water resources, mainly in the case of predominantly urban hydrographical basins.

These attributions have recently been strengthened upon approval of the Brazilian Statute of the City. This is a Federal Act, established in 2001, which proposes standards of public and social interest to govern the use of the urban property in favour of the collectivity safety and welfare, as well as the environmental balance. The urban policy established aims to organise the fulfilment of the social functions of the city and of the urban property by the application of a set of general guidelines, from which the following topics are detached:

- the guarantee of the right to sustainable cities, meaning the right to urban land, housing, environmental sanitation, urban infrastructure, transport and public services, work and leisure for present and future generations;
- the democratic management through people's participation representing segments of the community in the formulation, implementation and monitoring of plans, programs and projects for urban development;
- the planning of city development to prevent and correct the distortions of urban growth and its negative effects on the environment;
- the supply of urban infrastructure and community equipments, transport and public services to serve the interests and needs of the population;
- the protection, preservation and restoration of the natural and built environment, besides cultural, historical, artistic and landscape heritages.

Several important urban management tools were made available in the context of the Statute of the City and the Urban Master Plan is considered to be the basic instrument for the urban developing policy.

The possibility of achieving a sustainable water resource management must necessarily pass through a clear articulation with land use plans. What is observed in Brazil, however, is the disarticulation between instruments of water resource management and land use planning, reflecting, perhaps, the lack of legitimacy of planning and urban legislation in Brazilian cities, marked by a high degree of informality, and even illegality, in land use occupation. According to Tucci [4], the greatest difficulty for the implementation of integrated planning arises from the limited institutional capacity of municipalities in facing complex interdisciplinary problems, and in the sectoral ways in which local administrations are organized.

Here, however, it is worth stressing the differences among municipalities: while in large cities, mainly metropolitan cores, it is possible to find efficient administrations, with good capacity to access information and with relatively modern legislation, in other minor cities, like peripheral municipalities in metropolitan areas, a total obsolescence in the legislation is verified. This is aggravated by the absence of reliable general data and information about the processes of urban structuring and also by the small number and low qualification of the technical staff [5].

This inequality in the municipal scale presents a great obstacle for a greater effectiveness of water resource management structures and for the cooperation among the different hierarchical levels of government.

FLOOD CONTROL IN THE BAIXADA FLUMINENSE LOWLAND

Baixada Fluminense lowland is located in the western portion of the Guanabara Bay basin, in one of the most critical regions of Rio de Janeiro State, in terms of urban flooding. It is particularly interesting as an empirical study, considering the following aspects:

*   its location is in the metropolitan periphery;
*   there are areas with consolidated urban and industrial growth;
*   there rural areas in a process of urban development
*   the basin also contains rural areas still protected from urbanisation;
*   several areas present land use patterns that do not ensure minimal standards of living, especially those of poor drainage;
*   consequently, several serious flooding problems occur in the watershed plain areas;
*   water sources found in the basin area are used for complementing the Metropolitan Region drinking water supply;
*   Tinguá Biological Reserve, the main remnant of the Atlantic Forest in Rio de Janeiro State, is situated in this territory;
*   organised social movements, congregating federations of residents associations and entities involved in matters of environment, sanitation, housing, among others, are present in the basin, what demonstrates the great organisation capacity of its population vis-à-vis the questions related to citizenship and quality of life;
*   local administrations are becoming more committed to efficiency in public affairs, although in a still timid process;
*   the presence of major private and public investments in infrastructure will lead to significant transformations in the present urban configuration of the region.

Physical and socio-economic characteristics of the basin

The Iguaçu-Sarapuí River basin is situated in Baixada Fluminense lowlands. Its drainage area covers around 727 km², all of which is situated in the Rio de Janeiro Metropolitan Region. Iguaçu River springs in Serra do Tinguá massif, at an altitude of 1,600m. Its course runs southeast for approximately 43 km, until it reaches the outfall at Guanabara Bay. Its main tributaries from the left margins are Tinguá, Pati and Capivari Rivers, and, from the right margins, Botas and Sarapuí Rivers.

The physiography of Iguaçu-Sarapuí river basin is characterized by two main elements: the Serra do Mar Mountains and Baixada Fluminense lowlands, with a marked difference in altitude. The climate in the basin is hot and humid with a rainy season in the summer, the average annual precipitation being around 1,700mm, and the mean annual temperature approximately 22o C. The rivers run down the mountains in torrents with great erosive force, losing speed after reaching the plains, often overflowing their banks into large wetlands.

The basin fully encompasses the municipalities of Belford Roxo and Mesquita, also hosting part of the municipalities of Rio de Janeiro (covering the neighbourhoods of Bangu, Padre Miguel and Senador Câmara), Nilópolis, São João de Meriti, Nova Iguaçu and Duque de Caxias (Figure 1). According to the 2010 Brazilian census, the population of these municipalities reached 9,225,557 habitants (Table 1). However, just two of these municipalities are totally inserted in the basin.

Table 1: Municipal population, total municipal area, and insertion in Iguaçu-Sarapuí River Basin

City	Municipal Population			Total Area[1] (ha)	Area inside the basin[2] (ha)	% (*)
	Urban	Rural	Total			
Belford Roxo	469.332	-	469.332	7.350	7.350	10
Duque de Caxias	852.138	2.910	855.048	46.570	27.359	38
Nilópolis	157.425	-	157.425	1.920	1.042	1
Mesquita	168.376	-	168.376	3.477	3.477	5
Nova Iguaçu	787.563	8.694	796.257	53.183	27.894	38
Rio de Janeiro	6.320.446	-	6.320.446	126.420	3.290	5
São João de Meriti	458.673	-	458.673	3.490	2.293	3
Total	9.213.953	8.694	9.225.557	242.410	72.705	100

Source: (1) demographic census of 2010, with the territorial division of 2001, (2) Adapted from the Iguaçu Project; (*) percentage of the municipal area in relation to basin area.

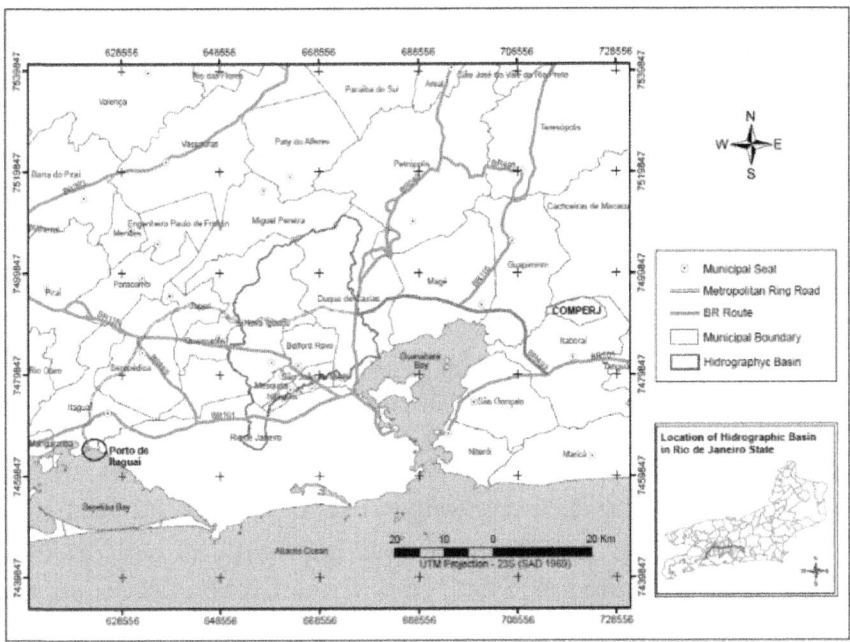

Figure 1: Iguaçu-Sarapuí River Basin

Table 2: Poverty and inequality map – Brazilian Municipalities, 2003- Poverty incidence in Baixada Fluminense Lowlands

Municipality	%
Belford Roxo	60,06
Duque de Caxias	53,53
Mesquita	-
Nilópolis	32,48
Nova Iguaçu	54,15
Rio de Janeiro	23,85
São João de Meriti	47,00

It is in the lower parts of the basin, with elevations near the medium sea level, where it is concentrated mostly of the urban area, with something about 1.5 million people living there. Calculations from IBGE, the Brazilian Institute of Geography and Statistics, show that the incidence of poverty in these municipalities is quite significant, especially in Belford Roxo, Nova Iguaçu and Duque de Caxias, affecting more than half of their populations (Table 2).

The structural analysis of per capita income and the capability to finance investments by municipalities in the region, according to the Observatory of the Metropolis [6], demonstrate the strong differences between the municipalities belonging to the Metropolitan Region of Rio de Janeiro. Such differences constitute obstacles to cooperation in solving common problems. Moreover, the fragile financial structure, coupled with the shortage of technical capacity, particularly in the areas of planning and budget, strengthen the uncertainty, discouraging long-term partnerships in infrastructure projects that could be used to promote social and economic development for the region.

After a century of intense population growth, Brazil has entered the new millennium with quite modest rates of population growth. As shown by the data of the last Census, the Brazilian population grew at an average rate of 1.6% per year in the 1990s, following a decline trend after the strong growth happened from the 1950 to 1970. Projections developed recently estimated that the Brazilian population is growing at rates below 1.3% per year.

The city of Rio de Janeiro has been the centre of services for the Metropolitan region, although this characteristic has not reflected in a high degree of attractiveness for population in recent times. The region remained with the lowest population growth rate among large Brazilian cities. It should be noted, however, that in absolute terms, there was a warming of migration in the last decade towards Rio de Janeiro. Between 1980 and 1991 the total number of migrants towards the metropolitan area of Rio de Janeiro was around 570,000 people, while between 1995 and 2000 (just in five years) the total migration reached 330,000 people. The capital of the state remained the main pole centre, receiving these migration flows and housing 195,000 migrants, i.e. 62% of the total [6].

According to Britto and Bessa [7], historical investments were made in the region by different state governors, like the one of the 1980s, with an amount up to R$ 3 billion, without, however, effectively guaranteeing universal access to environmental sanitation, housing and a better quality of life. Explanations for this are related with: (i) the lack of a profound diagnosis of the dimension of the problem in the region to correctly orient the profile of the interventions; (ii) the discontinuity and non-integration among the programs and projects implemented throughout these years; (iii) the political disputes in the region often decharacterised the projects, again lacking continuity; (iv) the fragility of social control in the process, once the format of the implemented programs have not provided an effective participation of the population (although this component existed in various of these projects); (v) the lack of institutional capacity, allied to the centralizing culture of the state governors in relation to sanitation management; (vi) the strong clientelistic culture in the municipal administrations; (vii) the growing demobilisation of organized social movements, which need members qualification for following up the policies implementation.

Flood control in the Baixada Fluminense lowlands

Floods in the Iguaçu-Sarapuí River Basin are aggravated basically by the f inadequate land use occupation, in the particular conditions of the lowlands of Baixada Fluminense. In this process, the most important factors are: lack of adequate urban infrastructure; deficiency of the sewage services and solid waste collection; uncontrolled exploitation of mineral deposits, mainly sand for construction purposes; disorderly, illegal occupation of river banks and floodplains; lack of adequate treatment for public roadways pavements; obstruction or strangulation of drainage due to structures built without the proper concerns (railway and road bridges, and water pipelines interferences), as well as walls and even buildings that partially obstruct river channels. At the heart of these problems one always finds either inadequate legislation regarding land use, or, in the great majority of cases, non-compliance with the existing legislation.

It is estimated that floods in the basin directly affect 189,000 people. However, the damage caused and the total number of people indirectly affected by floods are both difficult to estimate. Included in this latter category there are, for example, employees who cannot reach their workplaces and the interruption of traffic and commerce along the flooded roadways or nearby areas that become inaccessible.

In this context, in order to properly discuss the adequate possible planning actions for mitigating these problems, and to figure out the cause-effect process related to future scenarios, a mathematical model will be applied as an aiding tool. The case study alternatives are then introduced in order to allow the development of the discussions in practical terms, using examples of what may happen in the future without the proper concerns. The aim of hydrodynamic modelling was to evaluate the possible impacts of the expansion of urbanisation towards the interior of the basin without the adequate planning process and considering the construction of Metropolitan Ring Road, which is being taken as an urban development inductor factor. Another objective of the modelling consisted of evaluating the impact of an average rise in mean sea level, regarding the drainage system conditions, according to forecasts made by the Intergovernmental Panel on Climate Change (IPCC) [8]. In both situations, which may critically combine effects, planning actions are required in order to control future negative effects, otherwise the human and material losses could become irremediable.

Brief Description of MODCEL

In order to proceed with the proposed analysis, it was necessary to choose a mathematical model to support the simulations. With this aim, a hydrodynamic model for representing rural and urban floods – MODCEL [9, 10 e 11] was used. The construction of MODCEL, based on the concept of flow cells [12] intended to provide an alternative tool for integrated flood solution design and research. MODCEL is a model that integrates a hydrologic model, applied to each cell in the modelled area, with a hydrodynamic looped model, in a spatial representation that links surface flow, channel flow and underground pipe flow, This arrangement can be interpreted as a hydrologic-hydrodynamic pseudo 3D-model, although all mathematical relations written are one-dimensional. Pseudo 3D representation may be materialised by a vertical hydraulic link used to communicate two different layers of flow: a superficial one, corresponding to free surface channels and flooded areas; and a subterranean one, related to free surface or surcharged flow in storm drains.

The representation of urban surfaces by cells, acting as homogeneous compartments, in which rainfall run-off transformations are performed, allows the integration of all the basin area. The cells interact through hydraulic laws, represented by cell links capable to model different possible flow patterns. Different types of cells and links give versatility to the model. The cells, considered individually as units or taken in pre-arranged sets, are capable to represent the watershed landscape, composing more complex structures. Therefore, the task related to the topographic and hydraulic modelling is an important phase of the process. In large floodable areas, when leaving the drainage network, the water can follow any path, dictated by the topography and by the urban built patterns. Marginal sidewalks may become weirs for the spilling waters from the rivers, the streets may act as canals and the buildings, parks or squares may act like reservoirs. In this situation, it is perceived that overflowed waters may have an independent behaviour from the drainage network, generating their own flow patterns. These characteristics are adequately represented in MODCEL.

The modelled area of Iguaçu-Sarapuí River basin extended from Guanabara Bay to Botas River confluence. The upstream reaches of the basin, which were not divided in cells, had their flows determined through a hydrological model called HIDRO-FLU [13].

Simulation criteria

The main objective of the modelling of the lower and middle reaches of the Iguaçu River was to evaluate impacts caused by the expansion of uncontrolled urbanisation towards the middle/upper basin, arising from the development expected from the construction of the Metropolitan Ring Road, an important axial roadway. The effective rainfall calculation method used was that of the SCS [Soil Conservation Service] of the Department of Agriculture of the USA - USDA. The Curve Number (CN), the main hydrological parameter of this method, varied for each of the simulated scenarios in accordance with different stages of urbanisation, as described below:

- Past situation: the CN values were defined based on soil types and land use mapping from 1994 (LANDSAT satellite images) [14].

- Present situation: the CN values were determined by land use mapping, made on the basis of images from the 2006 Aster sensor [14].

- Future situation: assumed that the flat, still rural areas of the sub-basins of the Rivers Iguaçu (upper reach), Botas, Capivari, Pilar and Calombé, and the Outeiro canal will suffer a disorderly process of urbanisation, following the trend of peripherisation in progress in Baixada Fluminense lowlands. This future scenario corresponds to a horizon of approximately 20 years (2030).

- Controlled future situation: assumed an alteration in the current pattern of urbanisation of these areas, with the introduction of land use control by means of urban planning actions and adoption of more sustainable urban drainage techniques.

- Each modelled cell in the basin representation had an individualised CN, depending on its particular characteristics.

- Another objective of the modelling consisted of evaluating the impact of the mean sea level rise, as forecasted by the IPCC, on the drainage conditions of the hydrographical basin. The proposed scenarios tested the isolated and/or associated effect of the following variables:

- different hydro-meteorological conditions, alternating typical tidal situations and the effect of meteorological tide;

- variation in the soil impervious rates arising from the behaviour of future urbanisation, considering the maintenance of the current rates (without any increase in new urban areas); an increase in the impervious rates due to unplanned urban expansion; and a moderate increase in the rates due to planned control of urban expansion. For each of the simulated scenarios, CN values were adopted as presented in Table 3.

It is important to stress that this paper does not intend to look for final solutions in order to minimise present flood conditions (although this discussion will be considered conceptually in the context of this study, in a nest topic). The main aim refers to the possibility of discussing future conditions worsening due to the inadequate planning process that take place today.

Table 3: Curve Number (CN) used in each simulated scenario

Basin	Past CN	Current CN	Future CN	
			without control	with control
Iguaçu	65	66	77	72
Botas	81	81	82	81
Capivari	67.5	65	77.9	72
Outeiro	72	84	84	84
Pilar	75	76	78.2	76
Calombé	68	79	79.8	79

The return period considered for the design rainfall was 20 years. The hydrologic parameters and rainfall information adopted were based on the Iguaçu Project [15] calculations. Regarding to the impacts caused by alterations in mean sea level, a local tide table was used as the base information. This table was produced by the Diretoria de Hidrografia e Navegação da Marinha do Brasil (Hydrography and Shipping Directorate of the Brazilian Navy), with values ranging from 0.09 to 0.90m, representing the tidal variation on the Rio de Janeiro coast. The meteorological tides were considered to influence this value with a majoring of 0.80m. Besides, a possible increment of 0.60m in the mean sea level was also considered (IPCC forecast), due to climate change expectative. With the values mentioned, the proposed scenarios were simulated, considering the tidal variations, the dynamics of urbanisation, the rise in the mean sea level, and combinations among these variables.

Results obtained in the modeling

Figure 2 represents the areas susceptible to flooding for the former conditions of urbanisation (at the time of Iguaçu Project [15]), in the 90's, without taking into account the meteorological tides and the effects of climate change. It is, therefore, a condition of reference for the current and future scenarios comparison, referring to flooding conditions of more than 15 years ago.

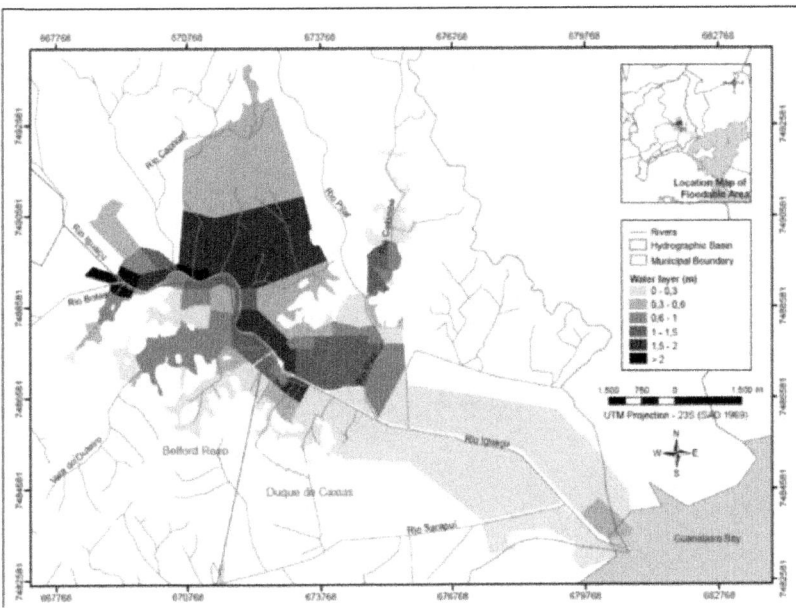

Figure 2: Reference flood map for former urban condition

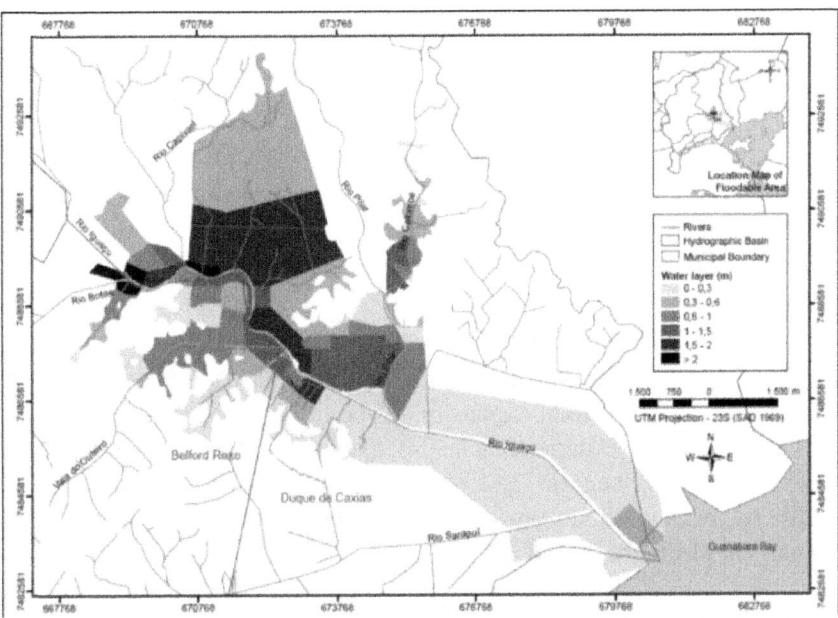

Figure 3: Flood map obtained for the present condition - Scenario 1

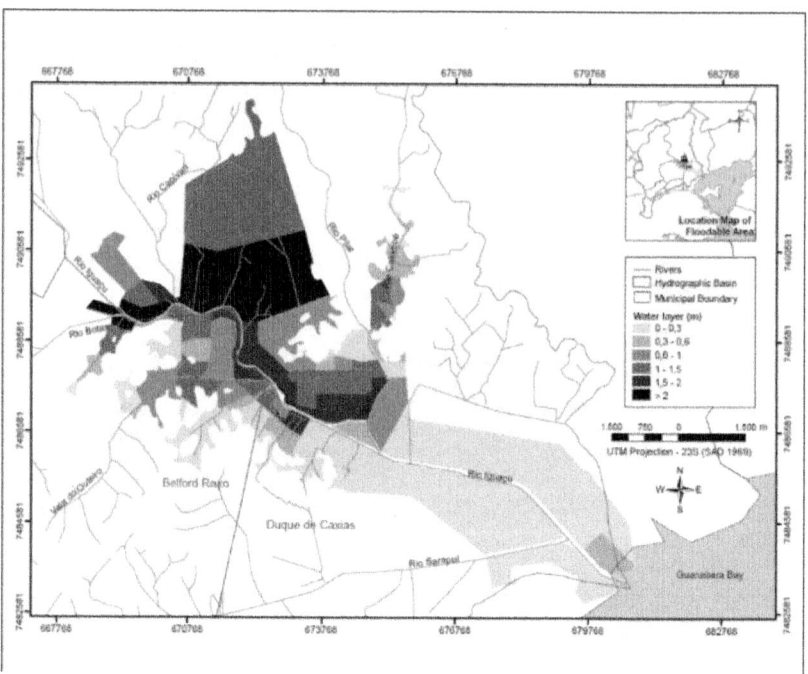

Figure 4: Flood map obtained for future condition - Scenario 2.

It is observed that there are significant differences in floods in past conditions from those in the present scenario. The alteration already occurred in the land occupation in the upper reaches of the basin in the period justifies this result. The flood maps presented in Figures 3 and 4, respectively, were obtained through the following conditions: current situation of urbanisation in the basin, without considering meteorological tides and the effects of climate change (Scenario 1); and future condition of the basin urbanisation, considering disorderly urban expansion, typical tides and without the effects of climate change (Scenario 2).

The comparison among these three scenarios allows the assessment of the isolated effect of the urban expansion in the flooding aggravation. When the CN is altered for the upper reaches of the drainage area, in the simulation corresponding to Scenario 2, a significant worsening is noticed in flood conditions, even without any other worsening factor acting, as seen in the comparison of Figures 3 and 4

If effective measures were implemented for land use development control, in order to prevent disorderly occupation in the middle and upper reaches of the basin, it can be seen, in Figure 5 (Scenario 3), that it is possible to avoid the worsening of floods in the referred sub-basins. It is perceived a reduction in the water levels in the densely urbanized areas, when compared with the previous development situation, without any control over land occupation.

The figures 6 and 7, presented in sequence, correspond to the following scenarios:

•   Figure 6: Flood map obtained for the future conditions of basin urbanisation with urban expansion without control over land use; meteorological tide of 80 cm and a 60 cm rise in the mean sea level due to climate changes (Scenario 4);

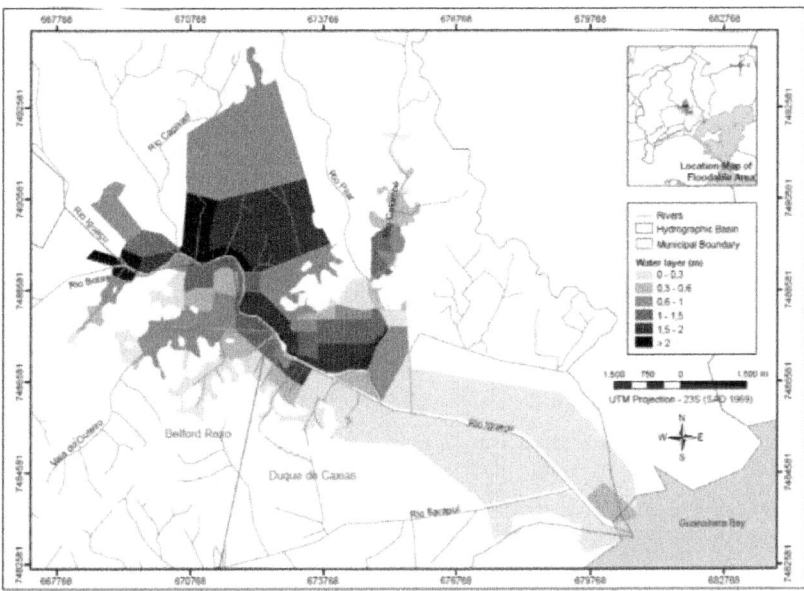

Figure 5: Flood map obtained for controlled future condition - Scenario 3

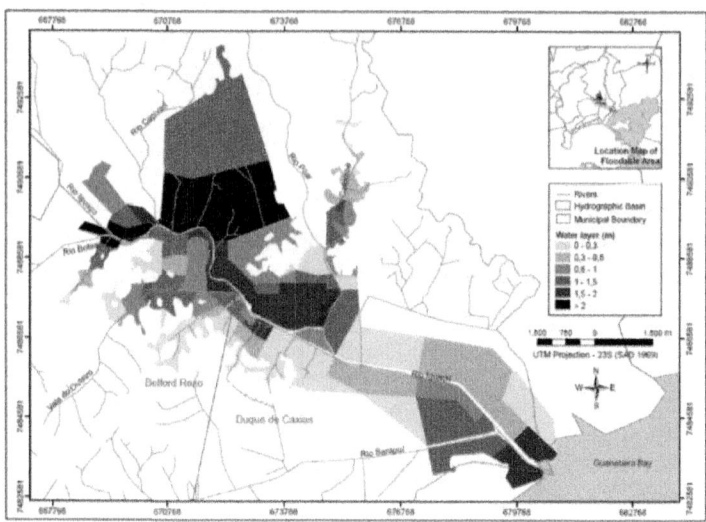

Figure 6: Flood map obtained for future condition, in the context of climate changes - Scenario 4

- Figure 7: Flood map obtained for the future conditions of basin urbanisation, with control over the land use development; meteorological tide of 80cm and climate change effects, with a 60 cm rise in mean sea level (Scenario 5).

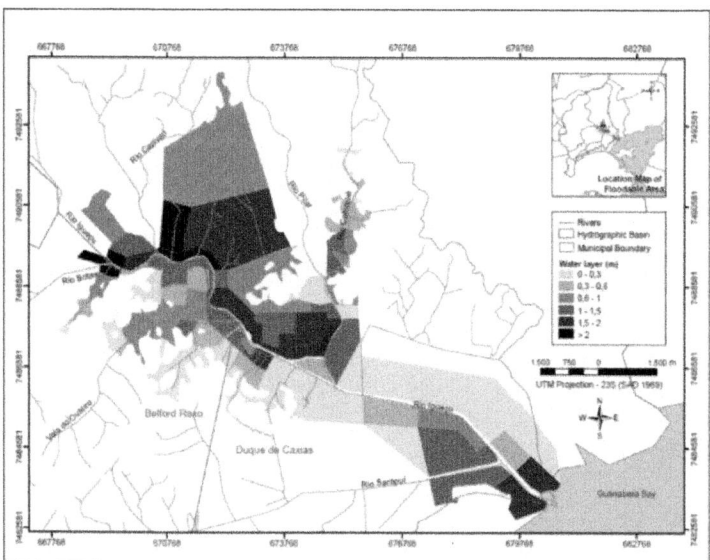

Figure 7: Flood map obtained for controlled future condition, in the context of climate changes- Scenario 5

These two scenarios test the conjugated effect of the three variables considered in the simulations: urbanisation of the upper basin, presence of meteorological tide and mean sea level rise. Based on these scenarios, it is possible to conclude that the disorderly urbanisation of the upper basin causes flooding aggravation in the downstream urban areas already consolidated, while the tidal variations cause even greater floods in the lower reaches (under tidal influence). The sea level rise will worsen the floods in the urban areas situated at low elevations, near the Iguaçu River estuary.

Both the urban expansion and the sea level rise are going to cause great impacts on the urban areas of the basin. Despite having their causes explained by independent variables, these factors, if combined, would lead to serious impacts on the population resident in the basin. If planning measures are not taken in advance, it will be very difficult to mitigate their impacts later.

CONCEPTUAL DISCUSSION

The urban drainage system includes two major subsystems: micro-drainage and macro-drainage. The micro-drainage system consists of the paving of streets, gutters, gullies, stormdrains and channels of small dimensions, intending to collect the runoff and conduct it to the macro-drainage net. Macro-drainage generally consists of natural or built channels of larger dimensions, receiving the input from micro-drainage, concentrating flows and discharging in the receiving water body. A complementary set of structures also take part in drainage systems, among which is possible to mention: reservoirs, protective dikes, and pumping stations. All these structures are arranged and designed to work in an n integrated way, intercepting, conveying, possibly infiltrating or temporarily storing and discharging the generated runoff. Ultimately, the receiving water body is the sea and this is the case of Iguaçu-Sarapuí Rivers.

The urban flooding process, by its turn, is directly associated with the failure of these subsystems, due to lack of maintenance, obsolescence, disordered urban growth or, as stated in recent discussions, due to the possibility of climate changes worsening flow conditions. Specifically to the drainage systems, the negative effects that may arise from the situation of climate changes refer to the increase of extreme rainfall events intensity, and to the restriction imposed by the expected sea level rise at the basin outfall. Evaluating this context, the increase in the mean sea level causes a reduction in the discharge capacity of the system, causing the drainage net to lose efficiency. The worsening of the extreme rainfall events intensity works in the other part of the problem, generating greater volumes to be drained by a system whose discharge capability diminished because of the new outfall restrictions. In this situation, in a context of already serious urban flooding problems, the effects generated by the possibility of climate changes can dramatically increase flooding areas, causing them to reach locations not previously affected by floods, increasing inundation depths and residence times, making the situation even worse.

Understanding how urbanisation affects floods is very important for urban flood control design. In general, it is possible to say that the urban flood control conjugates the adoption of structural measures that change the landscape of the basin, introducing interventions inside and outside the drainage network, to act directly in minimising the problem, and non-structural measures, associated with land use planning, environmental education and several possible other measures that allow a more harmonious coexistence with the phenomenon of flooding. The combination of structural and non-structural measures, in a context of planning integrated with urban growth, allows a composition capable of solving the problem of flooding in a harmonious and sustainable way. This approach, which is relatively recent, is being considered more appropriate to treat the urban flooding problem, by treating the problem in a systemic way and proposing actions that seek to minimize the impacts of urbanization.

This trend, though not motivated by the possibility of climate change, also goes toward this theme, with the possibility of reaching effective results, in opposition to the traditional approach that basically considers propositions of rectifying and canalising water courses. In this perspective, the traditional approach treats the consequence of the problem, related to the generation of exceeding superficial flows. The possibility of the mean sea level rising, however, limits the discharge capacity of the system and makes the traditional approach to fail. Thus, in this context, it is necessary to treat the problem of flow generation, acting in the causes of flooding, while trying to introduce infiltration and storage measures spread over the urban basin landscape in order to reduce and delay flood peaks, allow groundwater recharge and seek to restore the approximate natural flow conditions. This approach introduces the sustainable urbanization concept, proposing that the flood should not be transferred in space or time. This way, storage and infiltration measures may be important measures for sustaining adequate drainage conditions. Storage measures should consider detention or retention reservoirs, acting in-line with rivers or in the base of hill slopes, or combined in multifunctional landscapes in parks and public squares, or even in the plot level, as an on-source control option. By its turn, infiltration measures may involve reforestation actions, the use of pervious paving, or infiltration trenches, among others. All these measures, properly designed in an integrated manner, might be able to work preventively or correctively, if necessary, modifying the spatial and temporal distribution of flows, to face the new challenges.

The storage measures, because of their applicability and diversity of use, in different combinations with the drainage net configuration, are highlighted in this conceptual discussion. The reservoirs are able to attack the problem of flooding worsening, both from the point of view of the uncontrolled urban growth, as well as from the point of view of possible climate changes. The storage capacity of these reservoirs allows facing the larger volumes and to control surface runoff released to the network, minimising chances of system failure, with a time of response that matches the velocity of the critical superficial processes that generates floods. Infiltration measures are very important, because they are able to reduce flow volumes, but infiltration process takes more time and, in this case, time may be a critical factor when trying to control floods. So, infiltration measures are desirable, but may usually they do not prescind from storage measures.

PROPOSED SOLUTIONS FOR IGUAÇU-SARAPUÍ RIVER BASIN

The Iguaçu Project, related to the first Water Resources Management Master Plan, was the reference scenario used in this study. After more than one decade, the revision of the Water Resources Management Master Plan for Iguaçu-Sarapuí River Basin started in 2007 and finished in 2009. Lack of an adequate urban land use control and unplanned city growth led to several problems, as discussed previously. In the newer version of The Master Plan, the original set of proposed measures was reviewed. Part of these measures was maintained, especially in consolidated areas; however, whenever possible, new concepts on sustainable urban drainage were introduced. The basin was considered in an integrated way and environmental recovery concerns were added to the new plan. Irregular occupations of risky areas, subjected to frequent flooding, and especially riverbanks occupations, were considered not appropriated and people living in these houses without proper safe conditions needed to be relocated.

Both structural and non-structural measures were proposed for flood control purposes, ranging from short to long-term actions. Some of the proposed actions aiming to give more sustainable solutions considered:

- the maintenance of natural spaces free from urbanisation, preventing vegetation removal and the aggravation of flooding at the consolidated urban areas;
- the recovery of lost vegetated areas;
- a land use regulation and control, by means of the establishment of formal Environmental Preservation Areas;

- the implementation of urban parks;
- the creation of public consortiums for integrated planning of policies for multi-counties interests (recognising the importance of the metropolitan planning);
- the revision and adaptation of the municipalities urban planning instruments.
- In terms of flood control, riverbanks protection and natural vegetation preservation, three types of parks were proposed, as basic measures to be reproduced in a distributed way over the basin, encompassing the following functions (figure 8):
- Fluvial Urban Park – longitudinal parks along rivers, with the purpose of protecting river banks from irregular occupation by low income population.
- Flooding Urban Park – longitudinal parks implemented in low elevation areas to allow frequent inundations, with a storage function intending to help in damping flood peaks.
- Environmental Urban Park – parks with greater dimensions, flat or not, with the purpose of environmental preservation and land use valuing, aiming to minimise runoff generation and maintaining a buffer of pervious surfaces.

The figures 9 and 10 show two more detailed examples of the proposed parks, in practical conditions, being one for Sarapuí River, and another for Iguaçu River.

Complementary actions held by the State include the articulation with every Municipality in the basin, in order to implement the proposed measures, create local conditions for urban land use control and develop environmental education campaigns, with the financing of the Federal Government, through a specific Program of Developing Acceleration (PAC, in Portuguese). Besides, a habitation program is also being conducted in the basin, in order to support and allow people relocation from risky areas to safer near areas.

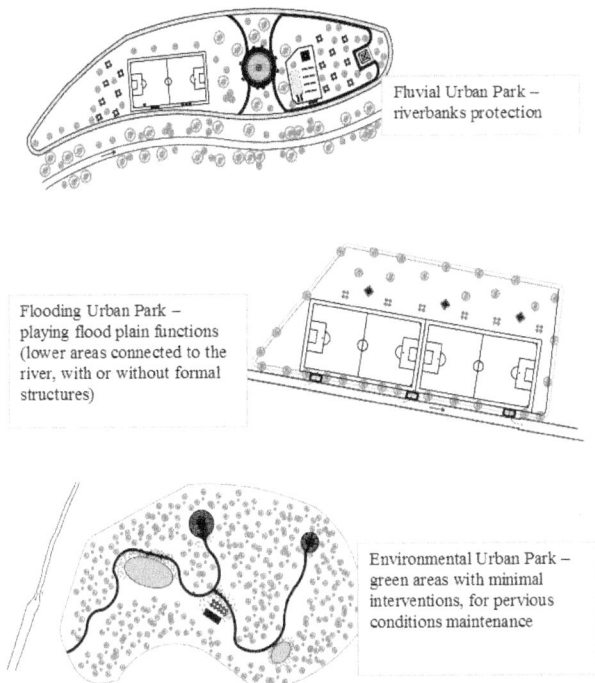

Fluvial Urban Park – riverbanks protection

Flooding Urban Park – playing flood plain functions (lower areas connected to the river, with or without formal structures)

Environmental Urban Park – green areas with minimal interventions, for pervious conditions maintenance

Figure 8: Fluvial parks typology – proposed distributed measures for flood control and environmental recovery

Flooding Urban Park of Gomes Freire polder

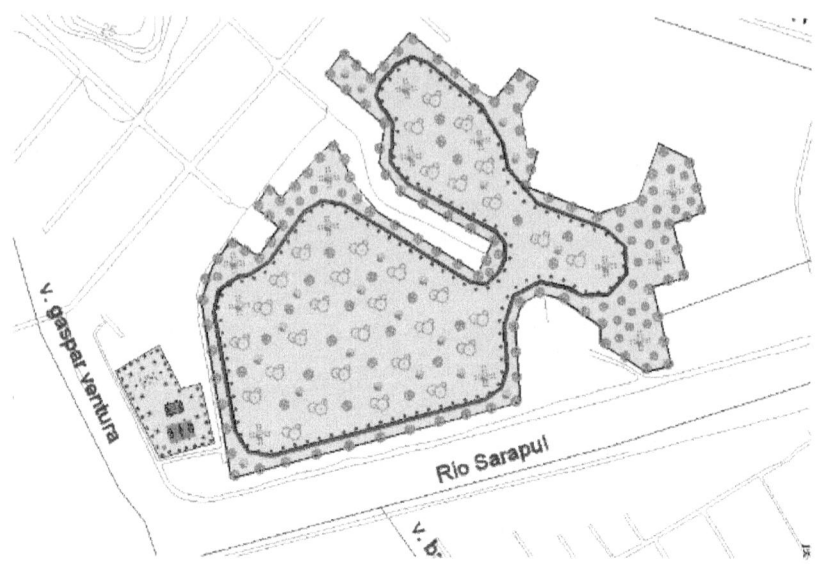

Figure 9: Flooding Urban Park examples – Sarapuí River

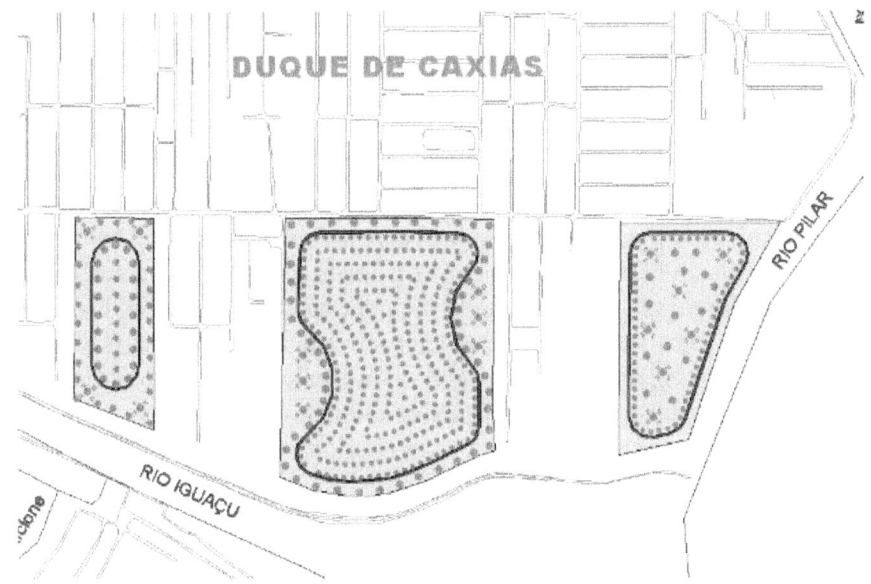

Figure 10: Flooding Urban Park examples – Iguaçu River

CONCLUSION

- Promoting integration of public policies that interact with the water resources is probably the most urgent and complex task on the agenda of public administrators who are really committed to a sustainable future for the metropolitan areas.

- There are reasons to believe that the new institutional arrangements in place in the country offer alternatives for the shared responsibilities involving states and municipalities, mainly in the large urban agglomerations. Specifically, in relation to municipalities, there is a vast spectrum of possibilities to be pursued within the Statute of the City. The new Master Plans can and must incorporate more effective mechanisms for land use management, using a greater range of legal, economic and fiscal instruments focused on urban development on a sustainable basis. However, master plans for urban development still lack mechanisms of inter-municipal coordination and regional agreements orientations that may prevent eventual unintended consequences of land use regulations, from one municipality to another.

- The Iguaçu-Sarapuí River basin still embodies conditions favourable to planning for urban flooding, albeit devised to apply for the long term.

A significant part of its territory remains in the form of areas still not incorporated into the urban fabric – notably the areas situated between the mountains that rise abruptly and the lowland itself. This enables the maintenance of areas with high soil pervious rates, provided that the urban fabric does not expand to those areas.

- The disorderly occupation in Baixada Fluminense lowlands is going to increase the frequency and intensity of the urban floods, causing major damage to the already urbanized areas. The main limiting factor for the expansion of the urban perimeter is the lack of highway connection and regular mass transport lines in the upper parts of the basin, maintaining low occupation rates and rural activities in these areas. It is also worth highlighting the lack of preparation of local administrations to deal with the probable resulting impacts of climate change, above all in urban areas situated at low elevations in relation to the sea level.

- Some of the actions proposed by this study were:

- maintenance of spaces free from urbanisation, preventing the aggravation of flooding at the consolidated urban areas;

- land use regulation and control, by means of the establishment of formal Environmental Preservation Areas;

- implementation of urban parks, mainly for storage purposes, minimising flooding impacts and preparing the basin for future worse climatic conditions;

- creation of public consortiums for integrated planning of policies for multi-counties interests (recognizing the importance of the metropolitan planning);

- revision and adaptation of the urban planning instruments for the municipalities.

- Complementary actions of state responsibility include articulation with every Municipality in the basin, in order to implement the proposed measures, create local conditions for urban land use control and develop environmental education campaigns about the risks of worsening the floods.

ACKNOWLEDGEMENT

The first author is grateful to the Support Program Postdoctoral CAPES/ FAPERJ. The second author acknowledges CNPq for his research support.

REFERENCES

1. Dourojeanni, Axel, & Jouravlev, Andrei. Gestión de cuencas y ríos vinculados con.centros urbanos. C E P A L - Comisión Económica para América Latina y el Caribe,1999.

2. Low-Beer, Jacqueline Doris, Cornejo, Ione Koseki. Instrumento de gestão integrada da.água em áreas urbanas. Subsídios ao Programa Nacional de Despoluição das Bacias.Hidrográficas e estudo exploratório de um.programa nacional de apoio à gestão.integrada. Relatório de Andamento. Extrato de resultados preliminares de pesquisa.(módulo Institucional). Convênio FINEPCT.-HIDRO 23.01.0547.00. Universidade de São.Paulo, Núcleo de Pesquisa em Informações Urbanas, 2002.

3. .Jouravlev, Andrei. Los municipios y la gestión de los recursos hídricos. Serie Recursos.Naturales e Infraestructura. CEPAL - Comisi.ón Económica para América Latina y el.Caribe, n° 66, 2003.

4. Tucci, Carlos E. M (2004). Gerenciamento inte.grado das inundações urbanas no Brasil..Rega/Global Water Partnership South Améric.a. Vol. 1, n° 1 Santiago: GWP/South.América, jan./jun.,2004.

5. .Ibge. Pesquisa de informações básicas munici.pais – suplemento de meio ambiente, 2002.

6. Observatório Das Metrópoles. Como andam.as metrópoles. Relatório Final – 21 de.dezembro de 2005. Disponível em www.ippu.r.ufrj.br/. Acesso em 8/08/2008. ABRH,.João Pessoa, 2005.

7. Britto, Ana Lucia Nogueira de Paiva, & BESSA, Eliane da Silva. Possibilidades de.Mudanças no Ambiente Cons.truído: o saneamento nos novos planos diretores da.Baixada Fluminense. ANAIS do IV Encontro. Nacional da ANPPAS..Brasília, DF, 2008.

8. Parry, M.L., Canziani, O.F., Palutikof, J.P., Van der Linden P.J. and Hanson, C.E. (eds)..Contribution of Working Group II to th.e Fourth Assessment Report of the.Intergovernmental Panel on Climate Change.. Cambridge University Press, Cambridge,.United Kingdom and New. York, NY, USA, 2007.

9. Miguez, M. G. Modelo Matemático de Células de Escoamento para Bacias Urbanas. Tese .(Doutorado em Engenharia Civil), COPPE / UFRJ, Rio de Janeiro, 2001.

10. Mascarenhas, F.C.B.. & Miguez, M.G. Urban Fl.ood Control through a Mathematical Cell .Model. In: Water International. .Vol. 27, n° 2, p. 208-218, 2002.

11. Mascarenhas, F.C.B., & Miguez, M.G. Math.ematical Modelling of Rural

and Urban .Floods: a hydraulic approach. .In: Flood Risk Simulation. .Wit Press, Gateshead, 2005.

12. Zanobetti, D.; Lorgeré, H.; Preissman, A.; Cunge, J. A. , Mekong Delta Mathematical .Program Construction. Journal of the Waterwa.ys and Harbours Division, ASCE, v.96, n. .WW2, pp. 181-199, 1970.

13. Magalhães, L. P. C., Magalhães, P. C., Mascarenhas, F. C. B., Miguez, M. G., Colonese, B. .L., & Bastos, E. T. Sistema Hidro-Flu para .Apoio a Projetos de Dr.enagem. XVI Simpósio .Brasileiro de Recursos Hídricos.

14. Carneiro, Paulo Roberto Ferreira. Controle .de Inundações em Bacias Metropolitanas, .Considerando a Integração do Planejamento do Uso Solo à Gestão dos Recursos .Hídricos. Estudo de caso: bacia dos rios Igua.çu/Sarapuí na Região Metropolitana do Rio .de Janeiro. 2008. IX, 296 p. (Doutorado em Engenharia Civil) Coordenação dos .Programas de Pós-Graduação de Engenharia da.Universidade Federal do Rio de Janeiro .(COPPE/UFRJ), Rio de Janeiro.

15. Laboratório De Hidrologia E Estudo Do Meio Ambiente Coppe/UFRJ - PNUD. Plano .Diretor de Recursos Hídricos da Bacia dos Rios Iguaçu/ Sarapuí, com Ênfase no Controle .de Inundações. Rio de .Janeiro: SERLA, 1996.

Chapter 7

THE ROLE OF TRADABLE PLANNING PERMITS IN ENVIRONMENTAL LAND USE PLANNING: A STOCKTAKE OF THE GERMAN DISCUSSION

Dirk Loehr[1]

[1]Trier University of Applied Sciences, Environmental Campus Birkenfeld, Germany

INTRODUCTION

The idea of tradable planning permits is subject to broad discussion in some developed countries such as Switzerland (for example, see [1]), but particularly in Germany (for example, see [2]).

The German federal government intends to reduce the daily land consumption to 30 ha per day in 2020 [3]. In 13 years between 1993 and 2010, land consumption in Germany was significantly higher than 100 ha per day. In the other 5 years, the undershooting of the 100 ha mark has been mostly due to lower economic growth rates or an economic slump [4]. Particularly rural areas were affected by excessive land consumption. Almost 50% of the converted land is sealed [5].

In order to achieve the 30 ha target, there is a broad consensus about the necessity to support planning by means of economic instruments. In this discussion, tradable planning permits turned out to be the instrument of choice, at least among the scientists. In Germany, a lot of research has been underway on this issue for years now (for example, see [6]). Among others, a pilot project is also in preparation [7, 8], as it was planned in the coalition agreement of the current federal government [9].

The idea of tradable planning permits stems from the concept of tradable CO_2 rights, more accurately the cap and trade system. Within the cap and trade regime, pollution rights should be limited in quantity and made tradable. Due to the cap on the pollution possible, the system is considered ecologically effective. If the mechanism is applied to the field of land use planning, the communal development plans are only legally valid if they are backed by planning permits, which have to be held by the communes. The communes – as the planning authorities – and not the land owners are the holders of the planning permits. This is an important difference from "tradable development rights", where private-sector actors are the sellers and buyers (for example, see [10]).

Due to the trade, the scheme is also regarded as being efficient because only those actors with the lowest marginal abatement costs reduce the emissions. The permits can be bought and sold by the communes on an organized trading platform.

The cap on the permits helps to circumvent rationality traps (game theory) which otherwise would appear. In Germany, for instance, communes competed against each other to attract new inhabitants and industries in order to get more tax revenues and higher shares out of the financial equalization scheme. This competition was a race to the bottom in many cases. Among others, the results in many cases have been almost empty residential or commercial areas and high infrastructure costs. However, if a community waives the preparation of new building areas, the neighbouring municipality takes the chance.

Within the cap and trade scheme, such rationality traps might be broken up [11]: Due to the costs of the permits, only such communes whose benefits of land development exceed the costs of the permits will buy planning permits and carry out land development. If land conversion can be avoided at costs below the costs of the planning permits, communes waive the right to further development. Maybe they can also reconvert the land into a natural state. Hence, if there is no need for holding permits, such communes will sell them to other communes (for example, see [12]). If they do not sell the "free" rights, they will suffer opportunity costs. This means that communes with high marginal abatement costs (tax revenues, jobs etc.) are the buyers of the rights, while communes with low marginal abatement costs are the sellers. In the end, all marginal abatement costs equalize at the price of the tradable permits. In the trading planning permits scheme, the secondary market is the institutional heart of the mechanism.

Although it sounds quite appealing at first glance, we want to show that the application of the cap and trade scheme to land use planning is anything but self-evident and not a promising approach per se.

HYPOTHESIS: NO MAGIC BULLET

Tradable planning permits are considered to be a sort of magic bullet. On the one hand, the cap on development permits makes the system effective. On the other hand, only those communes with the highest benefits (additional taxes and shares from financial equalization schemes) carry out the development. Communes with low opportunity costs waive the right to development. Hence the scheme is also efficient, because the planning rights are used at the locations with the highest benefits.

However, contrary to what intuition would suggest, we want to show that effectiveness and efficiency don't harmonize if the concept is applied to land use planning. In contrast, the cap and trade approach cannot meet the goals of efficiency and effectiveness at the same time ("incompatibility thesis") [13]. The argument is based on the following two statements:

- In order to be efficient, a cap and trade system needs wide system boundaries. At least in small or medium-sized countries, such wide system boundaries go hand in hand with a unified planning permit and a unit price.
- In contrast to CO_2, effective land use planning doesn't require control of a scale, but of a structure. A land use structure cannot be controlled effectively by a single planning permission with a single price.

THEORETICAL ISSUES AND REVIEW OF LITERATURE

There is a central difference between the cap and trade on CO_2 and the cap and trade on planning permits. Considering the consequences for global warming, it does not matter where the CO_2 is emitted due to the diffusion characteristics of the greenhouse gas. Hence the task is to control a scale (maximal CO_2 emissions anywhere) by capping the quantity of emissions. However, regarding land use planning, not only the scale but also the structure of land use has to be controlled. The quantity of land as a whole can hardly be extended. Instead, the relevant issue relates to changes in the structure of land use, which is for instance forestry, agriculture, industry, settlements etc. It is of central importance where the land use takes place and for which purpose. Hence CO_2 permits are a homogenous good, but land use rights shouldn't be.

CONSENSUS: PRIMACY OF PLANNING

At present, the structure of land use is controlled by the planning system. According to the proponents of the tradable development rights idea, land use planning should not be substituted but supported by the economic tool ("primacy of planning") [14].

Planning is necessary to break up a possible Nash equilibrium [11] caused by the behaviour of land owners: If, in the absence of any planning, only the willingness to pay decides about land use patterns, a spatial disaster may result and people may run into a rationality trap. If, for example, German people were allowed to realize the favoured model of the detached one-family house in green surroundings, urban sprawl would happen, with negative ecological, economic and social impacts.

At the same time, planning is necessary to protect such forms with weak financial endowments which cause important positive external effects. If no plan provided public spaces e.g. for kindergartens and schools, such forms would have to compete with actors with a high willingness to pay (e.g. banks). Hence they could not be realized. However, without such facilities, the value of the area would often be lower than with them. Good planning should consider the variety of functions of land (e.g. ecological, spiritual). Such forms of land use that move beyond efficiency and profitability are not only important for the cohesion of the social system, but in many cases also for the resilience of the ecological system (for example, see [15]). Planning has to balance the competing demands of various stakeholders, including groups with low budgets and the protection of nature.

Thesis: Efficiency needs wide system boundaries

Although the primacy of planning is wide consensus, in recent debates it has been argued that tradable planning permits may counteract land use planning [13]. The "incompatibility thesis" is based on the required design of a cap and trade regime. A major justification of the system is its efficiency. The efficiency of the cap and trade system is caused by differences in marginal abatement costs:

- Those communes with high marginal abatement costs buy planning permits on the market at the lower market price. The difference is the benefits from the cap and trade system.

- On the other hand, such communes with low marginal abatement costs reduce their harmful activities and sell the free certificates on the market. The difference between market price and marginal abatement costs is the profit from abatement.

In the end, the marginal abatement costs of all the actors equal the market price of the planning permits. The higher the differences of marginal abatement costs of the acting communes, the higher the efficiency potential of the regime will be.

However, high differences in abatement costs can be achieved by a wide design in terms of space, time, participants and the objects of trade:

- In categorical terms, diverse spatial categories (living, commerce, mixed use, traffic etc.) have to be gathered in one single planning permit ("universal" certificate). This is any land for human settlement and transport infrastructure without regard to its different components;

- Regarding the market participants, the discussion is about also including individuals or NGOs instead of only permitting communes as traders;

- In spatial terms, scientists agree that the trade boundaries have to be as wide as possible (e.g. whole of Germany, no single states);

- Considering the time dimension, banking and borrowing is also discussed in order to use differences in the marginal abatement costs over the timeline.

In Germany, in the last decade the preferred design is characterized by

- A country-wide regime (although the pilot project mentioned above will only comprise selected communes) which incorporates the administrative support of the different states of the federation [16].

- A universal certificate which comprises the whole area for settlement and traffic [16].

- Banking should be allowed (at least the transfer into the following provisional period), in order to allow long-term development strategies for the communes. In contrast, there is much scepticism with regard to providing the opportunity to use the rights before owning them (borrowing) [17].

- Regarding the market participants, an extension of participants beyond the communes has not been discussed seriously so far.

Hence the efficiency potential can only be exhausted if the target is "scale" instead of "structure" (which would make sub-markets necessary). The scale target has to fix wide system boundaries (in contrast to other tradable development rights schemes; for example, see [13]). The scale target and the wide system boundaries are mutually dependent.

There is also another reason why a working trade system would not be possible without wide boundaries: Narrow markets cause high price volatility of the permits. The higher the price volatility, the more insecure the economic success of abatement activities and the less abatement activities will take place. Thus the target is to set the condition for organized trade of the permits.

Thesis: Controlling structure by economic tools needs tightly segmented sub-markets

Having shown the necessity for wide system boundaries in a cap and trade model, the next question is whether a land use structure – not a scale – can be controlled within such a system. We want to illustrate the problem within figure 1 below. The land use plan sets the allowed land use A (e.g. industry) at the maximum of CA. The maximal land use B (e.g. housing) is limited by CB. The marginal abatement costs (MAC) are MACA for land use type A and MACB for land use type B. For simplification purposes, the illustration doesn't include more land use types.

If a commune waives the right to development of additional sites, it has to suffer marginal abatement costs. Such marginal abatement costs are mainly opportunity costs. If, for instance, a residential area is not realized, a German municipality gets lower shares of the income tax revenues, lower property tax revenues and lower revenues out of the fiscal equalization scheme. If an industrial site cannot be realized, the opportunity costs also comprise lost business tax revenues. Also indirect effects have to be considered, such as income multiplier effects which otherwise would have been initiated by construction activities. All these interrelations and effects are quite complex and include feedbacks within the system. Basically, the scale of opportunity costs is not quite clear. Fiscal impact analysis could provide for more cost transparency, but it is in an early stage. Some fiscal impact tools used so far for residential areas turned out to have quite different performance; for industrial areas no reliable fiscal impact tool is available so far. Also within the above-mentioned pilot model of tradable development rights the development of reliable fiscal impact analysis tools is acknowledged to be important in order to get a better idea about the marginal abatement costs.

However, in the subsequent figure we assume, contrary to the facts, that there is an accurate idea about the volume of the marginal abatement costs of land use type A and B. Hence we can derive a mathematical function of MAC, being dependent on the scale of land use of the different types.

First, let's assume that the caps for land use type A (CA) and B (CB) are set according to the land use plans. We assume that the land use planning also properly computes the marginal damages (MD), which are illustrated with the dotted line for both types of land use. With this theoretical "trick" we can take into account that planners care for quantity as well as for quality of land use (for example, see [17]). Therefore, the planning target (CA and CB) corresponds perfectly with the intersection of marginal abatement costs and marginal damage of land use type A (EA) and B (EB).

Moreover, theoretically the marginal damage of land use type B can be expressed in equivalents of the marginal damage of land use type A. Such equivalents are useful for the definition of a universal cap. Analogous equivalents are also used in the greenhouse gas emission permit schemes. In the Kyoto regime for instance, global warming potentials (GWPs) are used in order to express the global warming potential of the other greenhouse gases in relation to CO_2, whose GWP is standardized to 1. Hence, if caps for different land use types should reflect such equivalents (for simplification purposes a linear function is used below), we get for instance a function such as:

$$C_B = e \times C_A$$

(1)

Options The total cap results by aggregation of the caps of the individual land use types:

$$C_{A+B} = C_A \times e + C_A$$

(2)

Options It is important to note that the equivalents only reflect an average consideration. However, different communes may have different structures of land use; thus the equivalents don't represent their individual situation.

In the subsequent diagram, the added marginal abatement costs (MAC_{A+B}) show the aggregate demand curve, and the aggregated cap (C_{A+B}) shows the aggregate supply curve of all land used for settlement and traffic, set by the planning authorities. The intersection point determines the unit price of the universal development right P* [13]. The illustration holds true for an individual municipality as well as for the aggregation of municipalities.

The figure shows why a unit price causes economic incentives to violate the land use plans:

- Regarding land use type A, the mayor in charge will extend the land use until the intersection point (IA) of the unit price P* and the marginal abatement costs MAC_A. From this point on, the costs for additional planning permits exceed the benefits from additional land use. However, regarding the land use plan (and the marginal damage), more development of land use type A would be possible (up to E_A and C_A respectively). Insofar there is a loss of welfare, indicated by the gap between M_A and C_A. In order to support the land use planning, price P_A would be necessary instead of the unit price P*.

- Looking at land use type B, the mayor increases land development until his/her benefits of additional development MAC_B equal the unit price of the development P* (in point I_B or M_B respectively). However, this is much more than the land use plans have fixed (C_B). If the caps reflect the intersection point EB of marginal damage function MD_B and marginal abatement costs MAC_B, this point I_B is also far beyond efficiency (in E_B). In order to get an effective and efficient result, price P_B would have been necessary.

From the figure above we may derive two important results:

- The trading model, which is based on wide system boundaries and a single tradable planning permit, is not able to support land use planning. With a unit price, it is only able to control scale but not structure.

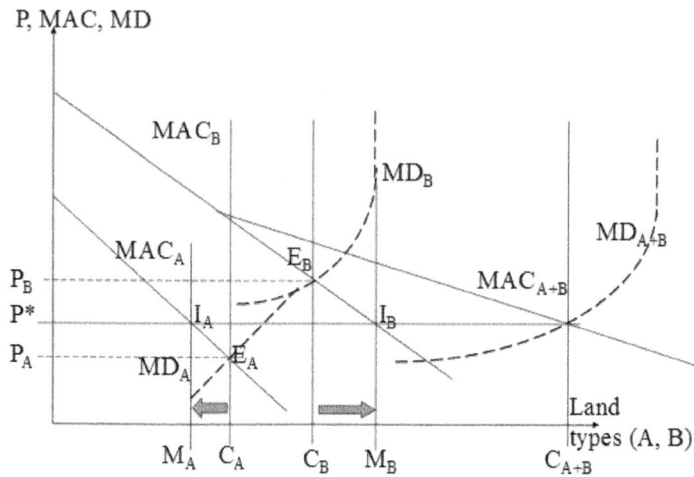

Figure 1: Aberrations with a universal planning permit (adapted from [13])

- Nonetheless, the supporters assert effectiveness. Obviously, they refer to the CO_2 blueprint and only consider the control of scale, but ignore the necessity to control structure. There is no assessment of the welfare losses which are caused by overshooting and undershooting of the planning targets so far (differences between $M_{A,B}$ and $C_{A,B}$). Therefore, also no proper statement about the net efficiency gains (efficiency gains minus the welfare losses, due to overshooting and undershooting) of the cap and trade system can be provided.

Sometimes, supporters suppose that the deviations and aberrations are negligible. Depending on the price of the permits, the position of the caps and the marginal abatement costs, the aberrations and welfare losses can be randomly quite high or low. There is no evidence for a correction mechanism, which might be able to reduce such aberrations systematically. Hence, overshootings and undershootings probably don't equalize in aggregation. The system is as effective as a poor marksman who is currently missing the target in different directions. Aggregating the shooting errors doesn't turn the poor marksman into a champion.

The "incompatibility thesis"

Against this background, we can describe the central system conflict as follows:

- The rationale of the cap and trade regime is efficiency. However, efficiency can only be achieved within a wide design of the system, among others based on a universal certificate (including all categories of land for human settlement and transport infrastructure);

- In contrast, the required primacy of planning can only be maintained within a tightly structured system. In contrast to the controlling of the scale of CO_2 emissions, the settlement structure cannot be controlled by one cap (certificate) and one price, but needs a diversity of prices with a diversity of certificates.

- If a structure is "treated" with a single price, welfare losses have to be expected due to overshooting and undershooting of the planning targets. Hence the net efficiency gains and net welfare effects of a cap and trade regime for land use policy are not clear at all.

The supporters of the tradable planning permits claim that planning should impose a correcting action if necessary. However, our argument is that precisely due to the counteracting economic signals, planning cannot impose such an action.

The problem could theoretically be solved by a variety of tightly segmented sub-markets with diverse price settings. Sub-markets which adapt the categories of land use planning (market segmentation in categorical terms) have been discussed e.g. by Henger and Schröter-Schlaack [17]. Spatial boundaries which limit the trading rights to regions with similar protection status (market segmentation in spatial terms) have been addressed e.g. by Williams [18], Walz et al. [16], Henger and Bizer [19]. However, in most countries such markets would be too small and thus inefficient [17]. At least in the German discussion, the conflict of goals which appears in the cap and trade scheme was decided in favour of efficiency and against market segmentation. Within a system with wide boundaries, such an incompatibility could perhaps be avoided in a few countries with high population and centralized land use planning systems, such as China. However, so far there is no sound research about the minimum size of such markets and the requirements for the land use planning system.

Addendum: Initial distribution

Within a cap and trade system, the initial distribution might be done by auction or by providing the permits to the communes without costs (according to alternative allotment formulas). Simulation experiments with German municipalities related to the cap and trade scheme also showed that the initial distribution of rights is quite a critical issue, which may endanger acceptance [20]. In order to guarantee the acquis of the communes, discussions have so far favoured "grandfathering" schemes, in which the status quo of land consumption is not touched. However, such grandfathering schemes are probably less efficient than auction schemes [13, 17].

RESULTS AND DISCUSSION

Basically, the required primacy of planning is not compatible with the system of tradable planning permits. If the basic idea of capping planning permits should be kept, a redesign of the regime is necessary.

Cap and auction

In order to support the planning system ("primacy of planning"), the planning permits should best be defined in a tight manner which is in line with the categories of the planning system. For instance, if planners are thinking in categories such as residential, industrial or mixed areas, traffic etc., the planning permits should also follow these categories. This also facilitates the handling of the system. Planning permits should basically be mandatory for all sorts of developed areas. For instance, recreational space may have also negative ecological impacts.

The caps could be administered on different administrative levels, e.g. at the level of the states, even at the level of regions (or in other countries: at county level). However, the administration needs a certain human capacity. In Germany, the 30 ha cap might be broken down into lower administrative levels without problems.

However, within a tight definition of a variety of planning permits at a low administration level (e.g. region), the "markets" for each right would be quite narrow. Hence the allocation mechanism shouldn't be based on the secondary market (trade – "horizontal coordination") but on the primary market, namely auctions ("vertical coordination"). The auction of planning permits to the communes could be done periodically. Giving the focus to auction doesn't mean a complete ban on trade but a reduction of its significance. Due to the tight design of the sub-markets, organized trade wouldn't be possible anyway. Instead, over-the-counter trade would be feasible, regulated by the planning permits administration. Within the proposed design, different types of planning rights (residential, industrial etc.) were auctioned and traded at different prices.

The administration should also guarantee that it will take back the planning permits for a fixed price (based on the auction price). Thus communes could also think about changing the land use plans and the redevelopment of shrinking areas into the natural state. Deconstruction would be encouraged, because the communes could be sure about the compensation.

A system which is based on auction might be designed tightly with regard to space, time, participants and objects of the design. The system may work with only a few participants. It may work even at regional level. Within the auction, the development rights are allocated to those communes that can make the best use of them. Moreover, without the overshooting and undershooting of the cap and trade regime, welfare losses also are avoided.

Despite the segregation of sub-markets, the system is also efficient. However, in terms of efficiency it is not clear if such a cap and auction system will or will not compete with a cap and trade system, which is based on a single, universal development right. Nonetheless, if in doubt, recognizing the primacy of planning within the conflicts of goals means subordination of efficiency. From a system theory point of view, economic efficiency shouldn't be a guiding value [15] of superior significance anyway, at least not in land use management. Instead, the guiding value of efficiency has to be balanced with other guiding values.

Completion by means of a financial equalization scheme

The regime sketched out above has to face some serious counterarguments relating to political viability: In an auction, the powerful communes with a high willingness to pay will prevail. Moreover, the financial situation of the communes would be even more strained and the municipalities' acquis would be encroached (see section 3.5.). Hence the question is how to increase the acceptance of a cap and auction model.

On the one hand, certain transition regulations such as free development rights for existing settlements would certainly be helpful, but not nearly enough on their own.

On the other hand, the view of the discussion about climate policy might be promising. Within the Kyoto regime, it has not been possible to put in place effective caps, mainly because the problems of distribution and "climate justice" have not been solved yet [21]. The regime was based on "grandfathering". Hence those countries with the most aggressive occupancy of the atmosphere and most responsibility for the climate problem got most rights. This was considered as being unjust by countries with developing and emerging economies. Basically, the same holds true in terms of land use permits.

However, many of the objections mentioned above could be countered by establishing a redistribution mechanism. Within such a redistribution mechanism, the money paid by the communes in the auction could be firstly collected in a fund ("land trust"), which is administered by the affected communes. Second, the money is redistributed to the communes, preferably according to the number of their inhabitants (other redistribution keys are also possible). In this respect, all inhabitants are considered as "co-owners" of the planning permits, and thus they should participate in equal shares from the revenues of the auction. Considering the CO_2 emission trading schemes, this idea has been popularized by Peter Barnes [22]. Applied to land use planning, a similar redistribution scheme has already been suggested by Krumm [23]. However, his proposal was based on a price-steering basis: Basically, communes should be charged for any new land conversion using a fixed rate per square meter. The money should be pooled in a fund and redistributed to the communes, preferably according to the number of citizens.

If redistribution to the communes were carried out according to the population, the payments into the "land trust" would be according to the land used per capita, whereas the redistribution would be according to the average use per capita. Hence, besides the cap, an additional incentive for a sustainable land use is implemented:

- If the actual land use per capita is higher than the average land use, the commune in charge is a net contributor to the "land trust";
- If the actual land use per capita is lower than the average, the responsible commune is a net beneficiary;
- If the actual land use corresponds to average land use, there is no difference compared with the status quo.

Because every commune tries to get net benefits out of the land trust, there will be a current dynamic incentive to carry out efficient and effective land use management. In terms of microeconomics, the dynamic incentive is pushed by the substitution effect, whereas the income effect is eliminated by the redistribution scheme ("Slutsky equation", see [24]).

Moreover, within this redistribution mechanism, an average access to the planning permits is granted, also for financially weak communes. The redistribution mechanism serves as an ecological financial equalization scheme between the municipalities. Not unlike a lease mechanism, communes with land consumption rates above average pay to communes with land consumption rates below average.

The effects of the redistribution system are far reaching. To mention just some of them:

- Currently, for instance, some German communes can take some "fiscal rents" due to their location, at the expense of other municipalities. This holds true particularly for the communes in the wealthy commuter belt of bigger cities ("Speckgürtel"). They benefit from the migration out of the bigger cities (e.g. young families), which are "bleeding". In such peripheral communes, land prices are often lower and the environmental conditions are often better than in the big cities. However, a great deal of the attractiveness is caused by uncompensated spillovers. According to the central locations principle [25] the bigger cities provide a variety of public goods at the commuter belt's benefit. Thus urban sprawl is fuelled, and the financial performance of bigger cities gets weaker and weaker. However, in the proposed regime, the whole fiscal surplus will be skimmed off dependent on the type of auction. The willingness to pay of the commuter belt's communes includes the expected fiscal rents (from spillovers). The fiscal rents are redistributed to all communes according to the number of the people, also to the bigger cities. Due to the higher density of population, the redistribution scheme will compensate the bigger cities for their efforts.

- The model is applicable in situations of growth as well as in shrinking areas. In aging societies such as Germany, in particular rural regions are affected by shrinking. However, land conversion and land consumption is highest ex urbia. Urban sprawl turns out to be luxury which is increasingly difficult to finance. The redistribution model may stimulate migration to more compact settlements, with a higher supply of public goods. The system would provide an incentive for renaturation measures in rural communes. This would have positive side effects, considering e.g. vacancy rates and the value of existing properties.

- By skimming off the rents from certain types of land use, communes get more indifferent towards land use alternatives. On a regional level, coordination between municipalities and the allotment of certain functions (industry, tourism etc.) towards different communes is easier than today. Thus, integrated approaches of regional development might be put in place without high resistance of the communes affected.

With regard to the technical implementation of the system, some minor problems have to be solved. For instance, in order to create equal conditions in the auction, the communes should pay into the land trust in the same "logical second" as the redistribution happens. This means the communes are only charged or rewarded by the balances (net position of pay-in and pay-out). Moreover, it has to be figured out on which administrative level the system should be applied. Basically, the redistribution mechanism should be tied to the scope of the cap and auction scheme.

One should be clear about the fact that no money for natural protection would be raised within the redistribution scheme. However, modifications are possible: If, for instance, a natural park as a common public good has to be financed, the redistribution could be carried out after first deducting the expenses for covering the park. Such decisions depend on the land trust and the planning authorities. A legal basis for the cooperation arrangement is necessary.

The proposed model may be appealing, but it is not a "silver bullet". The framework has to be completed by other instruments. For instance, the price of real estate may rise due to successful capping of planning rights. Thus, access problems for socially weak groups might be caused. Hence a suitable land taxation system which transfers shares of land rents and land value to the community would be desirable for example.

CONCLUSIONS

The concept of tradable planning permits transforms the idea of the CO_2 cap and trade regime to spatial planning. Analyzing the tool, we have at least to refer to effectiveness (planning, ecology), allocation (economy) and distributional aspects (social).

Regarding effectiveness, there is a broad consensus about the primacy of planning. Any economic tool should support planning instead of substituting it. However, planning land is not the same as planning the maximum permissible load of CO_2 in the atmosphere. The former requires a planning of structure, the latter a planning of scale, since it is irrelevant where the emission takes place.

Planning the structure of land use cannot be supported by a unit price, as a result of a universal certificate for all types of land use (for settlement and traffic). In contrast, a variety of sub-markets are necessary, with a different price setting. Meanwhile, more and more planners are also becoming sceptical about the supporting effects of a cap and trade regime.

Supporting the planning of a structure within a variety of sub-markets may be inferior compared with the efficiency of a cap and trade system with wide system boundaries. On the other hand, efficiency losses due to overshooting or undershooting might be avoided. The efficiency losses might be minimized by auctioning the permits to the needy communes on the primary market ("vertical allocation"). Although trade shouldn't be forbidden, an organized secondary market is dispensable (subordination of a "horizontal allocation mechanism"). Moreover, both systems would have to prevent strategic acquisitions of permits (impediment of development in other communes by an artificial shortage of supply), e.g. by a current devaluation of the permits.

Regarding the blueprint of CO_2 trade, a comprehensive arrangement on a global scale has so far failed due to distribution disputes. Also a cap and auction system for planning permits wouldn't be acceptable particularly for communes with a weak financial endowment if there were no correction. This is the reason why the cap and auction regime should be completed by a redistribution mechanism which is based on equal stakes in the scarce land use opportunities.

However, more research is necessary in order to deal with the details of the counter-proposal outlined in this article. So far, in Germany politics has supported the cap and trade approach; as has the allocation of research funds. Critics who pointed out the incompatibility between effectiveness and efficiency in the cap and trade approach have been pushed aside. This also holds true for the combination of caps, auction and redistribution, which couldn't be assessed so far. However, in experimental simulations the acceptance of the cap and trade regime among the practitioners was obviously not very high. Among others, the results of the cap and trade game turned out to be quite sensitive in terms of an increase of the complexity of the framework [20]. In contrast, at least the redistribution approach of Krumm was highly accepted (here, basically, also no fiscal impact assessment is necessary) [26]. Maybe it is time to widen the scope of the research paradigm to extend beyond the cap and trade regime.

REFERENCES

1. Zollinger F, Seidl I (2005) Flächenzertifikate.für eine nachhaltige Raumentwicklung? – .Ein Konzept für Baden-Württemberg und Erkenntnisse aus der Übertragung auf die .Schweiz. Informationen zur Raumentwicklung 4/5: 273-280.

2. Kriese U (2005) Handelbare Flächenfestset.zungskontingente – Anforderungen an ein .Mittel zur Beendigung des Landschaftsverbrauchs. Informationen zur .Raumentwicklung 4/5: 297-306.

3. Federal Ministry for the Environment, Natu.ral Conservation and Nuclear Safety (1998) .Nachhaltige Entwicklung in Deutschland.– Entwurf eines umweltpolitischen .Schwerpunktprogramms, Berlin, 147 p.

4. Penn-Bressel G (2011) Flächenneuinans.pruchnahme – Wirkungen auf Umwelt, .Städtebau und Ökonomie. Wirtschaftsdienst 11: 800-802.

5. Frerichs S, Lieber M, Preuß T (2010) Fläc.hen- und Standortinformationen erheben und .bewerten – Methoden und Konzepte für ei.n nachhaltiges Flächenmanagement. In: .Frerichs, S, Lieber, M, Preuß, T, editor.s. Flächen- und Standortbewertung für ein .nachhaltiges Flächenmanagemen.t. Berlin: Difu pp. 11-27.

6. Homepage of the "Refina" research programme, funded by the federal

government: .http://www.refina-info.de. Accessed: 2012, Mar 15.

7. Federal Ministry for the Environment, Natu.ral Conservation and Nuclear Safety (2010) .Environment research plan (Umweltforsch.ungsplan) 2010. Project Z 6 – 91 054/84, .Project Number 3710 16 106.

8. Homepage of the project "Forum Flächenzertifikate". Available: .http://www.ufz.de/index.php?de=21103 Accessed 2012 Mar 15

9. CDU, CSU, FDP (2009) Wachstum, Bildun.g, Zusammenhalt, Coalition Agreement. .Berlin.

10. LeJava, J P (2009) Transfer of Development Ri.ghts in New Jersey – A background paper, .New Jersey. Available: http://www.dvrpc.org/TDR/pdf/2009-10_LeJava_.Background_Paper.pdf Accessed 2012 Mar 15.

11. Nash J F (1950) Non-cooperative Games; Ph.D.. Thesis; Princeton University: Princeton, .USA. Available: http://www.prince.ton.edu/mudd/news/faq/topics/Non-.Cooperative_Games_Nash.pdf. Accessed 2012 Mar 15.

12. Krumm R (2004) Nachhaltigkeitskonforme Flächennutzungspolitik – Ökonomische .Steuerungsinstrumente und deren gesellschaf.tliche Akzeptanz, IAW research report, .Tübingen: IAW. 136 p.

13. Loehr D (2006) Handelbare Flächenausweis.ungskontingente: Eine gute Idee auf .Abwegen, in: Zeitschrift für Umweltpolitik und Umweltrecht 4: 529-544.

14. Bovet J (2006) Handelbare Flächenausweisun.gsrechte aus Steuerungsinstrument zur .Reduzierung der Fläche.ninanspruchnahme. Natur und Recht 6: 473-479.

15. Bossel H (1998) Globale Wende – Wege zu .einem gesellschaftlichen und ökologischen .Strukturwandel, Munich: Droemer. 464 p.

16. Walz R et al. (2005) Gestaltung.eines Modells handelbarer .Flächenausweisungskontingente unter Berück.sichtigung ökologi.scher, ökonomischer, .rechtlicher und sozialer Aspekt.e, Final Report. Research Project, funded by the Federal .Environmental Office, Project Number 203 16 .123/03, Dessau-Roßlau. 170 p. Available: .http://opus.kobv.de/zlb/volltexte/2009/.7870/pdf/3839.pdf. Accessed 2012 Mar 15.

17. Henger R, Schröter-Schlaack C (2008) .Designoptionen für den Handel mit .Flächenausweisungsrechten .in Deutschland, Land Use .Economics and Planning – .Discussion Paper, No. 08-02, University.of Göttingen, September. Available: .http://www.uni-goettingen.de/en/80714.html Accessed 2012 Mar 15.

18. Williams R C (2003) Cost-Effectiveness vs. Hot .Spots: Determining the

optimal size of .emissions permit trading zones. University of Texas at Austin Working Paper.

19. Henger R, Bizer K (2008). Tradable Planning .Permits for Land-use Control in Germany. .Land Use Economics and Planning.– Discussion Paper No. 08-01.

20. Küpfer C et al. (2010) Handelbare Fläc.henausweisungszertifikate – Experiment .Spiel.Raum: Ergebnisse einer Simulation in 14 Kommunen. Naturschutz und .Landschaftsplanung 42 (2): 39-47.

21. Wicke L (2006) Das Versagen des Kyoto-Protok.olls in seiner jetzigen Form und seine .strukturelle Weiterentwicklung. Zeit.schrift für Sozialökonomie 150: 3-9.

22. [22].Barnes P, Pomerance R (2000): Pie in the Sky .– The Battle for Atmospheric Scarcity Rent, .Washington. Available: http://community-w. ealth.org/_pdfs/articles-publications/   .commons/   paper-barnes-pomerance.pdf Accessed 2012 Mar 20.

23. Krumm R (2002) Die Baulandausweis.ungsumlage als ökonomisches .Steuerungsinstrument einer nachhaltigen Fl.ächenpolitik. IAW Discussion Papers 7. .Available: http://iaw.edu/i.aw/De:Publikationen:IAW-Reih. en:IAW-Diskussionspapiere: .2002 Accessed 2012 Mar 20.

24. Varian H (1992) Microeconomic Analysis, 3r.d ed., New York: W.W. Norton & Company .Inc. 506 p.

25. Christaller W (1933) : Die zentralen Orte in.Süddeutschland. 2nd ed. 1968. Darmstadt: .Wissenschaftliche Buch.gesellschaft. 331 p.

26. Preuß T et al. (2007) Perspektive Flächenkreis.laufwirtschaft, Vol. 3: Neue Instrumente .für neue Ziele. Berlin / Bonn: Difu. 109 p

Chapter 8

POLICY ARRANGEMENT FOR WASTE MANAGEMENT IN EAST AFRICA'S URBAN CENTRES

Christine Majale-Liyala[1]

[1]Environmental Planning and Management, Kenyatta University, Nairobi, Kenya

INTRODUCTION

Today waste management in developing countries and particularly in East Africa is characterised by the involvement of both state and non state actors. The types of arrangement for service provision range from self-provision through collective action independent of external agencies to indirect state provision through sub-contracting to other agencies – NGOs, private for profit companies, user groups among others. Generally, there is much agreement that monopolistic provision realized entirely through state agencies is unfeasible, undesirable, or simply rather old fashioned [1, 2, 3]. However, there is little consensus on the alternative. According to arguments presented by Joshi and Moore in [4], there is need to look beyond new discourses like New Public management and Public Private Partnerships indicating that the trend now is towards pragmatism, pluralism and adaptation to specific circumstances because the reality in such developing countries is highly diverse. Some services, it is argued, cannot be effectively delivered to the ultimate recipients by state agencies because the environment is too complex or variable, and the costs of interacting with very large numbers of poor households are too high. In such cases, users become involved in an organized way at the local level. There are arrangements therefore that do not fit into standard categories. Some of these unorthodox arrangements are of recent origin, and are seen to constitute (smart) adaptations to prevailing local circumstances. They are widespread in developing countries but they raise many issues. This chapter looks at these arrangements through the lens of policy arrangement approach [5] to help discern which arrangement results in better waste management.

The urban centres studied are Mwanza, Kisumu, and Jinja which are considered primary urban centres coming just after the capital cities in their respective countries. The three centres lie within the lake Victoria basin (See Figure 1) and therefore their individual efforts in waste management contribute to sustainable management of the lake by among other things, reducing the pollutant load into the Lake. These three were chosen because of certain similarities (and differences) but more so because they are all found on the shores of the lake basin as mentioned earlier and they are all primary urban centers which makes them comparable in urban status.

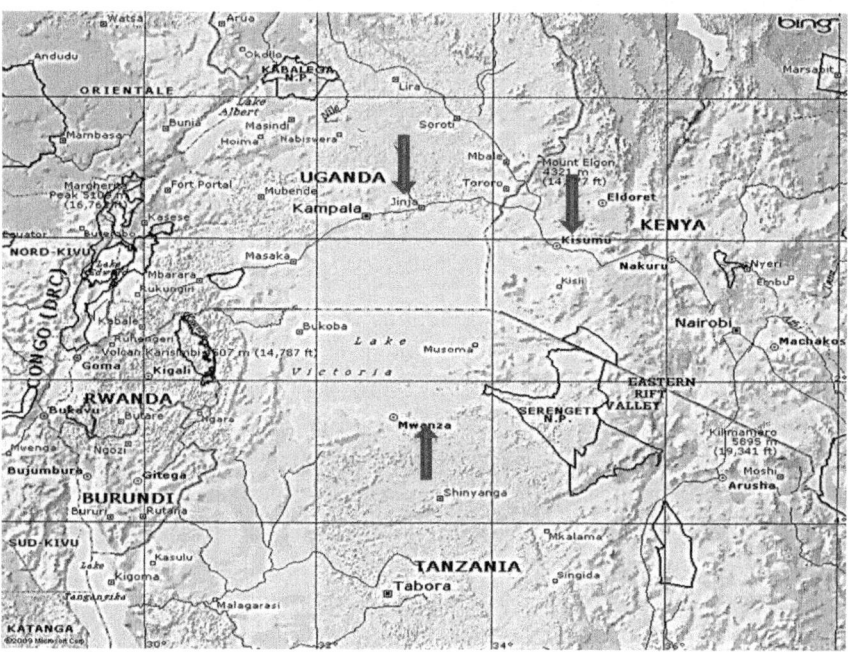

Figure 1: Map of The Lake Basin Showing Location of the Three Areas under Study in their respective countries –Marked With Arrows.

PROBLEM STATEMENT

Solid waste management is particularly a problem for urban centres in developing countries. Growing economies and swelling population numbers from both in-migration and natural growth are continually increasing the urban centers' sizes. These large and growing population is one of the main forces driving the centres overwhelming environmental challenges including solid waste management. Key interventions to addressing the solid waste management challenges could lie in the policy arrangements. This chapter therefore seeks to compare the policy arrangement of three urban centres in East Africa in order to conclude on which arrangement(s) presents the most flexible, robust and sustainable option for solid waste management. Flexible to include both state and non-state actors, robust to keep on running or operating under changing national circumstances (like the economy) and sustainable to contribute to the improvement of environmental conditions.

Profile Of The Urban Centres Studies

Kisumu is the third largest urban center in Kenya after Nairobi and Mombasa. It is located in Nyanza province in the Western part of Kenya. Geologically it sits on the arm of tertiary lava, which extends southwards overlooking the plains to the East and Winam gulf of Lake Victoria to the West. The Lava formation is attributed to the tectonomagnetic activities associated with the Kano-Rift valley system. As a result the city is curved into a trough with the walls of the Nandi escarpment to the East dropping onto the floor of the Kano flood plains and gently flowing to the Dunga wetlands at the shores of the Lake Victoria. Kisumu covers an area of $297km^2$ of land mass and $120km^2$ under the lake. The population of the council has been increasing rapidly, and at a growth rate of 2.8% per annum it was estimated at about 500,000 in 2007 from 322,734 people in 1999.

Jinja is the second largest urban center in Uganda after Kampala city. It is located 81km East of Kampala. It is situated just north of the equator, on the northern shores of Lake Victoria and at the source of the Nile River. The town lies on a tapering plateau with an average altitude of 1230 meters above sea level. The municipality has an extensive shoreline in the east, south and west of both Lake Victoria and the voluminous waters of the Victoria Nile. It occupies an area of $28km^2$. It has a resident population of about 86,512 people (population census 2002) with a day population that doubles that figure due to peri-urban migrant labor. At a growth rate of 2.4% per annum, the population as of 2007 is estimated at 95,121 people, see [6].

Mwanza is the second largest urban center in Tanzania after the city of Dar-es salaam. It covers an area of 1325 km^2 of which $425km^2$ is dry land and 900 km^2 is covered by water. Of the $425km^2$ dry land area, approximately 86.8 km^2 is urbanized while the remaining area consists of forested land, valleys, cultivated plains, grassy and undulating rocky hill areas. According to the 2002 National Census, Mwanza City had 476,646 people. With an annual natural growth rate of 3.2%, the population as of 2007, is estimated at 714,060 people, see [7].

Research Methodology

The study makes use of both secondary and primary data. The collection of secondary data involved a review of literature on the concept of policy arrangement; network,state and market governance as well as Bylaws and other legislative pieces in Kenya, Uganda and Tanzania. The study obtained primary data from interviews with urban authority officials (Public health Officers and Director of Environment at Jinja Municipal Council, Public health officers and solid waste manager at Mwanza City Council and Environmental Officers at Kisumu City Council). Non-state service providers both formal and informal in the three urban centres were also interviewed. In addition, about 600 questionnaires were administered to households and given the differences inherent in the three towns, different approaches were taken to select samples for the household survey as follows:

For Mwanza

In Mwanza, there are two districts and each has urban and rural wards. Urban wards (14 in number) were purposively selected because they receive solid waste management (SWM) services. Every urban ward receives services either from a Community Based Organization (CBO) or a private company working under a contract so all urban areas irrespective of income levels receive SWM services [8]. For the survey, stratified sampling was used. It was a disproportionate stratified sampling because for all the 14 urban wards (which were the strata), on average 13-14 households were interviewed irrespective of population figures in each ward. The goal was to have each ward represented by a minimum number of 10 households and to arrive at a maximum total sample of 200 households. Using the population census data and with the help of the ward leaders and the public health officers posted to each ward, households were randomly picked from a list and questionnaires administered.

For Kisumu:

In Kisumu, SWM is patterned much more along income levels. Previous empirical work, see [9], indicated that the council had not officially permitted non-state actors to operate but all SWM activities by these actors went on unofficially. SWM service providers (the non-state actors) defined their clients on the basis of their income. Community self-help groups were common in low income areas while private companies dominated high and middle income areas. Therefore a list of low, middle and high income estates was made, and then a few estates from each of the three categories were randomly picked

NB: Just like in Mwanza, Kisumu has civic wards, some of these wards are estates in themselves for instance, Nyalenda is a ward and an estate at the same time, while in other wards, there is more than one estate. There are in total 17 civic wards covering 41 estates. There are about 41 estates recognized by the council (11 high income, 17 middle income and 13 low income). The council provides waste management services in only 12 of these estates. The study aimed to administer 200 questionnaires just like in Mwanza. About half the number of estates in each income category mentioned earlier was randomly selected. The study ended up administering questionnaires in 6 high income, 9 middle income and 7 low income estates. About 10 questionnaires were administered in each of the estates selected. The number of households per estate varies from about 3,200 to about 12,000. Selecting households within the selected estates was done differently depending on the kind of estate arrangement. In planned estates like the Railway estates, the houses are numbered and organized in a certain pattern so it was easy to do systematic sampling, selecting every fourth household. In the informal estates like Nyalenda, the houses are not numbered or arranged in any particular order,the researchers were guided by the village names within Nyalenda which were listed and then one household from each village was randomly selected. To avoid covering a village more than once, the study used one research assistant per estate.

The income categorization used to obtain the sample reflects the general pattern of service provision in the town. The estates were classified according to one of the three distinct income categories. From each group of estates about half was randomly selected. Within each selected estate about 10 households were randomly selected. Therefore, based on the argument of inferential statistics, the data can be considered to reflect the situation of service provision among the different income categories in the town as a whole

For Jinja

In Jinja, there are three divisions [6]. Within each division there are parishes but solid waste management services have been contracted out per division and the work is given to two contractors. The divisions are:

- Central Division
- Walukuba Division
- Mpumudde Division

One contractor serves both Central and Walukuba divisions while the other contractor serves Mpumudde division. Service is therefore not structured along income levels as in Kisumu neither per ward as is the case in Mwanza. Waste is collected from skips (collection points) and not directly from the households. Contractors are paid per emptied skip. There are 119 collection points (skips) in Central, 10 in Walukuba and 20 in Mpumudde. Central and Mpumudde Divisions were picked for the study in order to show the differences (if any) in service provided by the two contractors. Frequency of questionnaires administered was higher in Central with about 180 questionnaires randomly distributed and 38 randomly distributed in Mpumudde. These frequencies were more or else in line with the distribution of collection points which are many in Central division. In the end, in total 218 household questionnaires were administered, so the aim to get a total minimum of 200 questionnaires, as for the other two towns, was reached.

With a list of street names (also referred to as roads/avenues/zones which are equivalent of the villages in Kisumu), households (numbering up to 10 in certain streets) were randomly selected from each street.

Theoretical Framework

In East Africa, the national state in each country has been and is the predominant actor when it comes to service provision. Recently though, the influence of market and civil society stakeholders is evident particularly in service provision at the local level. In this paper these developments are analysed using the policy arrangement approach developed by Arts and others [5, 10, 11]. As an approach, it has been mainly used in studies conducted in Western countries but can also provide a framework against which solid waste management is analysed in East Africa given the entry of non-state actors. Policy arrangement is defined as the temporary stabilization of the content and organization of a policy domain at a specific level of policy making. It is temporary because arrangements are under pressure of constant change either by policy innovations on the ground or by processes of political modernization. It has four main dimensions namely:

- The actors and their coalitions involved in the policy domain; and
- The division of power and influence between these actors, where power refers to the mobilization, division and deployment of resources, and influence to who determines policy outcomes and how;
- The rules of the game currently in operation, specifically the formal procedures for pursuit of policy and decision-making; and
- The current policy discourses where the concept of discourse refers to the views and narratives of the actors involved—in terms of norms and values, definitions of problems and approaches to solutions—in this case, the study captures network governance

These four dimensions of a policy arrangement are inextricably interwoven. This means that any change on one of the dimensions induces change on other dimensions. This relationship is symbolized by the tetrahedron, in which each of the corners represents one dimension (Figure. 2).

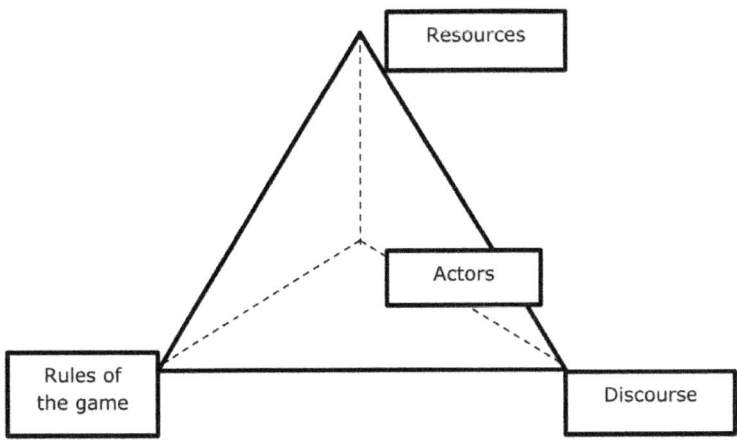

Figure 2: Tetrahedron showing Relation between Dimensions of Policy Arrangement

An analysis of an existing policy arrangement, including its problems or sticking points, concerns all four dimensions of the concept. The methods for mapping out the relevant actors, their coalitions and oppositions are familiar from network analysis. Methods are also available for assessing power relationships. Then existing rules of the game in the arrangement have to be reconstructed: Who decides on the agenda? Who participates in the policy game? Who is excluded? Who takes the decisions? Discourse analysis provides systematic instructions for analyzing the fourth dimension: What are the main concepts in policy discourse and the policy programme? What are the basic assumptions of the policy? What do relevant policy documents contain? How do the various players in the field interpret the policy concepts and basic assumptions?[12].

According to [11], policy innovations can be initiated from each of the dimensions. Policy agents may decide: (1) to allow more or new actors to participate in policy making or in coalition formation; (2) to reshape power relations, for example by adding to or withdrawing resources from a policy arrangement; (3) to reformulate the rules of the game on the basis of which policies are made; and (4) to reformulate the policy discourse concerned, for example by redefining its core concepts. However, innovations in one dimension tend to have consequences for other dimensions, and even for the arrangement as a whole. In other words, in some cases changes have been initiated by new coalitions (e.g. the participation of citizen groups), whereas in other cases they are provoked by innovative discourses, or reinforced by rules and resources, setting off a chain reaction of changes in all aspects. Finally, this chain may lead to the change of *entire* policy arrangements. The approach of policy arrangements helps to analyse such changes.

Changing Political Dynamics In Service Provision

The existence and performance of local authorities in East Africa in service provision, has a historical component depicting changes that have occurred dating back to the 1960's when the three East African countries attained independence. Olowu [13] explains that when these countries attained political independence with formal structures of democratic, representative government, political leaders in their bid to consolidate political power then opted for highly centralized modes of governance. This centralized mode of governance was reinforced by a culture of politics of patrimony in which all powers and resources flow from one source of power ('the father of the nation') to clients to shore up the regime. This pattern of power and resource distribution was strongly supported by both domestic and external actors until the late 1980s. The reasons adduced for adopting this approach included –rapid economic and social development actualized through centralized planning, unity and national integration, containment of corruption and political stability. In fact the argument was that if decentralization would be necessary at all it must be in the form of administrative decentralization or deconcentration—the sharing of responsibilities between central and local administrations which do not exercise any discretionary authority nor dispose of resources. Yet the 1990s marked an era of political and democratic approaches wherein decentralization was progressively being seen (by governments, external actors and the increasingly influential civil society lobbies) as a means of enhancing democracy and citizen participation and (by governments and external actors) as a way of reducing the role, and in particular the expenditures, of the central government [14]. Over time, these changes have necessitated governmental reconfigurations, many of which have a powerful 'local' governance orientation. They include resurgent regional organizations, public private partnerships in infrastructure creation and maintenance and service delivery, decentralisation, devolution and deconcentration of expertise and accountabilities within government departments, and contractual relationships between government and community providers, among others [15]. Non-state actors are increasingly getting involved in service provision. To date though, none of the three East African countries has any municipal service that is completely privatized as yet neither are there distinct policies on privatization of service provision within local authorities. The services most experimented with so far are solid waste management and water supply, but the former more than the latter.

Existing Solid Waste Management Arrangements

The arrangements of SWM in the three urban centres differ. This is captured within the four dimensions of policy arrangement:

Rules

Jinja

Jinja has three administrative divisions: Central, Walukuba and Mpumudde. Due to the efforts to move towards privatization in the country, the council contracted out solid waste management through open bidding, specifically collection and transfer of waste to the disposal site. The arrangement is an annual contract between the municipal council and the private entrepreneurs and in the financial year 2008/2009 two contractors won the tenders [16]. One serves two divisions: Central and Walukuba, while the other contractor provides service to Mpumudde. The legitimacy of the contractors is realized in a number of ways. First the Jinja Solid Waste Management By-Laws 2005, in their objectives recognize the role of private companies in the collection and disposal of waste when this is practiced in a sustainable manner and at a fee. The two contractors serving then were thus officially recognized by the council and in turn by a number of the households they served. Secondly, the tenders for contracts in SWM are advertised through the media and as earlier mentioned, there is open bidding. Contracts for companies that were involved previously can only be renewed on the basis of their performance. As far as decision making is concerned, the municipal authority is still at the helm of SWM, making policies and seeing to their implementation. The contractors do not attend council meetings and are therefore unable to make or influence decision-making. From an interview with one of the contractors however, it became clear that they are free to voice their opinions directly to the town clerk, which may or may not be taken into account when formulating policies.

MWANZA

In Mwanza, there are 21 wards, out of which only 14 wards receive solid waste management services. These are the wards in the urban sections of Mwanza city. Privatization of solid waste management resulted in the council awarding contracts to groups and in the financial year 2008/2009, contracts were awarded to Community Based Organizations and two private companies that serve the wards in the Central Business District [7, 16]. Every other ward is served by one or two Community Based Organizations (CBOs). In terms of legitimacy, it should be noted that first, the CBOs and the two private companies are legitimate organizations, officially recognized by the council and the people they serve as revealed by the interviews and household survey. They are awarded formal contracts after having won through a democratic process. The groups undergo registration as solid waste management service providers and pay a registration fee of Tshs25,000 (USD17.85). The private companies pay taxes to the Tanzania Revenue Authority. Secondly these contractors are well known to the people they serve because the members of these CBOs are local and belong to/are residents in the wards they serve.

KISUMU

In Kisumu, even with the municipality as the central locus of authority, see [17], legitimacy remains a key concern. This is so because there are questions regarding the legal mandate accorded to groups providing SWM informally. Most of these informal groups are registered by the ministry in charge of community development, they are however not formally recognized by law as actors in the domain of solid waste management. The presence and activities of these groups are nevertheless known by the council and some of these groups (see table 1 below) even responded to be operating through some form of 'franchise'[1] - in areas allocated to them by the municipal authority. With no legal papers to show the arrangements they are part of, most if not all of these groups are not legitimate in the SWM arena. This impinges on a number of issues, for instance seeking legal redress in case of payment defaults becomes a problem. Getting donor assistance also becomes a problem because questions will arise as concerning ties to the public, transparency and adherence to the mission of a group, representative status and the relationship between the group and the community served. On the other hand, in terms of community support, openness of information, democratic decision-making, these groups can be considered more legitimate than some official actors are.

Table 1: Legitimacy of groups involved in SWM services Source: Author based on field work

Form of arrangement	Numbers
'Franchise'	11
Quasi contract	1
Partnership	1
Unwritten authority to operate	2
Pay rent to council	1
None	15
TOTAL	31

ACTORS AND COALITIONS

Table 2: Summation of SWM Indicators in the three urban centres Source: Author based on field work

Town/SWM Status	Jinja	Mwanza	Kisumu
Waste collection arrangement	Private collectors formally contracted and the municipality	CBOs and 2 private collectors formally contracted and the municipality	CBOs, private companies and operating informally and the municipality
% of households receiving SWM service	60% (n=218)	82.5% (n=200)	46.5% (n= 200)
% of households that pay for waste collection	N/A	96.4% (n=165)	79.4% (n=93)
Satisfaction of SWM amongst households	62.6% (n=130)	51% (n=165)	70.6% (n=93)
% of waste collected in the towns	40-60%	88%	35-45% (municipal and non-municipal)

Jinja

Looking at the three urban centres, Jinja's SWM arrangement is a close representation of the model of market and networks. This is because of the presence of economic actors in the SWM arena. It is a close representation because these actors are not exactly the determining players in the field of SWM, yet they provide most of the SWM services to the council and its people. The economic actors are private entities in form of companies with several employees as casuals contracted to provide services. The casuals do the collection and sweeping. There are no coalitions or relations between the companies contracted. Field interviews with these contractors revealed that market competition prevents them from having any form of cooperation. As their contracts have an annual nature, they need to stay on top of the game to win the contract the coming year. Within this arrangement, the percentage of households receiving service stands at 60% (where n=218) (see table 2 above).

The networks in this arrangement refer to other actors with shared interests in SWM though the degree of cohesion varies between them. There is the involvement of women and youth groups in the road sweeping and clean-up activities which are done occasionally and mostly on a voluntary basis. There are also environmental groups that are actively involved, including National Environmental Management Authority (NEMA) as a wing of the government which has established pedagogic centers to showcase exemplary activities and is helping to source additional skips to be used in the council. Also involved are international institutions like Lake Victoria Region Local Authorities Cooperation (LVRLAC) who promote exchange of practices amongst the councils member of the organization. International Labour Organization (ILO) and the Lake Victoria Basin Commission (LVBC) have also actively taken part in capacity building.

MWANZA

Mwanza's SWM arrangements comes close to that which can be described as communities and networks. They come 'close' to this model because there are questions on extent to which the CBOs involved can exercise power. Although much of the SWM arena is dominated by CBOs (14 in number and 2 private firms) and these CBOs are the major implementers of the SWM policies, the local authority still dictates these policies, awards the contracts and generally steers everything that has to do with SWM. Unlike the situation in Jinja, as far as the contractors relating with each other is concerned, all the 16 groups interviewed in Mwanza belong to an association called the Mwanza Solid Waste Management Association (MASMA). MASMA meets once every month to share ideas on problem solving and opportunities that can be explored further. Apart from the association, neighboring CBOs (that is CBOs working in neighboring wards) work together in sharing experiences and sometimes even the use of equipment in case the workload is more than expected. This arrangement has resulted to 82.5 % (where n= 200) of households receiving services. As an arrangement it also promotes social sustainability having includes local communities as service providers within their own jurisdictions.

Networks in the Mwanza arrangement are visible in the different actors involved in SWM albeit to different degrees. Apart from the local authority, CBOs and the private companies, just like in Jinja, there is NEMA, different government ministries and regional organizations, in particular, LVRLAC to which Mwanza is a member and the LVBC. ILO has been very instrumental in training the CBOs to earn their income from the waste collection and also urging them to form an association.

KISUMU

The situation in Kisumu comes close to the model of hierarchy and networks. This is the case because the local authority is still solely responsible for solid waste management and the management style is actually still of the command-and-control type. The Department of Environment receives its directions and authority from the line ministry of Local Government and implements them at the local level. Unlike Mwanza and Jinja, Kisumu has no formal/official arrangement that involves non-state actors in collecting and transporting waste or in sweeping the roads. The local authority does the road sweeping itself, as well as the collection, transfer and disposal of waste, but these local authority services are concentrated in the Central Business District and only a few residential areas also benefit from them. Non-state actors (see table 3 below) provide service to most of the other residential areas in an unofficial manner. This is likely to affect the robustness of the system as the non-state actors cannot be held accountable for the SWM services they provide.

Table 3: SWM Groups in Kisumu City Council and Those Interviewed Source: Field work in Kisumu

SWM	Existing Numbers	Numbers Interviewed
Recyclers	23	6
Groups: CBOs and youth groups	27	17
Private companies/individuals	18	8
TOTAL	68	31

In terms of relationships between the non-state actors themselves, the study revealed that they work together during cleanups and some even share their working equipment. Like in Mwanza, they have also formed an association called the Kisumu Waste Managers Association. This arrangement has resulted to 46.5% (where n=200) of households receiving services (see table 4 below for the percentage of service coverage provided by the different actors).

Table 4: Service providers to Households in Kisumu N=200 Source: Household survey in Kisumu.

Service provider	Frequency	Percentage
CBO	21	10.5
Municipality	18	9.0
Private Company	54	27.0
Others	107	53.5
	200	100.0

Just like the other two towns, the networks refers to occasional involvement of other actors like NEMA, LVRLAC, LVBC, UN-Habitat, Practical Action amongst others.

RESOURCES

JINJA

In Jinja, the interviews with public health officers revealed that the municipal authority has a budget for SWM which is mostly funded by the central government. There are also fees collected from business premises (Ugshs20,000 - USD14 per business)[2] - and a dumping fee at the disposal site but all these go to the central reserve at the council where they tend to be absorbed by overall council expenditures. The Director of Environment informed that the authority actually has relative autonomy in putting up its own budget, see [18]. Yet it is evident that because most of the funds come from the central government and they are conditional, it is unlikely that adequate resources would be set aside for waste management compared to other 'important' municipal services. For the year 2008, the budgeted expenditure for waste management was Ugshs69,600,000 (USD49,714) but the actual expenditure came to Ugshs120,000,000 (USD85,714) reflecting a deficit of about Ugshs50,000,000 (USD35,714) on the budget.

Payments from the local authority to the contractors are made as per the number of skips emptied to the disposal grounds. The contractor earns Ugshs28,000 (USD20) per small skip (3 tons) emptied and Ugshs30,000 (USD21.4) for a bigger skip (3.5 tons) emptied. On average the contractor serving the two divisions empties 18 skips per day while the one serving Mpumudde division empties on average 7 skips per day. In addition, the contractors use local authority vehicles for transporting the skips to the dumpsite. They hire the vehicles at Ugshs100,000 (USD71.43) per truck per month. They cover the costs for minor and major repairs and fuel as well. The drivers of the trucks are however employed by the local authority and not by the contractors themselves. This is to allow the authority to control and monitor the disposal of waste because the contractor is paid as per the number of skips emptied. One outstanding aspect of the Jinja arrangement which is not the case for Mwanza and Kisumu (and which is also contrary to the 'markets' arrangement), is that households do not pay for the service. The reticence to pay for SWM is because households are convinced that it is the responsibility of the council to provide the service and at no cost. This conviction remains very strong so that attempts to introduce a fee of Ugshs.2000 (USD1.42) per household per month were not successful. This is likely to affect the robustness of service provision as a fee on waste could probably have been ring-fenced and used on SWM instead of entirely relying on transfers from central government. given that the Service satisfaction from the household survey stands at 62.6% (where n=130).

MWANZA

In Mwanza, the study learned that the council also gets most of its SWM funds from the central government. There are additional funds from the fee charged at the dumping site and also from the fee charged to CBOs to have their waste transferred to the dumping ground but like in Jinja, these funds end up at the central reserve in the local authority. The solid waste manager informed the study that SWM is not properly defined in the authority's overall budget but that their total expenditures for the year 2007 went up to Tshs210,900,000 (USD150,643). This figure like the one in Jinja is small compared to the expenditures in Kisumu and this could be linked to the fact that the costs incurred in Jinja and Mwanza are shared by the council and the private sector contracted to provide SWM services.

The contractors in Mwanza charge different rates for SWM for different land uses but all households pay a standard fee of Tshs400 (USD0.28) per household per month. From the survey 96.4% (where n=165) of households pay for waste collection. Interestingly though only 51% of these are satisfied with service provision. Possible reasons for this could be that the largest percentage of service providers are CBOs who, as it emerged from the field interviews, have no incentive to invest and improve SWM given that it is an annual contract. This could impact on their level of professionalism. It is also possible that the population has very high levels of expectation about the performance of the contractors. A number of the households gave recommendations in line with improving the skills of CBOs, improving the infrastructure used for collection, showing that they expect more than they are receiving.

The CBOs pay Tshs8000 (USD5.7) per trip to the local authority for transferring waste to the disposal grounds. The private companies however take their own waste to the disposal grounds and pay for its disposal. In addition, the CBOs and the private companies that have been awarded the SWM contracts are paid by the city council at a rate of USD1.2 for every 300m length of tarmac road that is cleaned daily.

KISUMU

For Kisumu, SWM receives its funds from a conservancy fee of USD 0.67 charged per household per month through the water bill, but this includes only those households that have metered water connection (that is connected to the central grid). There are also other funds originating from the dumping fee charged at USD1.4 per load of a pick-up truck. Businesses, particularly the markets and other commercial areas, are charged for SWM through their business license. The other percentage of funds for SWM comes from what is transferred from the national government to the local authority - the Local Authority Transfer Fund (LATF). All these contributions combined are however, not sufficient to adequately run the SWM system for the municipality as a whole. In the financial year 2006/07, the annual income from solid waste management was USD70,000 against an annual expenditure of about USD420,000.

Kisumu's scenario differs from the other two urban centres when it comes to the payment systems. The areas that are served by the local authority have their costs taken care of in the water bill and some of the households responded that they pay for waste as part of their house rent. The private companies are operating in open competition and work purely on a willing-buyer-willing-seller basis. From the survey, their services are mostly offered in high and middle income estates. Payments are made at the end of the month as per a verbal agreement with the household. CBOs who operate mostly in middle and low income areas also charge fees agreed upon with each household. Given the high number of informal operators, fees charged for waste collection varies but the average fees for different residential areas are revealed through the household survey as shown in table 5 below.

Table 5: Payment Rates for SWM services. Source: Household Field Survey in Kisumu

Residential area	Payment rates (Kshs./Month)
Low income areas	40.00 – 100.00 (USD)
Middle income areas	150.00 – 250.00 (USD)
High income areas	250.00 – 500.00 (USD)

The survey revealed that 79.4 % (where n=93) pay for waste collection and the service satisfaction from these households stands at 70.6%.

DISCOURSE

When collecting primary data, there was a lot of hype from the local authority officers regarding privatization of solid waste management services. This was particularly the case in Mwanza and Jinja where non-state actors have been formally involved in service provision. The term privatization has been applied to three different methods of increasing the activity of the private sector in providing public services: 1) private sector choice, financing, and production of a service;2) public-sector choice and financing with private sector production of the service selected; 3) and deregulation of private firms providing services. In the first case, the entire responsibility for a service is transferred from the public sector to the private sector, and individual consumers select and purchase the amount of services they desire from private providers. For example, solid-waste collection is provided by private firms in some communities. The second version of privatization refers to joint activity of the public and private sectors in providing services. In this case, consumers select and pay for the quantity and type of service desired through government, which then contracts with private firms to produce the desired amount and category of service. Although the government provides for the service, a private firm carries out the actual execution of it. The government determines the service level and pays the amount specified in the contract, but leaves decisions about production decisions to the private firm. The third form of privatization means that government reduces or eliminates the regulatory restrictions imposed on private firms providing specific services. The cases studied are neither here nor there as far as privatization is concerned and therefore based on discussions presented in [3], the paper settles on network governance to describe the desirable direction for the arrangements in the three urban centres. A network approach to governance is decentring the state as the unique organ for governance and replacing it with pluricentric forms of governance. Networks permit inter-organizational interactions of exchange, concerted action, and joint production in a formal or informal manner. These networks vary in composition from domain to domain, but they are likely to consist of government agencies, key legislators, pressure groups, relevant private companies and civil society organisations such as NGOs and CBOs [19]. Also citizens themselves may be engaged in such network arrangements but this has to be achieved through constructive reciprocities, as they can not be forced through formal state interventions. This network perspective shows many similarities with the popular notion of partnerships in governance seeking to determine the respective roles of public as well as private actors in collaborating to improved public services [20]. Network governance arrangements intend to achieve their objectives through the combined efforts of these different sets of actors, but the respective roles and responsibilities of the actors involved remain distinct while the state is no longer the sole locus of authority. This is however not the case in the three urban centres where the state is still the locus of authority. Yet given the outcome of the surveys and interviews, network arrangement is a desirable option because its flexibility allows diverse social actors to engage actively in finding concrete options for providing sustainable solid waste management services and does not force them to wait until national and urban state authorities are willing and able to engage.

CONCLUSION

The policy arrangements for waste management in the three urban centres are different. The one in Jinja typifies an arrangement in which the market dominates. Here, private collectors are formally contracted to provide service with some assistance from the municipal authority. The arrangement in Mwanza is community dominated with more than 14 CBOs and 2 private collectors formally contracted to provide service with some assistance from the city authority. In Kisumu, the urban authority itself solely provides waste management services. The involvement of non- state actors in Kisumu is unofficial and informal. Jinja and Mwanza arrangements are not privatized per se but looking at the official involvement of non-state actors and the outcome of the household survey in terms of quantities of waste collected, percentages of households covered and SWM satisfaction amongst households in these two towns, the chapter concludes that a balanced arrangement is needed where all societal actors can play their role. It is clear that involving non-state actors as in a network governance arrangement is truly plausible and these actors, both formal and informal, need to work under an effective and strengthened government in order to afford all income groups access to solid waste management services and to ensure flexibility and robustness of the services provided. This means solid waste management services cannot be left to the state actors only. The collective effort by all actors in therefore is likely to ultimately contribute to sustainable management of the Lake Basin. Since the three towns border the Lake Victoria, proper waste management by both the state and non-state actors in respective towns is expected to contribute to the total reduction of the pollution load into the lake waters.

NOTES

[1] - The word franchise is in quotation marks because there are no legal papers to show for it and the arrangement is only franchise by name but not in actual sense.

[2] - At an exchange rate of Kshs 70 to USD 1 at the time the study was conducted. Kshs 1 to Ugshs 20, Kshs1 to Tshs 20

REFERENCES

1.   Callaghy, T. 1993. Political Passions and Economic Interests; Economic Reform and Political Structure in Africa. In: Mkandawire, T. Thinking About Developmental States in Africa. A paper Presented at the UNU-AERC Workshop on Institution as and Development in Africa. Tokyo-Japan: UNU Headquarters; 1998.

2. Lewis, P. Economic Reform and Political Transition in Africa: The Quest for a Politics of Development. World Politics 1996;49 (1) 92-129.

3. Oosterveer, P. Urban environmental services and the state in East Africa; between neo-developmental and network governance approaches. Geoforum 2009; 40 (6) 1061-1068.

4. Joshi, A. and Moore, M. Institutionalised Co-production: Unorthodox Public Service Delivery in Challenging Environments. The Journal of Development Studies 2004; 40(4) 31-49.

5. Arts, B. and Tatenhove , J. Environmental Policy Arrangements: A New Concept. In: Goverde, Henri, (eds.) Global and European Polity? Organizations, policies, contexts. Aldershot: Ashgate; 2000. p 223-237.

6. Jinja Municipal Council. 2007.Brief Profile of the Municipality. Jinja Municipal Council, Uganda.

7. Mwanza City Council. 2007. Mwanza city Brief profile. Mwanza City Council, Tanzania.

8. Mwanza City Council-Refuse Collection and Disposal By-laws of 2000. Mwanza City Council, Tanzania.

9. Majale, C. L. Modernising Solid Waste Management at Municipal Level: Institutional Arrangements In Urban Centres of East Africa. PhD thesis. Wageningen University The Netherlands; 2011.

10. Tatenhove, J., Arts, B. and Leroy, P. editors. Political Modernisation and the Environment. The Renewal of Environmental Policy Arrangements. Dordrecht/Boston/London, Kluwer Academic Publishers; 2000.

11. Tatenhove, J. and Pieter Leroy, P. Environment and participation in a Context of Political Modernisation. Environmental Values 2003; 12 (2003) 155–74

12. Bas, A., Leroy. P and Tatenhove, J. Political Modernisation and Policy Arrangements: A Framework for Understanding Environmental Policy Change. Public Organization Rev 2006 ;6: 93–106

13. Olowu, D. Governance, Institutional Reforms and Policy Processes in Africa In: Olowu D. and Sako S. (eds.) Better Governance and Public policy: Capacity Building for democratic Renewal in Africa. Bloomfield: Kumarian Press ; 2002. p 53

14. Conyers, D. Decentralization and Service Delivery: Lessons from Sub-Saharan Africa. IDS Bulletin 2007; 38 (1) 18-32.

15. Olukoshi, A. Changing Patterns of Politics in Africa. In: Boron, A.; Lechini, G. (eds). Politics and Social Movements in a Hegemonic World: Lessons from Africa, Asia and Latin America. Buenos Aires: CLACSO;

2005.p177-201. Retrieved from- http://bibliotecavirtual.clacso.org.ar/ar/libros/sursur/politics/Olukoshi.rtf

16. Mwanza City Council. Mwanza City Tender Document 2008/2009. Mwanza City Council, Tanzania

17. Republic of Kenya, 1998 - Local Government Act Cap 265. Revised Edition. Nairobi: Government Printers.

18. Government of Uganda- Local Government Act Cap 243. Kampala: LDC Publishers.

19. Keeley, J. and I. Scoones. Understanding environmental policy processes; a review. IDS Working Paper 89. Surrey: Institute of Development Studies University of Sussex; 1999.

20. Baud, I., Post, J., and Furedy, C. editors. Solid Waste Management and Recycling. Actors, Partnerships and Policies in Hyderabad, India and Nairobi, Kenya. Dordrecht, Boston, London: Kluwer Academic Publishers; 2004

Chapter 9

AN INTEGRATED LAND-USE SYSTEM MODEL FOR THE JORDAN RIVER REGION

Jennifer Koch, Florian Wimmer, Rüdiger Schaldach and Janina Onigkeit

Center for Environmental Systems Research, University of Kassel, Germany

INTRODUCTION

The Jordan River region (Israel, Jordan, and the Palestinian National Authority (PA)) is one of.the most water scarce regions of the world. The total renewable water resource values in the.Jordan River region are 52 to 535 m.^{3}.per capita and year [15], which is far below the threshold.value of 1000 m.^{3}.per capita and year indicating chronic water scarcity [14]. On average,.water resources withdrawn for agricultural activities, such as irrigated crop production,.amount to two thirds of the total actual renewable water resources in the Jordan River region.[17]. This makes the agricultural sector the region's major water user and shows the strong.regional impact of agricultural land-use activities on water resources. Besides the effect on.water resources, land-use activities also have a considerable effect on other natural resources.[20]. Examples are desertification caused by maladjusted land management policies [1, 4],. biodiversity loss due to habitat destruction and fragmentation [41, 64], and salinization of land.induced by irrigation [22]. Current pressures on natural resources in the Jordan River region.are likely to aggravate in the future due to high projected population growth rates, economic.development, and changing climate conditions. This may cause a further degradation of the. region's ecosystems and reduce their capacity to provide ecosystem services in the long run..Hence, there is an urgent need for a better understanding of the complex relationships in these.human-environmental systems, in order to develop sustainable management strategies for the.use of natural resources in the Jordan River region..Water resources in the Jordan River region are largely transboundary and their distribution.between Israel, Jordan, and PA is a potential source of conflicts. Hence, strategies for.sustainable natural resource management in this region have to capture regulations at the.state level and have to be based on consistent assessment methods and collaboration between. the parties involved. This makes modeling approaches operating at the small scale or.approaches focusing solely on natural systems unsuitable. However, existing integrated.modeling approaches that cover the entire Jordan River region, such as presented in the Global.Environmental Outlook 4 [58], apply spatial resolutions that are too coarse to capture the.biophysical heterogeneity in the region, which is governed by a steep precipitation gradient.[13].. In order to gain a better scientific understanding of the linkages between natural resources,.land management, and ecosystem functioning in the Jordan River region, we developed the.integrated modeling system LandSHIFT.JR (Land Simulation to Harmonize and Integrate.Freshwater Availability and the Terrestrial Environment - Jordan River). LandSHIFT.JR is.based on the spatially explicit land-use model LandSHIFT [49] and covers Israel, Jordan, and.PA. It applies a cellular automata approach to calculate land-use changes and corresponding.irrigation water requirements under current and future climate conditions. LandSHIFT.JR.operates on a regular grid with a spatial resolution of 30 arc seconds. It was tailored.specifically to the environmental and socio-economic conditions in the Jordan River region.[28, 29]. Since scarce water resources, vegetation degradation due to overgrazing, detrimental. effects of climate change on crop yields and irrigation water requirements, and increased soil.salinity caused by irrigation are the major environmental issues in the Jordan River region,.LandSHIFT.JR explicitly addresses these issues. This distinguishes LandSHIFT.JR from other.integrated land-use modeling systems operating at a similar spatial resolution and scale, e.g..the CLUE(-S) model [61, 62]. LandSHIFT.JR simulates the spatial and temporal dynamics of.land-use systems in the Jordan River region and allows exploring the impact of alterations.in socio-economic and biophysical conditions on the spatial distribution and intensity of.land-use activities and the feedback of land-use changes on socio-economic conditions. The.modeling system's main field of application is the simulation of spatially explicit, mid-.to long-term scenarios of land-use change. These scenarios show trends in land use and.support the identification of hot spots of change and competition for land. Thus, spatially.explicit land-use change scenarios generated with LandSHIFT. JR provide scientific support.for evaluation and formulation of sustainable land-use planning and promote informed.decision making..The objective of this chapter is to provide a comprehensive description of the integrated. modeling system LandSHIFT.JR and of its validation. Moreover, we present an example of.an application of LandSHIFT.JR - a scenario-based assessment of land-use changes in the.Jordan River region. In section 2, a short description of the biophysical conditions and the.most important land characteristics of the

Jordan River region is provided. In sections 3 to 5,.a detailed description of LandSHIFT.JR is given. The description focuses on the underlying.concepts of the modeling system as well as on the distinctive features of LandSHIFT. JR. The.structure of these sections follows the "Overview, Design concepts, Details" protocol for.model descriptions as proposed by [23] and is based on a description of an earlier version.of LandSHIFT.JR [30]. Sections 6 and 7 provide overviews of the validation of LandSHIFT.JR.and the results of an application example, respectively. The chapter closes with a discussion.of and conclusions on the integrated modeling system, its validation, and simulation results.in section 8

The Jordan River region

LandSHIFT.JR was developed for Israel, Jordan, and PA (Fig. 1). The Jordan River region is.bordered by Lebanon and Syria in the North, by Iraq and Saudi Arabia in the East, by Egypt.in the Southwest, and by the Mediterranean Sea in the West. The region ranges from 34.22.∘.E,.29.19.∘.N to 39.30.∘.E, 33.38.∘.N. The terrain in the Jordan River region is very heterogeneous. The. Coastal Plain, stretching along the Mediterranean Sea, is flanked by the Negev desert in the Southeast and a mountainous region in the East and Northeast. In the North, the mountains.force the coastal air masses to rise and, as a result, induce relatively high precipitation amounts.[13]. A key physiographic feature of the Jordan River region is the Great Rift Valley in which.the Jordan River, Lake Tiberias, and the Dead Sea are located. With 407 m below sea level, the. Dead Sea marks the lowest point in the region and on the Earth's surface. The highland area.in the Western part of Jordan, located along the Great Rift Valley, rises to elevations of 1200.m above sea level and drops gradually in elevation towards the East, where it develops into.the Jordan desert plateau [13]. The point with the highest elevation in the Jordan River region.is the Jabal Umm ad Dami, located in the South Jordan desert, with about 1854 m above sea level.

The climate in the Northern, Central, and Western part of the Jordan River region is.Mediterranean, characterized by hot, dry summers and cool winters [13]. In the residual.part of the Jordan River region a semi-arid to arid climate predominates. A dominant feature.of the regional climatic conditions is the steep precipitation gradient, ranging from 900 mm.mean annual precipitation in the Northern tip of Israel to less than 50 mm in the desert areas.in the South of Israel and the South and Southeast of Jordan. Temperatures also exhibit a high.spatial variability across the Jordan River region with cold winters and hot summers in the.mountainous regions and more moderate extremes in the Rift Valley and the Coastal Plain.[13].

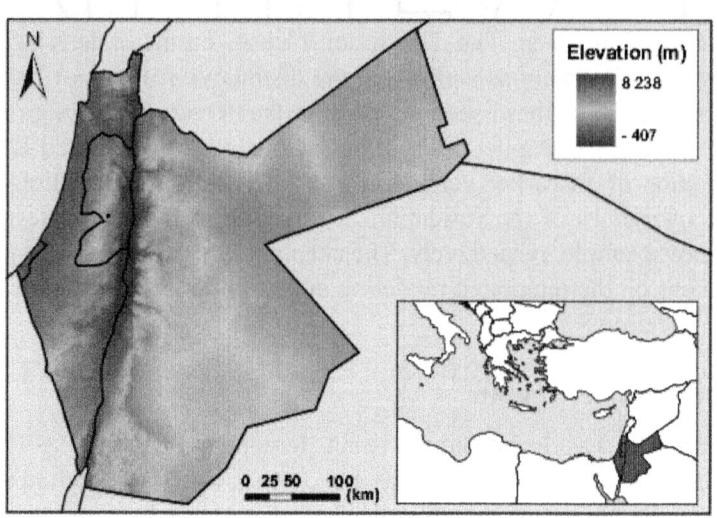

Figure 1: The study region covers Israel, Jordan, and the Palestinian National Authority

The Jordan River region covers about 116 thousand km.2.of land area and 1 thousand km.2.of.inland water area. Approximately 76.1% of the land area in the region is located in Jordan,.18.7% in Israel, and 5.2% in PA [18]. About 2600 km.2.in the region are forest area. Arable land.and permanent cropland sum up to about 9200 km.2.. Approximately 3000 km.2.in the Jordan. River region are equipped for irrigation. Thereof about two thirds are located in Israel. About.14 million people live in the Jordan River region [18]. The largest cities in the study region are.Amman, Jerusalem, Tel Aviv, and Gaza

OVERVIEW

Purpose

LandSHIFT.JR is a regional version of the integrated modeling system LandSHIFT [49]. It was.adjusted and further developed to specifically simulate the spatial and temporal dynamics.of land-use systems in the Jordan River region. LandSHIFT.JR was designed for exploring.the effects of changes in socio-economic, climatic, and biophysical conditions on the spatial.distribution and intensity of land-use activities. In addition, LandSHIFT.JR serves as a tool to.formalize knowledge on and gain new insights into the functioning of land-use systems in the.Jordan River region. It can be used to test hypotheses about processes and interlinkages within.land-use systems, promote the understanding of these systems by identifying key processes.and their interlinkages, and, as a result, reveal demands for future research activities.. LandSHIFT.JR's main field of application is the simulation of spatially explicit, mid- to.long-term future scenarios of land-use and land-cover change. These scenarios explore.possible trends in land use and visualize alternative land-use configurations. Main model.output are maps displaying changes in land-use patterns. These maps help to reveal hot spots.of land-use change and allow for the identification of priority areas for further research or.focus areas for alternative management strategies. By these means, spatially explicit land-use. change scenarios generated with LandSHIFT.JR provide scientific support for the evaluation.and formulation of sustainable land-use planning and promote informed decision making.

State variables and scales

The representation of land-use systems in LandSHIFT.JR is operationalized on interacting.spatial scale levels. On these scale levels, the state variables of the modeled land-use systems.are defined. In total, there are four different spatial scale levels:

- Macro level:.The spatial definition of the macro level is based on states. The state of a.macro-level entity (i.e. a state) is specified by the state variables.population.,.crop demand.,.livestock number.(goats and sheep),.yield change.driven by technological progress, and.fraction of irrigated crop production.in total crop production. The state variables. crop demand.,.yield change., and.fraction of irrigation area.are specified separately for each crop category..Changes in macro-level state variables constitute driving forces of land-use change in.LandSHIFT.JR..

- Intermediate level I:.The spatial scale hierarchy of LandSHIFT. JR includes a level based.on natural regions. This scale level allows including information on crop demands with.a higher spatial resolution such as the output from economic land-use models [26]. The.only state variable specified on this level is.crop demand.. This state variable is specified.separately for each crop category; a change in.crop demand. constitutes a driving force of.land-use change in LandSHIFT.JR..Crop demand.can only be specified on one spatial scale.level. In case it is specified on the macro level, it cannot be specified on the intermediate. level I and vice versa..

- Intermediate level II:.The spatial configuration of the intermediate level II is specified.by a regular grid with a spatial resolution of 0.02 decimal degrees (dd), which equals approximately 2.2 km at the equator. The state variables defined on this level include.potential.irrigated wheat yields., potential.rainfed wheat yields., and.net irrigation water.requirements.. Changes in potential yields are considered to be drivers of land-use change..

- Micro level:.The geographic area of each state is specified by the micro leve.l-ar.egular.grid with a uniform cell size of 30 arc seconds, which equals about 0.00833 dd or 1 km.at the equator. The state of a micro-level grid cell is specified by the state variables.dominant.land-use type.,.settlement area.,.population density.,.stocking density.for sheep and.goats,.net primary productivity.(NPP) of rangeland and natural vegetation,.relative human.appropriation of net primary production.(rel. HANPP [25, 29]), and.crop production..

The latter.is defined separately for each crop category. Furthermore, a set of quasi-static landscape.characteristics (e.g. slope) and land-use constraints (e.g. conservation areas) are defined.on the micro leve LandSHIFT.JR applies a 5-year time step. The length of the simulation period depends.on the research question the respective simulation experiment is supposed to answer and.typically ranges between 20 and 50 years. After each simulation step, LandSHIFT.JR writes the.simulation results to files. This output comprises micro-level maps displaying the dominant.land-use and land-cover type, population density, net irrigation water requirements, stocking.density, and rel. HANPP. Moreover, the output includes a set of indicators and area statistics. aggregated to the macro level

Process overview and scheduling

The processes implemented in LandSHIFT.JR are organized in three modules (Fig. 2),.which operate on the different spatial scale levels by modifying the scale-specific state.variables. The.Biophysics module., which describes the environmental subsystem, comprises.process representations for the calculation of potential irrigated and rainfed wheat yields, net.irrigation water requirements, and NPP of rangeland and natural vegetation. All the variables. provided by the Biophysics module are climate dependent and, hence, differ between climate.scenarios. This module operates on the intermediate level II and on the micro level. The.Socio-economy module.and the.Land Use Change module.(LUC module) represent the.human subsystem. The Socio-economy module provides information on population growth,.agricultural production and trade (implemented via the state variables crop demand and.livestock numbers), and yield change due to technological progress. The processes of this.module operate on the macro level and, in case crop demands are specified with a higher.spatial resolution, also on the intermediate level I. For each simulation step, the Biophysics.module and the Socio-economy module are executed and the corresponding state variables.are updated. Subsequently, the updated information is used by the LUC module to simulate.changes in land use and land cover. The processes of the LUC module operate on the.micro level. Bidirectional information exchange between the modules is implemented via.the exchange of the state variables..The LUC module calculates the extent and location of land-use and land-cover changes..Therefore, it implements four land-use activities: housing and infrastructure, irrigated crop.production, rainfed crop production, and livestock grazing. The processes representing.the different land-use activities are organized in submodules:.METRO.for housing and

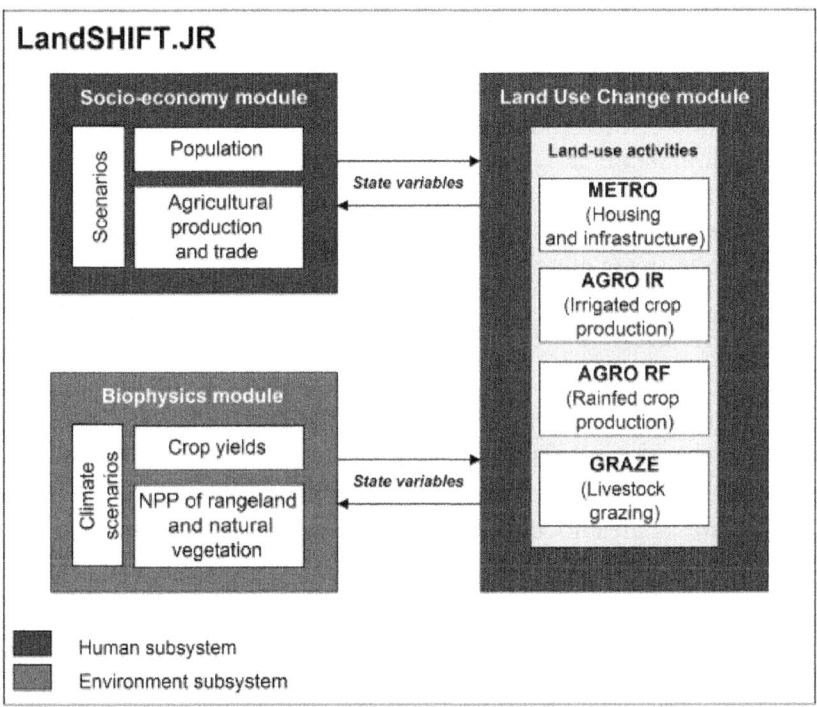

Figure 2: Conceptual structure of the integrated modeling system LandSHIFT.JR, adapted from [48]..infrastructure,.AGRO IR.for irrigated crop production,.AGRO RF.for rainfed crop.production, and.GRAZE.for livestock grazing.

The competition between these activities for.land is addressed by a ranking of the four activities, which defines the sequence of execution..The ranking can be defined flexibly based on the research question; a straightforward way.of ranking land-use activities is to follow their economic importance:

METRO.AGRO IR.AGRO RF.GRAZE. In one simulation step, cells occupied by a superordinate land-use.activity are unavailable for a subordinate land-use activity..In every simulation step, each land-use activity submodule executes the functional parts.demand processing.,.preference ranking., and. demand allocation.. This complies with the.generalized structure of spatially explicit land-use change models as presented by [63]..First, within the demand processing part, driving forces of land-use change are converted.to macro-level/intermediate level I demands for services (e.g. housing) and agricultural. commodities. Second, within the preference ranking part, the suitability of the micro-level.grid cells for the different land-use activities is assessed, resulting in suitability maps. The.grid cells are then ranked based on their suitability. Third, within the demand allocation part,.each land-use activity manipulates the dominant land-use type as well as the corresponding.state variable (population density for METRO, irrigated crop production for AGRO IR, rainfed.crop production for AGRO RF, stocking density for GRAZE) of the best-suited micro-level.grid cells, in order to meet the demand for the service or agricultural commodity under.consideration. The range and magnitude of change is constrained by the demand for the.service or agricultural commodity on the one hand and by the supply, i.e., the productivity on.the particular micro-level grid cells on the other hand

Design concepts

The choice of design concepts was guided by the purpose of the modeling system as described.in section 3.1. LandSHIFT.JR combines a set of different design concepts that can be specified.as a dynamic, integrated, process based, and spatially explicit..Research questions that land-use modeling typically addresses are related to the timing and.rate of land-use changes [35]. A prerequisite for the representation of the temporal behavior of.land-use systems is a dynamic modeling approach [63]. LandSHIFT.JR applies such a dynamic.modeling approach. It subdivides the simulation period into several time steps and, hence,.fulfills the basic requirements for the simulation of land-use change trajectories, feedbacks,.and path dependencies in the evolution of land-use systems..According to [2], integrated modeling systems have to include information from more than.one discipline, organize information in a modularized program structure, and link scientific.findings with policy analysis. LandSHIFT.JR was developed to bring together information.from different disciplines to support decision making and it is typically applied in the.context of scenario analyses with strong relevance for land-use planning and policy [29, 31]..Furthermore, it provides a framework for the combination of biophysical and socio-economic.information with geographic information in form of a modularized program structure..LandSHIFT.JR applies a process-based modeling approach in order to describe the land-use.systems of the Jordan River region as human-environmental systems and to explore the.interlinkages between their subsystems. The modeling system includes representations of.the key processes resulting in changes in human-environment systems. As pivotal process,.LandSHIFT.JR implements human decision making with regard to the extent, location, and.intensity of land-use activities. The process-based approach allows analyzing trajectories and.intermediate states of land-use and land-cover change [63]..Spatially explicit land-use models simulate changes in land use for individual spatial entities.[63]. In case of LandSHIFT. JR these spatial entities are cells of a regular grid. Spatially explicit.models, such as LandSHIFT.JR, are able to simulate the location and spatial variability of.land-use and land-cover changes and, as a result, enable the analysis of the interlinkages.between socio-economic and biophysical environments as well as variations in location and.quantity of land use.

DETAILS

Initialization

Since LandSHIFT.JR integrates data from different fields and sources, considerable effort.is required to synchronize the different datasets in an initial simulation step. In order.to harmonize the information on population density and the land-use/land-cover map.information on urban areas, LandSHIFT.JR initially reads the basic land-use/land-cover map.(derived from the MODIS global land cover dataset [21]). This information is then combined.with the micro-level information on population density [8]. On micro-level grid cells, at which.the land-use type is not "urban", but where the population density exceeds the upper limit for.population density in rural regions, the land-use type is changed to "urban". Furthermore,spatial information on population density is combined with information on per capita area.demands [12] in order to calculate the settlement area on non-urban grid cells; this area is not. available for land-use activities such as crop production or livestock grazing.. Available land-use/land-cover map products for the Jordan River region do not distinguish.between area under crop for different crops. Furthermore, grazing areas are not assigned.separately. Hence, an initial distribution of rangeland as well as area under crop for.the considered crop categories has to be generated artificially. In order to derive the.initial distribution of area under crop, LandSHIFT.JR distributes areas for the different crop.categories to the best suited micro-level grid cells (see section 5.3.3). These areas under crop. are derived from national statistics for Israel.1., Jordan.2., and PA.[3].for the year

2000. Based on.the MIRCA2000 dataset [44], double cropping is assumed for rainfed and irrigated vegetables.in all three states. The applied area values are displayed in Table 1. The best suited micro-level.grid cells, on which these areas are distributed, are preferably those cells that are categorized.as "cropland" in the underlying land-use/land-cover map. The distribution of areas under.crop is carried out separately for irrigated and rainfed crop production. In case grid cells.categorized as "cropland" in the underlying map are not categorized as one of the considered.crop categories during initialization, their land-use type is set to "other crops" and kept.static for the rest of the simulation run. This is based on the assumption that "cropland".in the original land-use/land-cover map also includes areas covered with crops that are not.contained in one of the considered categories and that for these crops no drivers are specified.as model input

Table 1: Cropland and rangeland areas derived from national statistics and FAOSTAT [18] used to initialize cropland and rangeland area distribution in LandSHIFT.JR

State	IR fruits [km²]	IR vegetables [km²]	IR cereals [km²]	RF fruits [km²]	RF vegetables [km²]	RF cereals [km²]	Rangeland [km²]
Israel	661.4	253.6	643.0	162.6	22.1	1206.9	1480
Jordan	348.2	155.3	110.3	521.3	9.1	1045.5	7910
PA	81.6	67.1	26.8	1092.9	19.9	440.4	1500

Based on the resulting land-use type distribution, LandSHIFT.JR relates the macro-level.production.pcens.c.for each of the irrigated and rainfed crop categories.c.(derived from.census data) to the sum of the local production on grid cells with that crop category in the.newly generated map. pcalc.c.. This is done in order to calculate a separate management.parameter. base.c.for each of these categories. The management parameter is defined as.base.c.=.pcens.c./.pcalc.c.. It accounts for inconsistencies between different data sources and.uncertainties due to agricultural management strategies (e.g. multiple cropping, fertilization).that affect the total production of a crop but are not explicitly considered in LandSHIFT.JR..Crop production values applied in this context are.4.: about 1.78 million tonnes of fruits (Israel:.1.304 million tonnes, Jordan: 0.232 million tonnes, PA: 0.246 million tonnes), about 3.01 million.tonnes of vegetables and melons (Israel: 1.643 million tonnes, Jordan: 0.825 million tonnes,.PA: 0.541 million tonnes), and about 0.27 million tonnes of cereals (Israel: 0.183 million tonnes, Jordan: 0.044 million tonnes, PA: 0.041 million tonnes). The crop specific management.parameter is evaluated for initial conditions and is applied in the following simulation steps.in order to adjust the yield values and, as a result, transfer potential crop yields into actual.yields. Based on irrigated area under crop, rainfed area under crop, and the adjusted crop.yields, the fraction of irrigated crop production in total crop production is calculated. This.parameter is also invariant..There are two different modes available in LandSHIFT.JR for calculating the initial distribution.of rangeland and the related stocking densities for small ruminants: a production-driven. approach and an area-driven approach. For the production-driven approach, the forage.demand is allocated to the best suited micro-level grid cells and the land-use type of these.cells is converted to "rangeland". The forage demand is calculated from livestock numbers.derived from statistical data and the forage demand per animal [40]. Livestock numbers used.in this context were derived from FAOSTAT [18] and amount to 0.4 million sheep and goats.in Israel, 2.2 million sheep and goats in Jordan, and 0.9 million sheep and goats in PA. For the.area-driven approach, rangeland area for the year 2000 (also derived from FAOSTAT, Table 1).is allocated to the best suited micro-level grid cells. The land-use type of these grid cells is.set to "rangeland". The stocking densities on the rangeland cells are then calculated from the.local NPP of rangeland and natural vegetation and the forage demand per sheep or goat.

INPUT

LandSHIFT.JR input comprises time series on population, crop demands, yield change due.to technological progress, livestock numbers as well as socio-economic information, e.g. on.environmental policies or regional planning. For the application example presented in this.chapter, this information is derived from the participatory scenario exercise of the GLOWA.Jordan River project.5.. An overview of the scenarios and the corresponding values for the.drivers of land-use change is given in [6]. Besides the above mentioned input specified on.the macro level and/or intermediate level I, LandSHIFT.JR requires data on landscape and.land-use characteristics specified on intermediate level II and on the micro level. This category.of input includes potential crop yields and NPP under current and future climate conditions.or landscape attributes such as slope or river network density. A detailed description of the.data input on landscape and land-use characteristics is given in section 5.3.4..

Submodels

The processes in LandSHIFT.JR are organized in the three submodels Biophysics module,.Socio-economy module, and LUC module. The details of these submodels are described in.this section

Biophysics module

In each simulation step, the Biophysics module updates the state variables potential irrigated.wheat yields, potential rainfed wheat yields, net irrigation water requirements, and NPP of.rangeland and natural vegetation. The updated information is then provided to the LUC

module. The calculation of wheat yields and irrigation water requirements is based on the.output of the dynamic, process-based crop growth model EPIC [67, 68]. In order to include.future progress in the agricultural sector such as new management methods or fertilizers into.the crop yield calculations, yields are corrected with a state-specific factor for yield change.due to technological progress. The calculation of the state variable NPP of rangeland and.natural vegetation is based on output of WADISCAPE [34]. In contrast to the wheat yield.calculations, no effect of technological change on productivity is taken into account. This.is based on the assumption that small ruminant grazing in the Jordan River region usually.takes place on largely unmanaged marginal lands. The impact of changing climate conditions.is considered for wheat yields, irrigation water requirements, and NPP. This is realized by.a correction for climate change based on a linear interpolation between the productivities or.water requirements calculated for current climate conditions and the respective productivities.or water requirements calculated for future climate conditions given by regional climate.simulations for the Jordan River region [53]

GEPIC

We applied GIS-based EPIC (GEPIC) [37], a combination of the crop growth model.EPIC [67, 68] with a GIS, to simulate wheat yields and crop water requirements under current.and future climate. EPIC has been used successfully to simulate crop yields under a wide.range of weather conditions, soil properties, and management schemes [37]. EPIC works on.a daily time step and considers the major processes in the soil-crop-atmosphere-management.system [67]. We used simulated potential yield under rainfed and optimal irrigated conditions.for wheat as a proxy crop type. In order to derive irrigated/rainfed yields for the crop.categories fruits, vegetables, and cereals from irrigated/rainfed wheat yield, an additional.processing step was required. We multiplied the grid cell values of potential wheat yield by.the ratio of mean actual yield for an irrigated/rainfed crop category to the mean potential.yield on irrigated/rainfed areas covered by the crop category. This step was based on values.for the year 2000. The actual yields stem from IMPACT model [45] calculations that were also.used to provide input on future crop production. By this means, we ensure the consistency.of yield values between the various model drivers and inputs. At the same time, we are able.to include spatial and temporal variability of the crop yield simulations with GEPIC in our.analysis

WADISCAPE

The WADISCAPE model [34] provides information on stocking capacities.6.as well as information on the relationship between stocking density with small ruminants.(goats and sheep) and productivity of green biomass.7.under current and future climate.conditions. WADISCAPE simulates the growth and dispersal of herbaceous plants and.dwarf shrubs in artificial, fractal wadi landscapes (wadiscapes) of 1.5 km.×.1.5 km. The.main exogenous driver of vegetation dynamics in WADISCAPE is water availability, which.is calculated from precipitation under consideration of topography. The simulation of.vegetation dynamics is based on validated small-scale models of annual plants [33, 34] and.dwarf shrubs [38]. WADISCAPE simulations were conducted for five climatic regions (arid.to mesic Mediterranean) and, in factorial combination, five varying slope categories (0.°.to.30.°.). In order to determine the stocking capacity of the vegetation, these simulations were.conducted for stocking densities ranging from 0 to 10 animals per hectare.

Socio-economy module

The Socio-economy module operates on the macro level and, if crop demand information.with a higher spatial resolution is included, additionally on the intermediate level I. The.module accounts for the organization and processing of the state variables population, crop.demand, livestock number, and changes in crop yield due to technological progress. For.historical periods, information on these state variables is derived from statistical databases.(e.g. FAOSTAT [18]). For future periods, this information is typically generated with. participatory scenario development, following the SAS approach [3] and the economic model.IMPACT [45]. An update of the state variables is carried out by the Socio-economy module.within each simulation step.

IMPACT.

The International Model for Policy Analysis of Agricultural Commodities and Trade.(IMPACT), a representation of a competitive global market for agricultural commodities [45],.was designed for the analysis of current conditions and possible future developments in food.demand, supply, trade, prices, and malnutrition outcomes. The model covers 32 commodities.and 36 countries/regions linked through trade and, hence, accounts for almost all of the.world's food production and consumption. IMPACT is based on a system of supply and.demand elasticities implemented via linear and nonlinear equations. It incorporates demand.as a function of price, income and population growth, and changes in crop production. The.changes in crop production are determined on the basis of crop prices and productivity.growth rates [45].

SAS

The SAS (Story And Simulation) approach to scenario analysis [3] combines both,.quantitative and qualitative aspects of scenarios. The combination of those two aspects makes.the resulting scenarios on the one hand generally understandable and on the other hand.suitable for planning purposes. Distinctive features of SAS are the iterative structure and.the intensive participation of experts and stakeholders. A detailed description of SAS is.provided by [3]; a description of the SAS application in the context of a scenario analysis.for the Jordan River region is given in [6]. Results from this scenario analysis were used to.derive information on future development of population and livestock numbers in the Jordan.River region for the application example (see section 7

Land Use Change module

The LUC module is the central component of LandSHIFT.JR. The module accomplishes the.simulation of the location and quantity of land-use and land-cover changes. This is realized by.a regionalization of the macro-level/ intermediate level I demands for area intensive services.and agricultural commodities to the micro level. The basic principle is to allocate the demands. to the most suitable micro-level grid cells by changing the land-use type, population density,.crop production or livestock density of as many cells as required to meet the demand..Each service or commodity is linked to a land-use type. The LUC module implements the.submodules METRO (housing and infrastructure), AGRO IR (irrigated crop production),.AGRO RF (rainfed crop production), and GRAZE (livestock grazing). In every simulation

step, these four submodules are executed subsequently and each of this submodules executes.the three functional parts demand processing, preference ranking, and demand allocation. In.the following, the general operating mode of the functional parts is described

Demand processing.

The functional part demand processing is responsible for the.transformation of the drivers of land-use change to macro-level/intermediate level I demands.for the implemented services and commodities

Preference ranking.

The functional part preference ranking operates on the micro level and.

serves for the identification and ranking of the preferred grid cells for the different land-use.activities and the corresponding land-use types. A method from the field of Multi Criteria.Analysis [11] is applied in order to calculate the preference values of the grid cells for the.different land-use types. The preference value.ψ.k.of a grid cell.k.is calculated as

$$\psi_k = \underbrace{\sum_{i=1}^{n} w_i f_i(p_{i,k})}_{\text{suitability}} \times \underbrace{\prod_{j=1}^{m} g_j(c_{j,k})}_{\text{constraints}}$$

(1)

with $\sum_i(w_i) = 1$ and $f_i(p_{i,k}), g_j(c_{j,k}) \in [0,1]$. The first part of the equation is the sum of the different weighted suitability factors.p., contributing to the suitability of a grid cell.k.for.a particular land-use type. The weights.w.determine the importance of a suitability factor in.the analysis. The factor weights were determined according to the CRITIC method [10]. This.method allows the calculation of "objective weights" on the basis of the contrast intensity.between the evaluation criteria, i.e., the standard deviation of normalized criteria values and.the inter-criteria correlation. The second term of the equation is the product of the land-use.constraints.c.. These constraints reflect important aspects of human decision making, e.g..land-use restrictions in conservation areas. One constraint implemented in LandSHIFT.JR.is the transition between the different land-use types: not all land-use and land-cover types.can be converted into each other. These conversion elasticities are a frequently used method.in the field of land-use modeling [62]. A summary of all possible conversions is given in Table

Table 2: Land-use transition matrix. Possible conversions are indicated by "+", impossible conversions are indicated by

From / To	Urban	IR cropland	RF cropland	Rangeland	Set aside	Natural veg.
Urban	+	-	-	-	-	-
IR cropland	+	+	-	-	+	-
RF cropland	+	+	+	-	+	-
Rangeland	+	+	+	+	-	-
Set aside	+	+	+	+	+	-
Natural veg.	+	+	+	+	-	+

The suitability factors, their weights, and the land-use constraints are specified on the macro.level and implemented as time-dependent variables. This enables the representation of.changing policies and environmental boundary conditions. Suitability factors and constraints.are normalized by factor-specific value functions.f.and constraint specific value functions.g...Value functions, based on logistic regression analysis, can be defined as positive or negative

relationships and are scaled by the range of the respective factor within a state in order to.account for the spatial heterogeneity..Suitability factors and constraints can be state variables, landscape attributes, zoning.regulations, or spatial neighborhood characteristics. The neighborhood of the micro-level.grid cells is analyzed in each simulation step in order to generate information about the.land-use/land-cover type of the adjacent cells. The neighborhood of a cell can be defined.by type and order, e.g. von Neumann or Moore neighborhood. Additionally, a (geographic).search radius can be specified. The set of relevant suitability factors and land-use constraints,.the types of value functions, and the factor weights can be derived either by data driven.procedures (e.g. geostatistical analysis) or by expert knowledge (e.g. by means of the.Analytical Hierarchy Process [46])..Demand allocation..The functional part demand allocation assigns the macro-level.and intermediate-level I demands for the implemented services and commodities to the.micro-level grid cells with the highest preference for the associated land-use type. For this.functional part, each land-use activity implements its own allocation strategy

METRO.

The submodule METRO simulates the spatial and temporal dynamics of area.for housing and infrastructure. Changes in quantity and location of this area are.driven by alterations in population numbers, specified on the macro-level. The demand.allocation procedure for METRO distinguishes between municipal regions and rural.regions. Depending on the category, a different kind of growth process is applied..Therefore, the micro-level grid cells are grouped into these two categories. A municipal.cell is defined as a cell that features the land-use type "urban" or has at least one grid cell.with the land-use type "urban" in its direct neighborhood. All other cells are defined as.rural cells. The growth of urban areas is implemented as urban encroachment process [69],.i.e., new area for housing and infrastructure is located at the edges of existing urban area.[54]..In order to allocate additional population, a three-step procedure is applied. First, a.parameter defines the fractions of the additional population that is assigned to municipal.and rural regions, respectively. Second, depending on the grid cell's actual population.density and suitability values, additional population is allocated. On cells with the.land-use type "urban", an upper threshold for population density is defined, which limits.the population amount that can be allocated to these grid cells.

Third, based on the.recalculated population densities, land-use conversions are calculated: rural cells feature a.threshold value for population density. In case, this population density value is exceeded,.the land-use type of the grid cells is changed to "urban"..In rural regions, each cell has a fraction of settlement area that is occupied by housing and.infrastructure. The amount of settlement area on a rural grid cell is calculated based on.population density and the per capita area demand [12]. The area not required for housing.and infrastructure is available for other land use or land cover that specifies the dominant.land-use type on rural grid cells. On grid cells with the dominant land-use type "urban",.all area is required for housing and infrastructure.

AGRO IR and AGRO RF.

The two AGRO submodules AGRO IR and AGRO RF are.separate submodules that are executed one after another (see section 3.3). AGRO IR is.responsible for the allocation of the crop categories irrigated fruits (excluding melons),. irrigated vegetables and melons, and irrigated cereals. AGRO RF allocates the crop.categories rainfed fruits (excluding melons), rainfed vegetables and melons, and rainfed.cereals. The crop category definition is based on the crop type aggregation of the FAOSTAT.database [18]. Both, AGRO IR and AGRO RF, follow the same general approach regarding.demand allocation, and are hence described jointly..The basic principle of the demand allocation part in AGRO is to formulate a."compromise-solution"-problem for the calculation of a quasi-optimum crop allocation, in.order to deal with the competition for land resources between the different crop categories..This is implemented as a modified version of the Multi-Objective Land Allocation (MOLA).heuristic [11]. This heuristic resolves emerging conflicts by a pair-wise comparison; cells.claimed by more than one crop category are allocated to the category with the higher.preference value. In LandSHIFT, the heuristic was modified in two ways [49]. First, the.modified version allocates crop demands instead of a given area. Second, pattern stability.is considered in the conflict resolution step, i.e., the land-use patterns remain constant if.no changes in crop demands occur..The amount of crop production on a micro-level grid cell is based on the local production.P.. The local production.P.for a crop category.c.at simulation step.t.for a particular grid.cell, is defined as

$$P_c(t) = base_c \times y_c(t) \times (1 + tech_c(t)) \times a_c(t)$$

(2)

P.c.(.t.).micro-level grid cell production of crop category.c.for simulation step.t.[Mg],.base._c.management parameter for crop category category.c.[-],.y._c.(.t.).micro-level grid cell yield for crop category.c.in simulation step.t.[Mg km^{-2}.],.tech.c.(.t.).technology-induced yield change for crop category.c.in simulation step.t.[-],.a.c.(.t.).available cell area for production of crop category.c.in simulation step.t.[km^2.]..The crop production.P.is computed by combining state variables from different spatial.scale levels (crop yield and yield change) and the cell area.a.that is not used as settlement.area. The local crop yield is updated in each simulation step by the Biophysics module in.order to include changes due to alterations in climatic conditions. The management factor.base.is a proxy for agricultural management characteristics (see section 5.1), which are.not directly taken into account by LandSHIFT.JR. If not enough suitable land resources.are available to allocate the crop demands, unmet demands are documented in a text.file. In case more cropland was allocation in a previous simulation step as required in.the following simulation step, the land-use type of dispensable cells is converted to "set.aside" (fallow)..

GRAZE..THE GRAZE

submodule accounts for the spatial and temporal dynamics of.livestock grazing. Changes in quantity and location of grazing area, which has the land-use.type "rangeland", are driven by alterations in livestock numbers (sheep and goats) given.in livestock units (LU), specified on the macro-level. Based on the livestock number, the.amount of required forage, which has to be provided by grazing land, is calculated. This.is done under consideration of the daily forage demand per LU and the share of grazing.in feed composition. The residual share in feed composition is assumed to be covered by.crops and crop residues and is considered indirectly in LandSHIFT.JR..The demand allocation part of GRAZE is based on a relationship between grazing.intensity (stocking density) and local biomass productivity (NPP of rangeland and natura...vegetation). This relationship is specified by non-linear correlation functions between.stocking density (number of small ruminants per hectare) and green biomass productivity. (tonnes per hectare), calculated with WADISCAPE [34]. The correlation functions were.generated for all combinations of five slope categories (0.°.to.< .5.°.,≥.5.°.to.<.12.5.°.,≥.12.5.°.to.<.17.5.°.,≥.17.5.°.to.<.25.°.,≥.25.°) with five categories of mean annual precipitation (Table.3). Areas with mean annual precipitation values, that are not covered by the WADISCAPE.simulations (values below 80 mm mean annual precipitation) are not suitable for livestock. grazing. Except for micro-level grid cells located in these areas, each micro-level grid cell.is attributed to one of the correlation functions depending on the grid-cell value for slope.and mean annual precipitation.

Table 3: Mean annual precipitation categories in WADISCAPE [34].

Category	Mean annual precipitation [mm]
Arid	≥ 80 to < 200
Semiarid	≥ 200 to < 400
Dry Mediterranean	≥ 400 to < 500
Typical Mediterranean	≥ 500 to < 700
Mesic Mediterranean	≥ 700 to < 960

Two different allocation routines are available for calculating the initial distribution of rangeland and the corresponding stocking densities:.1. Demand-driven approach: The forage demand is allocated to the preferred micro-level. grid cells and the land-use type of these cells is converted to "rangeland". The local.biomass productivity is calculated from the non-linear correlation function that is valid.for the respective grid cells, assuming no former grazing activity on these grid cells..Based on the available biomass productivity, the local stocking density is calculated.under consideration of the forage demand per animal..2. Area-driven approach: Instead of a forage demand, a certain amount of rangeland area.(Table 1) is allocated to the micro-level grid cells. The land-use type of these grid cells.is converted to "rangeland". The potential total biomass production on these grid cells.is calculated from the non-linear correlation functions, assuming no former use of these.cells as rangeland. Based on the potential biomass production on the resulting area, the.stocking density is adjusted and assigned to the grid cells, in order to meet the forage. demand..In order to calculate the local biomass productivity in the following simulation steps, the.cell's correlation function is chosen and combined with the stocking density set in the.initial simulation step. The actual stocking density is then calculated from this productivity.via the forage demand and assigned to the grid cell. In the subsequent simulation step,.this stocking density is then used to derive the new local productivity from the cell specific. correlation function. This procedure is repeated for each simulation step [29].. An important effect of this feedback between stocking density and biomass productivity is.the resulting self-regulation of the grazing system: the allocation of high stocking densities.in one simulation step results in reduced biomass productivity in the following simulation.step and, hence, lower stocking densities. In addition to the dynamic calculation of local.biomass productivity, change in biomass productivity due to climate change, also derived.from WADISCAPE calculations driven by regional climate simulations [53], is considered..The GRAZE demand allocation part features two different methods for rangeland.management: (1) sustainable rangeland management and (2) intensive rangeland.management [29]. These allocation modes use micro-level grid cell specific information.on stocking capacities calculated by WADISCAPE. The allocation modes apply different.procedures in case the local stocking density exceeds the stocking capacity (overgrazing)..In case of sustainable management, the local sustainable stocking capacity defines the. maximum possible stocking density at a grid cell. The sustainable stocking capacity is.a user defined fraction of the maximum stocking capacity. Each time the stocking density,.assessed from local biomass productivity, exceeds the sustainable stocking capacity of the.grid cell, the stocking density is set back to the sustainable stocking capacity, i.e., no.overgrazing is allowed. For intensive management, this limitation is not applied and.the stocking density is exclusively limited by the local biomass productivity. For both.managements, the upper limit for stocking density is 10 animals per hectare, given by the. range of the WADISCAPE simulations [34].

Besides the above mentioned land-use/land-cover types urban, irrigated fruits (excl. melons),.irrigated vegetables (incl. melons), irrigated cereals, rainfed fruits (excl. melons), rainfed.vegetables (incl. melons), rainfed cereals, other crops, set aside, and rangeland, a set of.other types exist. These are: forests, cropland/natural vegetation mosaic, grassland, shrub.land, woody savannah, barren land, water, and wetlands. Changes in those are not directly. simulated by LandSHIFT.JR but result from land-use conversions of the land-use types that.area covered by METRO, AGRO, and GRAZE.

Submodel parameterization

METRO.

For METRO, two suitability factors were considered: terrain slope [59] and travel.time to major cities [57]. In Table 4, all suitability factors and their weights for the.different land-use activities are displayed. As land-use constraint, conservation areas were.implemented. As a result, no new urban area can be allocated in conservation areas. Spatially.explicit information on national and international nature conservation area was derived from.the World Database on Protected Areas [66]. The basic principle of METRO is to convert. the population to a cell specific population density value. For this purpose, one part of.the population is allocated to urban areas; the residual part is allocated to rural areas. The.fraction of population allocation to urban areas is 65 % [58]. In case that the rural population.density exceeds 5000 people/km.2., or the area demand for housing and infrastructure on a.grid cell exceeds 80 % of the grid cell size, the land-cover type of the grid cell is changed to."urban". The maximum population density per grid cell is 26098 people/km^2., derived from. the population density map for the study region for the year 2000 [8]..AGRO

IR..For AGRO IR, six different suitability factors were considered. Besides terrain.slope and travel time to major cities, additionally area equipped for irrigation [52], irrigated.crop yields, population density, and river network density were considered. Irrigated crop.yields were calculated with GEPIC and vary with time based on changes in climate conditions..Population density for the year 2000 is derived from the CIESIN dataset [8] and is updated by

Table 4: Suitability factors and corresponding weights for the different land-use activities.

Activity	Suitability factor	Factor weight
METRO	Terrain slope	0.366
	Travel time to major cities	0.634
AGRO IR	Area equipped for irrigation	0.233
	Irrigated crop yield	0.145
	Population density	0.049
	River network density	0.262
	Terrain slope	0.147
	Travel time to major cities	0.164
AGRO RF	Population density	0.067
	Rainfed crop yield	0.311
	Terrain slope	0.258
	Travel time to major cities	0.364
GRAZE	Population density	0.044
	NPP of rangeland/nat. veg.	0.529
	River network density	0.256
	Terrain slope	0.170

LandSHIFT.JR over the course of the simulation. The river network density is calculated as.the line density of rivers per grid cell, based on the RWDB2 river-surface water body network.dataset [19]. As land-use constraints, conservation areas are considered. Furthermore, a.risk map on soil sensitivity to adverse effects of irrigation with treated wastewater [47] was.included and can be used for future studies.

AGRO RF..For AGRO RF, four suitability factors were considered. These are rainfed crop.yields, slope, travel time to major cities, and population density. Rainfed crop yields were.calculated with GEPIC and vary with time based on changes in climate conditions. The only.land-use constraint for this activity is conservation area..

GRAZE..For GRAZE, the four considered suitability factors are NPP on rangeland and natural.vegetation, slope, river network density, and population density. In conservation areas, the.use as rangeland is constrained. The information on NPP is derived from

WADISCAPE.calculations. To derive the forage demand from the livestock number, we assume one sheep.or goat to equal 0.125 LU [51]. In addition, we apply a regional factor for Israel (0.8) and.Jordan/PA (0.42) that considers the geographical variability in animal body size [51]. The.daily forage demand per goat or sheep is 1.35 kg dry matter [40] of which we assume 30 % to.be covered by grazing [4]. The consumable part of the aboveground green biomass is 75 %...

MODEL VALIDATION

We applied three different methods to validate our modeling system. First, we validated

the underlying assumptions of the suitability assessment with the Relative Operating

Characteristics (ROC) method [43]. Second, we used the MODIS land cover dataset for the years 2001 and 2005 to perform a map comparison analysis using version 2.0 of the Map.Comparison Kit.[8].. Third, we compared macro-level simulation results on area under crop for.the different irrigated and rainfed crop categories for the year 2005 with the corresponding.values from statistical databases...

Relative Operating Characteristics

The agreement of simulated and observed land-use change depends on the agreement of both.the quantity and location of change. Only if the simulated quantity of change equals the.observed quantity of change, the simulated land-use changes can agree perfectly with the.real land-use changes. On contrary, if the simulated quantity of change equals the observed.quantity of change the location of simulated change can still lead to disagreement of modeled.and real land-use change..The ROC method [43] allows assessing to what degree the model is capable to assess the.right location of change independently of the simulated quantity of change. Hence, the.ROC analysis can be used to validate the underlying preference ranking processes that guide.the location of changes in land use and land cover, represented by a suitability map. For. this purpose, an independent real-change map indicating observed land-use or land-cover.changes is necessary. Since the ROC method is only meaningful for testing the suitability map.for the conversion of any land-use/land-cover type to one single land-use/land-cover type.at a time, we applied the method to validate the suitability maps for each land-use activity.separately. Whereas the suitability map for a specific land-use activity is a direct model output.of LandSHIFT.JR, the categorical real-change map had to be constructed. For this purpose,.observed raster maps for two points in time were compared to each other. Grid cells that.feature a land-use or land-cover change between these two points in time were categorized as.change.cells whereas all other cells are categorized as.non-change.cells..The ROC method compares the real-change map to a sequence of virtual simulated land-use.change maps that result from a successively increasing quantity of change. The maps are. derived by assuming that land-use change occurs on cells where the suitability value exceeds a.certain threshold. Typically, the minimum, the deciles, and the maximum of the distribution of.the suitability values are used as thresholds to prepare a sequence of maps assuming land-use.change on 0 % to 100 % of all cells in 10 %-steps. In order to compare each of these maps to.the real-change map, the rates of true positives (TP) and false positives (FP) are calculated. A.cell is counted as a TP if real land-use change is modeled correctly. In contrast, if simulated.land-use change coincides with non-change in reality, the cell is counted as a FP. The rates of.TP and FP are computed as the ratio of the number of TPs and the number of possible TPs.and the ratio of the number of FPs and the number of possible FPs, respectively. Based on.the results of each comparison in the sequence, the ROC-diagram is constructed by plotting a.curve in a coordinate system with the FP-rate on the x-axis and the TP-rate on the y-axis. The.ROC curve starts at the point (FP = 0, TP = 0), resulting from the assumption of zero simulated.land-use change, and ends at the point (FP = 1, TP = 1), resulting from the assumption that.land-use change is simulated on all cells. The performance metric of ROC, the area under the.curve (AUC), is calculated by trapezoidal approximation. On average, a random suitability

assessment results in a value of AUC = 0.5. In contrast, a suitability map that assigns the.n.highest values to the.n.cells where real change occurs (the perfect suitability map) yields AUC.= 1. Hence, an AUC-value between 0.5 and 1 indicates that the suitability assessment explains.the location of change better than a random process..We performed three separate ROC analyses for the land-use activities METRO, AGRO, and.GRAZE. Therefore, we compiled three different real-change maps. For METRO and AGRO,.we used the MODIS land cover dataset for the years 2001 and 2005. All cells that were "urban".("cropland") in the 2005 map but not in the 2001 map are categorized as change for METRO.(AGRO). For GRAZE, the real change map was derived from the small ruminant density.(SRD) maps adjusted to match FAO totals for the years 2000 and 2005 [16]. We defined real.change from non-grazing to grazing if the small ruminant density increases by 25% and by a.minimum of 25 animals per km.^{2}.over the five year period. The ROC curves resulting from the.analyses are shown in Fig. 3.

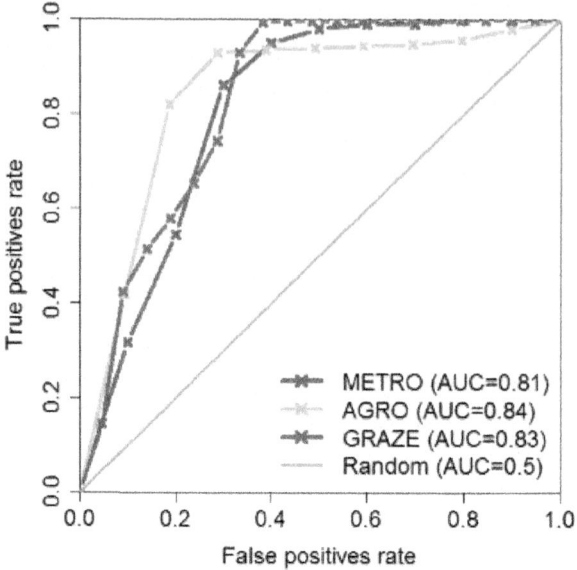

Figure 3: Relative Operating Characteristics (ROC) curves for the three land-use activities METRO,.AGRO, and GRAZE. The 45.°.line indicates the ROC curve for randomly distributed suitability values..The area under the curve (AUC) is the performance measure of ROC

Map comparison analysis

We carried out a map comparison analysis to validate the resulting land-use maps. For this.purpose, we compared the simulated land-use map.S.for the year 2005 with the MODIS land.cover map for the same year, which we considered the actual or reference land-use map.A.,by.calculating the kappa coefficient of agreement (.к.) [9, 42] and kappa simulation (.к.sim.)[60]..We applied.к.because it is commonly used for validation of simulated land-use maps. The.coefficient takes into account that the proportion of cells that are classified correctly by chance,.denoted as the expected proportion correct.p.$_e$, can be very large. The.p.$_e$ depends on the ..number of categories and the number of cells in each category in.S.and.A.. Based on the.observed proportion correct.p.$_o$ and.p.$_e$,.к.is defined as [60]:

$$\kappa = \frac{p_o - p_e}{1 - p_e}$$

(3)

Values for.к.range from -1 (indicating no agreement for any of the cells) to 1 (indicating perfect.agreement of.S.and.A.). If.p.$_o$ is equal to.p.$_e$, i.e., if the land-use types are allocated randomly,.к.is.equal to 0. The.к.coefficient tends to overestimate the performance of land-use change models,.which use an initial land-use map as a starting point, if the number of actually changing cells. is small compared to the number of cells with persistent land-use. In this case, a model that.randomly allocates a small quantity of change, or simulates no change at all, can reach.к.values close to 1. Hence, we also calculated the.к.$_{sim.}$ coefficient, which considers the number.of actual and simulated land-use transitions for the calculation of the expected proportion.correct.p.$_e$ ($\cdot_{transition.}$).. In order to calculate.p.$_e$ ($\cdot_{transition.}$)., additionally the initial land-use map.was considered. The value range for.к.$_{sim.}$ is similar to that of.к.and can be interpreted in the.same way. Similarly to the standard.к.,.к.$_{sim.}$ is then defined as [60]

$$\kappa_{sim} = \frac{p_o - p_{e(transition)}}{1 - p_{e(transition)}}$$

(4)

In order to calculate.к.and.к.$_{sim.}$, the land-use categories in the simulated land-use map.and the MODIS dataset were harmonized. For this purpose, the land-use categories that.LandSHIFT.JR simulates explicitly, i.e., "urban land", "cropland", and "rangeland", were.coded similarly in both maps. The remaining land-use types, e.g. "barren land", were lumped.together in the categories "natural land-cover" or "water". Rangeland is not classified as a.separate land-use type in the MODIS dataset. Therefore, we used the SRD map to derive the.extent of rangeland. We defined a cell as rangeland if the density of small ruminants was.87 animals per km.^{2} or higher and at the same time the land-use/land-cover type assigned.in the MODIS map was different from urban, cropland, and water. The threshold value of.small ruminant density was adjusted in order to maximize.к.. Since SRD is provided on a.different spatial resolution (0.05 dd) and the conversion of SRD to "real" grazing land is very.straightforward we consider the classification of rangeland to be rather inaccurate. Therefore,.we tested the model performance based on two different sets of land-use maps. In set "UCR".urban, cropland, and rangeland were considered; in set "UC" only urban and cropland were.considered as separate land-use categories..For the "UCR" set, the validation results for the map comparison were 0.6 and 0.12 for.к.and.к.$_{sim.}$, respectively. A value of.к.=0.6 indicates that the agreement of the simulated and observed.land-use map was significantly better than it can be expected for a random model. Compared. to other studies, which report.к.values from 0.6 to above 0.9 for land-use change modeling.[36, 65], the agreement of LandSHIFT.JR results and the reference map was relatively low..However, it is important to bear in mind that we did not calibrate LandSHIFT.JR in order to.maximize the agreement to observed datasets. The results are entirely based on parsimonious.assumptions and objective methods to derive model parameters, e.g. the suitability factor. weights. Hence, lower.κ.-values are to be expected..The.κ.coefficient can be interpreted as the gain in agreement of the model as compared to.a baseline assumption. For standard.κ.the baseline is a process that randomly allocates the.proportion of categories given be the model. For.$κ_{sim}$, the baseline is an improved random.process using the additional information that possible changes in land use are limited to a.certain, potentially very small, proportion of the cells, which is derived from the simulation.results and the reference map. Therefore, the expected proportion correct increases for.$κ_{sim}$ and the values are generally lower. Hence, a.$κ_{sim}$ of 0.12 still indicates that LandSHIFT. JR.explains the land-use changes in the study region significantly better than the improved.baseline process..When we used only the information originally given by the MODIS dataset (i.e. omitting.the land-use type rangeland and using the set "UC").κ.increased to 0.72 and.$κ_{sim}$ increased.to 0.22. This can partly be attributed to the inaccuracies induced by the simple approach to.derive the reference distribution of rangeland. Furthermore, it is important to consider that.the reference map is derived from a remote sensing product (MODIS) and the small ruminant.density dataset, which both are subject to classification and measurement errors. Additional.sources of error may by introduced by data preparation, e.g. spatial aggregation (MODIS) and.disaggregation (SRD)

Comparison with statistics

We compared the simulated area for rainfed and irrigated cropland for the year 2005 to.estimates of the national statistical agencies of Israel, Jordan, and PA (Table 5). Although the.model results for area under crops were in very good agreement for PA, the model simulated.considerably higher area demands in Israel and Jordan (Table 5). For Israel, the simulated area.demand for irrigated and rainfed cropland in 2005 was 48% and 66% higher than reported.by the statistics, respectively. According to the Central Bureau of Statistics in Israel, the.method to estimate the area under crops has changed starting in 2003. For that reason, a.comparison to earlier years is not possible. However, LandSHIFT. JR uses the estimates of area.under crops for the base year 2000 as an initial condition. Hence, the simulated area and the.area reported by the statistics cannot be compared directly. For Jordan, the simulated area.demand was overestimated by 41% and 57% for irrigated and rain-fed crops, respectively.. This discrepancy can partly be explained by the fact that, according to the state statistics, the.area under crops increased by only 4% while the production of agricultural products, which.is the main driver of LandSHIFT.JR, increased by 46% [18]. Assuming that high-quality land.resources are already in use for crop production, this is only possible if crop productivity.increases considerably due to massive changes in agricultural management, e.g., fertilizer.application or irrigation techniques. Currently, LandSHIFT.JR cannot simulate such effects. because of missing input data. According to the MODIS land cover dataset for 2005 the area.of cropland increased by about 63 %, which is more consistent with the relative increase in.crop production simulated with LandSHIFT.JR

APPLICATION EXAMPLE

In order to give an application example of LandSHIFT.JR, we set up a modeling exercise. As drivers for the model, we use the assumptions on the dynamics of population number agricultural production, livestock production, and yield change due to technological progress.as given by the GLOWA Jordan River. Modest Hopes.scenario [6]. Figure 4 shows the.LandSHIFT.JR results for land-use/land-cover distribution, population density, and livestock.density for the base year (2000) and the corresponding projections for the year 2050.

Table 5: Area under rainfed and irrigated crops in 2005 for Israel, Jordan, and the Palestinian National Authority (PA) as simulated with LandSHIFT.JR and estimated by the national statistical agencies.

	Rainfed cropland		Irrigated cropland		Total cropland	
	Statistics [km²]	LandSHIFT.JR [km²]	Statistics [km²]	LandSHIFT.JR [km²]	Statistics [km²]	LandSHIFT.JR [km²]
Israel	1283	2129 (+66%)	1298	1926 (+48%)	2581	4055 (+57%)
Jordan	1663	2613 (+57%)	610	858 (+41%)	2273	3471 (+53%)
PA	1545	1561 (+1%)	184	184 (0%)	1729	1745 (+1%)

A comparison of Fig. 4 (a) and (b) shows considerable increases in the area demands for.the main land-use activities. By 2050, the extent of urban land increases by about 56 %, while.irrigated cropland expands to more than twice, rainfed cropland to more than three times, and.grazing land to more than four times the area as compared to 2000. The figures for agricultural.area reflect the ranking of the four activities: the lower the priority of a land-use activity is the.lower is the productivity on the areas it is allocated to and, consequently, the larger is the area.expansion needed to fulfill the demands. The increasing population density between 2000.and 2050 is shown in Fig. 4 (c) and (d).

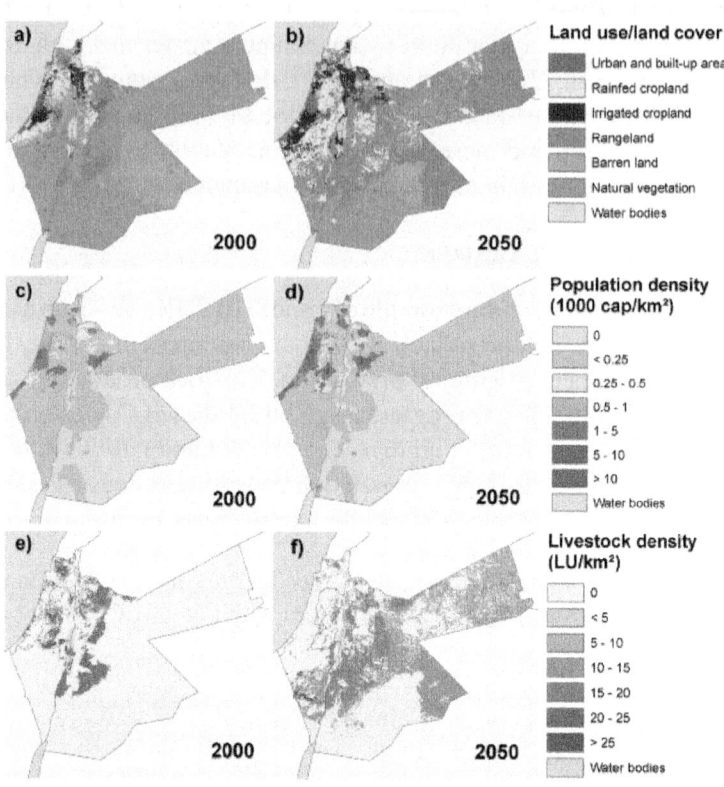

Figure 4: Maps of land-use and land-cover distribution, population density, and livestock density for the years 2000 and 2050 simulated with LandSHIFT.JR for the Modest Hopes scenario [6]

The maps show the typical differences between the.modeling approaches for rural and urban population growth. On the one hand, the urban.encroachment approach leads to relative fast growth of urban land (population density above.5000 people/km.^{2}) at the edges of existing cities or urban centers. On the other hand, rural.population density increases uniformly and proportional to the initial population density,.which is distributed homogeneously over administrative units. Hence, the outlines of these.districts can partly be recognized in the maps. The land-use activity with the lowest priority.is grazing. Therefore, rangeland is more and more displaced from areas with relatively high.productivity, where it is predominantly allocated in 2000 (Fig. 4 (e)), and shifted to less.productive areas (Fig. 4 (f)). This leads to a vast extent of rangeland with low stocking.densities in 2050. The expansion of irrigated cropland (Fig. 4 (a) and (b)) causes irrigation.water requirements to rise. Table 6 presents the simulated total irrigation water demand and.area specific irrigation water demand on state level. According to these figures, the projected.irrigation water demand almost doubles in PA and Jordan and is about threefold in Israel in.2050 as compared to the year 2000

Table 6: Total simulated (change between 2000 and 2050 in parenthesis) and average area specific irrigation water demand in 2000 and 2050 for Israel, Jordan, and the Palestinian National Authority (PA).

State	IR water demand [10^6 m³]		Avg. IR water demand [mm]	
	(2000)	(2050)	(2000)	(2050)
Israel	638	1477 (+132%)	31	72
Jordan	322	772 (+140%)	4	9
PA	86	162 (+88%)	14	26

DISCUSSION AND CONCLUSIONS

In this chapter, we introduce the integrated modeling system LandSHIFT.JR for the Jordan.River region. We give a detailed description of the modeling system, its parameterization,.and validation. We furthermore present a sample application of LandSHIFT.JR for the.Modest.Hopes.scenario, developed in the context of the GLOWA Jordan River scenario exercise [6]..Since vegetation degradation due to overgrazing is a major problem in the Jordan River region.[1] and since the intensity levels of grazing management strongly affect the environment via.different pathways (e.g. woody encroachment [7], biodiversity loss [1] or erosion [27]) we.developed a separate module for livestock grazing, that not only implements indicators for.grazing intensity, but also includes different rangeland management strategies [28, 29]. This.allows to consider the effect of rangeland management strategies in environmental impact.assessments..In contrast to earlier versions of LandSHIFT.JR, the current version includes the effect of.changing climate conditions on crop yields and productivity of natural vegetation, which.was shown to have a strong effect on land demand in the Jordan River region [32]. The.indirect effect of productivity on area demand for the different agricultural activities is.included indirectly by spatially explicit simulation models (WADISCAPE and GEPIC), driven.by high-resolution climate change simulations for the Jordan River region [53]. This allows.the inclusion of a high level of spatial detail into the simulations of land-use and land-cover.change, which is carried out on a grid with a spatial resolution of 30 arc seconds. This is of high. importance in a region with such high variability in biogeographic conditions as the Jordan.River region. Furthermore, it applies a consistent assessment method to the entire Jordan.River region and allows the combined assessment of socio-economic and climate impact on.the food production systems in the Jordan River region which is considered to be mandatory.[55, 56]..Another striking feature of the presented modeling system is the separate module for. irrigated crop production. This module allows to simulate spatial and temporal dynamics.of irrigated crop production and the resulting land-use patterns and intensities [32]. The.model also enables an assessment of climate dependent net irrigation water requirements.simulated with the GEPIC model. Hence, the modeling system can now be used the evaluate.the effect of changes in cropland extent (induced by changing climate conditions and/or.demands for agricultural commodities) on the net irrigation water requirements. However,.it has to be mentioned that the current LandSHIFT.JR version only evaluates the demand.and no connection to water supply is implemented so far. LandSHIFT. JR considers only.crop categories and does not differentiate between crop types. The net irrigation water.requirements for the different crop categories were inferred from GEPIC simulations for.wheat yields using a crop-specific adjustment parameter. This approach introduces some.inaccuracy into the simulation and, as a result, makes the simulation results more suitable.for the evaluation of changes in water requirements as compared to the absolute amounts..Furthermore, no information on conveyance efficiencies or irrigation efficiencies (e.g. drip.irrigation versus sprinkler irrigation) is included, which would be required to derive the gross.irrigation water requirements..In order to validate LandSHIFT.JR, three different validation methods were applied: (1) ROC.analysis [43], (2) map comparison using.κ.and.κ$_{sim}$ as performance measures [42, 60], and (3).a comparison of the quantity of simulated land-use changes with observed land-use changes..The results for the ROC analysis (AUC = 0.81 for METRO, AUC = 0.84 for AGRO, and.AUC = 0.83 for GRAZE) indicate that the suitability assessment in LandSHIFT.JR explains. the location of change to a high degree. The validation results for the map comparison.are at the lower range of values reported for land-use models, with 0.6 and 0.12 for.κ.and.κ$_{sim}$ (0.72 and 0.22 without rangeland), respectively. Bearing in mind that the modeling.approach of LandSHIFT.JR does not include a calibration step (e.g. [50, 65]), but is entirely.based on parsimonious assumptions and objective methods, we consider these values as.acceptable. The comparison of observed and simulated land-use changes shows an almost. perfect agreement for PA. Discrepancies resulting for Israel and Jordan might partly be.induced by inconsistencies in the reported values. Based on the validation results, we consider.LandSHIFT.JR suitable for the simulation of the location and quantity of land-use changes in.the Jordan River region.As shown for the application example, LandSHIFT.JR implements modules for the four.land-use activities infrastructure and housing, irrigated crop production, rainfed crop.production, and livestock grazing. For each land-use activity, besides the dominant land-use.types also an indicator of land-use intensity is allocated (population density, irrigated or.rainfed crop production amount, stocking density). Hence, the model concept implemented.in LandSHIFT. JR considers not only land-use patterns, but also the corresponding land-use. intensities. This makes LandSHIFT.JR land-use simulation results suitable for applications.focusing on natural resource management and environmental impact assessment [24, 39]..We see a potential for improvement regarding the validation process. The spatially explicit.validation of rangeland, net irrigation water requirements, and the separate validation of.irrigated and rainfed cropland was limited by insufficient data availability. This will be caught.up for, once suitable datasets are available. We encounter this validation issues by choosing a.straightforward modeling approach, based on logical assumptions and renunciation of model.calibration and consider this approach as second best to data..In addition to extensive sensitivity and uncertainty analyses to improve the scientific.knowledge and understanding of land-use systems in the Jordan River region, we see a strong.potential for future studies on the relationship between irrigation water supply (including.treated wastewater), net irrigation water requirements, and soil sensitivity towards the.irrigation with treated wastewater [47]. For this purpose, additional GEPIC simulations for.other crop types besides wheat would be required in order to be able to assess the irrigation.water requirements more accurately. This would allow for interesting analyses regarding the.potential of using treated wastewater for irrigation purposes, under consideration of possible.environmental problems associated with the use of treated wastewater for irrigation [5].

ACKNOWLEDGMENTS

This study is part of the GLOWA Jordan River project financed by the German Federal.Ministry of Education and Research (FKZ 01LW0502). We thank Katja Geissler (Potsdam.University, Research Group Plant Ecology and Nature Conservation) and Martin Köchy.(Johann Heinrich von Thünen Insitut, Braunschweig) for the provision of WADISCAPE model.output. Furthermore, we thank Gerhard Smiatek (IMK-IFU, Institute for Meteorology and.Climate Research-Atmospheric Environmental Research) for the provision of regional climate.simulation results.

REFERENCES

1.　Abahussain, A.A., Abdu, A.S., Al-Zubari, W.K., El-Deen, N.A. & Abdul-Raheem, M..(2002). Desertification in the Arab region: analysis of current status and trends,.Journal.of Arid Environments.51(4):521–545.

2.　Alcamo, J. (2002). Three issues for improving integrated models:

uncertainty, social.science, and legitimacy,.in.Gethmann, C.F. & Lingner, S. (eds.),.Integrative Modellierung.zum Globalen Wandel., Springer Verlag, Berlin, Heidelberg, Germany, pp. 3–14.

3.   Alcamo, J. (2009)..Environmental futures: the practice of environmental scenario analysis.,.Elsevier, Amsterdam, The Netherlands.

4.   Al-Jaloudy, M.A. (2001)..FAO country profiles: Jordan. Country pasture/ forage resource.profiles., Food and Agriculture Organization of the United Nations, Rome, Italy. URL:.http://www.fao.org/ag/AGP/AGPC/ doc/Counprof/Jordan/Jordan.htm

5.   Al-Nakshabandi, G.A., Saqqar, M.M., Shatanawi, M.R., Fayyad, M. & Al-Horani, H..(1997). Some environmental problems associated with the use of treated wastewater for.irrigation in Jordan,.Agricultural Water Management.34(1):81–94.

6.   Anonymous (2011)..Future management of the Jordan River basin's water and land resources.under climate change - a scenario analysis., Summary report, Center for Environmental.Systems Research, Kassel, Germany and Israel/Palestine Center for Research and.Information, Jerusalem, Israel.

7.   Asner, G.P., Elmore, A.J., Olander, L.P., Martin, R.E. & Harris, A.T. (2004). Grazing.systems, ecosystem responses, and global change,. Annual Review of Environment and.Resources.29:261–299.

8.   CIESIN (2004)..Global rural-urban mapping project (GRUMP): urban/ rural population grids.,.Center for International Earth Science Information Network, Columbia University,.International Food Policy Research Institute, the World Bank & Centro Internacional.de Agricultura Tropical. URL:.http://sedac.ciesin.columbia.edu/gpw/

9.   Cohen, J. (1960). A coefficient of agreement for nominal scales,. Educational and.Psychological Measurement.20(1):37–46.

10.   [10] Diakoulaki, D., Mavrotas, G. & Papayannakis, L. (1995). Determining objective weights.in multiple criteria problems: the CRITIC method,.Computers & Operations Research.22(7):763–770.

11.   Eastman, J.R., Jin, W., Kyem, P.A.K. & Toledano, J. (1995). Raster procedures for.multi-criteria/multi-objective decisions,.Photogrammetric Engineering & Remote Sensing.61(5):539–547.

12.   Erb, K.-H., Gaube, V., Krausmann, F., Plutzar, C., Bondeau, A. & Haberl, H. (2007). A.comprehensive global 5 min resolution land-use data set for the year 2000 consistent.with national census data,.Journal of Land Use Science.2(3):191–224.

13. EXACT (1998)..Overview of Middle East water resources,.Executive Action Team, Middle.East Water Data Banks Project. URL:.http://exact-me.org/overview/index..htm

14. Falkenmark, M. & Rockström, J. (2004)..Balancing water for humans and nature: the new.approach in ecohydrology., Earthscan, London, United Kingdom.

15. FAO (2003)..Review of world water resources by country., Water Reports 23, Food and.Agriculture Organization of the United Nations, Rome, Italy.

16. FAO (2007)..Gridded livestock of the world 2007., Food and Agriculture Organization of the.United Nations, Rome, Italy.

17. FAO (2012)..AQUASTAT: FAO's information system on water and agriculture., Food and.Agriculture Organization of the United Nations, Rome, Italy. URL:.http://www.fao..org/nr/water/aquastat/main/index.stm

18. FAO (2012)..FAO statistical database., Food and Agriculture Organization of the United.Nations, Rome, Italy. URL:.http://faostat.fao.org/

19. FIMA (2011)..RWDB2 River-surface water body network., Food and Agriculture.Organization of the United Nations, Aquaculture Management and Conservation.Service (FIMA), Rome, Italy. URL:. http://www.fao.org/geonetwork/srv/en/.main.home

20. Foley, J.A., DeFries, R., Asner, G.P., Barford, C., Bonan, G., Carpenter, S.R., Chapin,.F.S., Coe, M.T., Daily, G.C., Gibbs, H.K., Helkowski, J.H., Holloway, T., Howard, E.A.,.Kucharik, C.J., Monfreda, C., Patz, J.A., Prentice, I.C., Ramankutty, N. & Snyder, P.K..(2005). Global consequences of land use,.Science.309(5734):570–574.

21. Friedl, M.A., McIver, D.K., Hodges, J.C.F., Zhang, X.Y., Muchoney, D., Strahler, A.H.,.Woodcock, C.E., Gopal, S., Schneider, A., Cooper, A., Baccini, A., Gao, F. & Schaaf, C..(2002). Global land cover mapping from MODIS: algorithms and early results,.Remote.Sensing of Environment.83(1-2):287–302.

22. Ghasemi, F., Jakeman, A.J. & Nix, H.A. (1995)..Salinisation of land and water.resources: human causes, extent, management and case studies., Centre for Resources.and Environmental Studies, Australian National University, CAB International,.Wallingford, United Kingdom.

23. Grimm, V., Berger, U., Bastiansen, F., Eliassen, S., Ginot, V., Giske, J., Goss-Custard,.J., Grand, T., Heinz, S.K., Huse, G., Huth, A., Jepsen, J.U., Jørgensen, C., Mooij,.W.M., Müller, B., Pe'er, G., Piou, C., Railsback, S.F., Robbins, A.M., Robbins, M.M.,.Rossmanith, E., Rüger, N., Strand,

E., Souissi, S., Stillman, R.A., Vabø, R., Visser, U..& DeAngelis, D.L. (2006). A standard protocol for describing individual-based and.agent-based models,.Ecological Modelling.198(1-2):115–126.

24. Gunkel, A. & Lange, J. (2012). New insights into the natural variability of water.resources in the lower Jordan River basin,.Water Resources Management.26(4):963–980.

25. Haberl, H., Erb, K.-H., Krausmann, F., Gaube, V., Bondeau, A., Plutzar, C., Gingrich,.S., Lucht, W. & Fischer-Kowalski, M. (2007). Quantifying and mapping the human.appropriation of net primary production in earth's terrestrial ecosystems,.Proceedings.of the National Academy of Sciences of the United States of America.104(31):12942–12947.

26. Kan, I., Rapaport-Rom, M. & Shechter, M. (2007). Economic analysis of climate-change.impacts on agricultural profitability and land use: the case of Israel,.Proceedings of 15th.Annual Conference of the European Association of Environmental and Resource Economists.(EAERE)., Thessaloniki, Greece, pp. 14-17.

27. Khresat, S.A., Rawajfih, Z. & Mohammad, M. (1998). Land degradation in.north-western Jordan: causes and processes,.Journal of Arid Environments.39(4):623–629.

28. [28] Koch, J., Schaldach, R. & Köchy, M. (2008). Modeling the impacts of grazing land.management on land-use change for the Jordan River region,.Global and Planetary Change.64(3-4):177–187.

29. Koch, J., Schaldach, R. & Kölking, C. (2009). Modelling the impact of rangeland.management strategies on (semi-)natural vegetation in Jordan,. Proceedings of the.18th World IMACS Congress and MODSIM09 International Congress on Modelling and.Simulation., Cairns, Australia, pp. 1929–1935.

30. Koch, J. (2010)..Modeling the impacts of land-use change on ecosystems at the regional and.continental scale., Thesis, kassel university press GmbH, Kassel, Germany.

31. Koch, J., Onigkeit, J., Schaldach, R., Alcamo, J., Köchy, M., Wolff, H.-P. & Kan, I. (2011)..Land-use change scenarios for the Jordan River region,.International Journal of Sustainable.Water and Environmental Systems.3(1):25–31.

32. Koch, J., Wimmer, F., Schaldach, R., Onigkeit, J. & Folberth, C. (2012). Modelling.the impact of climate change on irrigation area demand in the Jordan River region,.in.Seppelt, R., Voinov, A.A., Lange, S. & Bankamp. D. (eds.),.Proceedings of the 2012.International Congress on Environmental Modelling and Software: Managing Resources of

a.Limited Planet (iEMSs 2012)., Leipzig, Germany.

33. Köchy, M. (2008). Effects of simulated daily precipitation patterns on annual plant.populations depend on life stage and climatic region,.BMC Ecology.8(4).

34. Köchy, M., Mathaj, M., Jeltsch, F. & Malkinson, D. (2008). Resilience of stocking capacity.to changing climate in arid to Mediterranean landscapes,.Regional Environmental Change.8(2):73–87.

35. Lambin, E.F., Rounsevell, M.D.A. & Geist, H.J. (2000). Are agricultural land-use models.able to predict changes in land-use intensity?. Agriculture, Ecosystems & Environment.82(1-3):321–331.

36. Lauf, S., Haase, D., Hostert, P., Lakes, T. & Kleinschmit, B. (2012). Uncovering.land-use dynamics driven by human decision-makin.g - a combined model approach.using cellular automata and system dynamics,. Environmental Modelling & Software.27-28:71–82.

37. Liu, J., Williams, J.R., Zehnder, A.J.B. & Yang, H. (2007). GEPIC - modelling wheat yield.and crop water productivity with high resolution on a global scale,.Agricultural Systems.94(2):478–493.

38. Malkinson, D. & Jeltsch, F. (2007). Intraspecific facilitation: a missing process along.increasing stress gradients - insights from simulated shrub populations,.Ecography.30(3):339–348.

39. Menzel, L., Koch, J., Onigkeit, J. & Schaldach, R. (2009). Modelling the effects of land-use.and land-cover change on water availability in the Jordan River region,.Advances in.Geosciences.21:73–80.

40. Perevolotsky, A., Landau, S., Kababia, D. & Ungar, E.D. (1998). Diet selection in.dairy goats grazing woody Mediterranean rangeland,. Applied Animal Behaviour Science.57(1-2):117–131.

41. Pimm, S.L. & Raven, P. (2000). Biodiversity: extinction by numbers,. Nature.403:843–845.

42. Pontius Jr., R.G. (2000). Quantification error versus location error in comparison of.categorical maps,.Photogrammetric Engineering & Remote Sensing.66(8):1011–1016.

43. Pontius Jr., R.G. & Schneider, L.C. (2001). Land-cover change model validation by an.ROC method for the Ipswich watershed, Massachusetts, USA,.Agriculture, Ecosystems &.Environment.85(1-3):239–248.

44. Portmann, F.T., Siebert, S. & Döll, P. (2010). MIRCA2000 - Global monthly irrigated and.rainfed crop areas around the year 2000: a new high-resolution data set for agricultural.and hydrological modeling,. Global Biogeochemical Cycles.24:GB1011.

45. Rosegrant, M.W., Meijer, S. & Cline, S.A. (2002)..International model for policy analysis.of agricultural commodities and trade (IMPACT): model description., International Food and.Policy Research Institute, Washington, DC, United States. URL:.http://www.ifpri..org/themes/impact/impactmodel.pdf

46. [Saaty, R.W. (1987). The analytic hierarchy process - what it is and how it is used,.Mathematical Modelling.9(3-5):161–176.

47. Schacht, K., Gönster, S., Jüschke, E., Chen, Y., Tarchitzky, J., Al-Bakri, J., Al-Karablieh, E..& Marschner, B. (2011). Evaluation of soil sensitivity towards the irrigation with treated.wastewater in the Jordan River region,. Water.3(4):1092–1111.

48. Schaldach, R. & Koch, J. (2009). Conceptual design and implementation of a model.for the integrated simulation of large-scale land-use systems,. in.Athanasiadis, I.N.,.Mitkas, P.A., Rizzoli, A.E. & Gómez, J.M. (eds.),. Information Technologies in Environmental.Engineering., Springer Verlag, Berlin, Heidelberg, Germany, pp. 425–438.

49. Schaldach, R., Alcamo, J., Koch, J., Kölking, C., Lapola, D.M., Schüngel, J. & Priess, J.A..(2011). An integrated approach to modelling land-use change on continental and global.scales,.Environmental Modelling & Software.26(8):1041–1051.

50. Schweitzer, C., Priess, J.A. & Das, S. (2011). A generic framework for land-use.modelling,.Environmental Modelling & Software.26(8):1052–1055.

51. Seré, C. & Steinfeld, H. (1996)..World livestock production systems - current status, issues.and trends., FAO Animal Production and Health Paper 127, Food and Agriculture.Organization of the United Nations, Rome, Italy.

52. Siebert, S., Döll, P., Feick, S., Hoogeveen, J. & Frenken, K. (2007).. Global Map of.Irrigation Areas version 4.0.1., Johann Wolfgang Goethe University, Frankfurt am Main,.Germany/Food and Agriculture Organization of the United Nations, Rome, Italy.

53. Smiatek, G., Kunstmann, H. & Heckl, A. (2011). High-resolution climate change.simulations for the Jordan River area,.Journal of Geophysical Research.116:D16111.

54. Solecki, W.D. & Oliveri, C. (2004). Downscaling climate change scenarios in an urban.land use change model,.Journal of Environmental Management.72(1-2):105–115.

55. Tubiello, F.N., Soussana, J.-F. & Howden, S.M. (2007). Crop and pasture response to

56. climate change,.Proceedings of the National Academy of Sciences of the United States of.America.104(50):19686–19690.

57. Tubiello, F.N., Amthor, J.S., Boote, K.J., Donatelli, M., Easterling, W., Fischer, G., Gifford,.R.M., Howden, M., Reilly, J. & Rosenzweig, C. (2007). Crop response to elevated CO_2.and world food supply: a comment on "Food for Thought ..." by Long et al., Science.312:1918–1921, 2006,. European Journal of Agronomy.26(3):215–223.

58. Uchida, H. & Nelson, A. (2008)..Agglomeration index: towards a new measure of urban.concentration., Background Paper, World Development Report 2009.

59. [58] UNEP (2007)..Global Environment Outlook - environment for development (GEO-4)., United.Nations Environment Programme, Nairobi, Kenya.

60. U.S. Geological Survey (1998).HYDRO1k: elevation derivative database., Earth Resources.Observation and Science (EROS) Center, Sioux Falls, SD, United States. URL:.http://eros.usgs.gov/#/Find_Data/ Products_and_Data_Available/.gtopo30/hydro

61. van Vliet, J., Bregt, A.K. & Hagen-Zanker, A. (2011). Revisiting Kappa to account for.change in the accuracy assessment of land-use change models,.Ecological Modelling.222(8):1367–1375.

62. Verburg, P.H., de Koning, G.H.J., Kok, K., Veldkamp, A. & Bouma, J. (1999). A spatial.explicit allocation procedure for modelling the pattern of land use change based upon.actual land use,.Ecological Modelling.116(1):45–61 30.Will-be-set-by-IN-TECH

63. Verburg, P.H., Soepboer, W., Veldkamp, A., Limpiada, R., Espaldon, V. & Mastura,.S.S.A. (2002). Modeling the spatial dynamics of regional land use: the CLUE-S model,.Environmental Management.30(3):391–405.

64. Verburg, P.H., Kok, K., Pontius, R.G. & Veldkamp, A. (2006). Modeling land-use and.land-cover change,.in.Lambin, E.F. & Geist, H.J. (eds.),. Land-use and land-cover change.- local processes and global impacts., Springer Verlag, Berlin, Heidelberg, Germany, pp..117–135.

65. Vitousek, P.M. (1994). Beyond global warming: ecology and global change,.Ecology.75(7):1861–1876.

66. Wang, F., Hasbani, J.-G., Wang, X. & Marceau, D.J. (2011). Identifying dominant factors.for the calibration of a land-use cellular automata model using Rough Set Theory,.Computers, Environment and Urban Systems.35(2):116–125.

67. WDPA Consortium (2004).Word Database on Protected Areas..URL:.

http://www.wdpa..org/.

68. Williams, J.R., Jones, C.A., Kiniry, J.R. & Spanel, D.A. (1989). The EPIC crop growth.model,.Transactions of the American Society of Agricultural Engineers.32(2):497–511.

69. Williams, J.R. (1995). The EPIC model,.in.Singh, V.P. (ed.),.Computer Models of Watershed.Hydrology., Water Resources Publications, Colorado, United States, pp. 909–1000.

70. [Wu, F. (1999). GIS-based simulation as an exploratory analysis for space-time processes,.Journal of Geographical Systems.1(3):199–218

Chapter 10

CONSERVATION AND SUSTAINABILITY OF MEXICAN CARIBBEAN CORAL REEFS AND THE THREATS OF A HUMAN-INDUCED PHASE-SHIFT

José D. Carriquiry[1], Linda M. Barranco-Servin[1], Julio A. Villaescusa[1], Victor F. Camacho-Ibar[1], Hector Reyes-Bonilla[2] and Amílcar L. Cupul-Magaña[3]

[1] Instituto de Investigaciones Oceanológicas, Universidad Autónoma de Baja California, Ensenada, Baja California, Mexico

[2] Universidad Autonoma de Baja California, Mexico

[3] Universidad de Guadalajara, Puerto Vallarta, Jal, Mexico

INTRODUCTION

Natural ecosystems around the world are continually changing, but in recent decades it has become increasingly evident that terrestrial and marine environments are degrading. It is considered that the main cause is the rapid human population growth and increasing demand of resources for our survival. In particular for coral reef ecosystems, they are experiencing a significant change as reflected in the decline of coral cover and diversity of species [1, 2].

Coral reefs are marine ecosystems of great ecological and economic importance to mankind. These ecosystems are characterized by high productivity and biodiversity [3] caused by the high diversity of habitats created by its complex, calcium carbonate three-dimensional structure that facilitates the diversification of niches and space availability for the establishment of a variety species. Additionally, they serve as important spawning areas for breeding and feeding of the organisms that are part of the ecosystem [4]. The interactions that exist between the species that inhabit coral reefs create an ecological balance that maintains ecosystem functioning, which is reflected in its ability to maintain high productivity in these groups of reef-building coral species and fish, and play key roles in the regulation of ecological processes

[2]. From the economic point of view, coral reefs supply mankind with such services as organisms for food, biochemical components, building materials, coastal protection against storms and waves, recreational opportunities and aesthetic and cultural benefits (see [4]). It is estimated that more than 100 countries have coastlines with coral reefs, and tens of millions of people depend on them as part of their livelihood or as part of their intake of protein [5]. It is noteworthy that the quantity and quality of these services depends on the health of coral reefs.

Unfortunately, many of these ecosystems are in serious state of degradation and it is considered that the health of coral reefs is in a worldwide crisis [6]. By 2008 it was estimated that coral reefs had effectively lost 19% of their original area, an additional 15% was seriously threatened and was considered to it will be lost within the next 10 to 20 years, while 20% are under threat with the possibility of being lost within 20 to 40 years. These estimates were made under a scenario of 'business as usual' and do not consider the threats posed by global climate change [7].

Paleoecologic work suggests that this pattern of degradation in various areas of the Caribbean is unprecedented within the past millennia. Also, there is no convincing evidence that global stressors (e.g. induced bleaching by temperature changes and reduced calcification rates via increasing levels of atmospheric CO_2) are responsible for the overall pattern of recent coral degradation, making it more likely that local stressors are responsible for the recent degradation occurred in the Caribbean [1]. In this sense, it is particularly evident that degradation of coral reefs occurs near densely populated areas, hence indicating that anthropogenic factors work synergistically against the stability of these ecosystems. Among these factors, the increased levels of nutrients and the over-exploitation of reef organisms are the best studied and considered to be responsible for the larger part of the impacts [8-9]. However, we cannot omit mentioning other impacts such as increased discharge of sediment and pollutants, uncontrolled tourism and introduction of new species (e.g., lionfish) and diseases [2,4, 10-11]. Added to all these factors is the threat of global climate change and the increased intensity of natural phenomena such as severe storms and hurricanes, and the development of the El Niño / La Niña - Southern Oscillation phenomenon --and their associated surface temperature change-- have contributed to the health degradation of coral reef ecosystems [12].

In many locations around the world the anthropogenic stress on coral reefs has exceeded the regenerative capacity of the ecosystems, causing dramatic changes in species composition and thus a severe economic loss [2]. This change in the structure and functioning of the ecosystem is known as "phase-shift" or alternative stable state; the most cited example is the shift from an original coral-dominated reef to one dominated by macroalgae[13-14], although several other transitions have been documented [2]. This phase-shift is a consequence of the loss of resilience, defined as the ability of an ecosystem to absorb perturbations, its resistance to change and its capability to regenerate after a natural or anthropogenic disturbance [15]. The degree to which the phase-shift --or alternative state-- is stable or reversible is poorly understood and represents one of the main challenges for research and management of coral reefs.

Given the nature of the problem and the huge importance of coral reefs to mankind, it is recognized that urgent action is needed to conserve and promote its sustainable use [7]. Thus, it is urgent to assess current management practices that focus mainly on safeguarding the biodiversity of coral reefs by delimiting marine protected areas (MPAs) where human activities are controlled, or prohibited in the case of fishing. This kind of management is increasingly prevalent. But if it succeeds by adequately fulfilling its objective, it would provide a spatial refuge for the organisms that are distributed within the area, allowing critical functional groups to persist, and so continue to build local resilience of the ecosystem. However, only a few marine protected areas meet its conservation and functionality goals. At the global scale, 18% of the coral reef area is within the boundaries of a marine protected area and only 1.6% of these are properly managed [16]. To adequately address the crisis requires that management efforts are based on a better understanding of the ecological processes that maintain the resilience of coral reefs. Managing for improved resilience, incorporating the role of human activities as shapers of ecosystems provides a basis for addressing the uncertainty of a changing environment [2].

THE CASE OF CORAL REEFS IN THE MEXICAN CARIBBEAN

The degradation of coral reefs in the Wider Caribbean is alarming. Population growth in the region has led to a combined effect of increased pollution and reduced herbivore populations as a result of overfishing and/or diseases [1,17]. These reefs are continually cited as examples of a phase-shift.

In the case of coral reefs in the Mexican Caribbean (from the Northern tip of the Yucatan peninsula to the southern international border with Belize) studies are scarce and the poor distribution of the information generated from implemented management programs limits our knowledge of these systems and the successes of the conservation programs. In this sense, the current status and health of the coral reefs of the Mexican Caribbean at the regional scale are not well known, but at the local scale, clear signs of deterioration have been reported in some reefs despite being designated and managed as protected areas. To achieve conservation and sustainable use of coral reefs requires a better understanding of the dynamics of these ecosystems and the processes that support or undermine resilience; we need reliable scientific information that can be used for management plans at local and regional levels. This study presents the ecological and hydrological characterization of thirteen coral reefs distributed within three National Parks in the Yucatan: Isla Mujeres-Cancun-Nizuc National Park (PNIMCN), the National Park of Puerto Morelos reefs (PNAPM) and Cozumel reefs National Park (PNAC) (Figure1) in order to present the current state of these reefs and identify possible causes of degradation, if present.

For the analysis of coral/benthic community structure we used the following benthic diversity indices: Margalef's D› index that measures species richness independently of the sample size, based on the ratio of the numberof species (S) and the total numberof individuals observed (N). Simpson's index of dominance (λ) that measures the probability from a non-repeated random draw of two organisms from a community, that they belong to the same species. The Shannon-Wiener's index of diversity (H›) measures the degree of uncertainty in predicting to which species belongs an individual chosen at random from asample of Sspecies and Nindividuals.

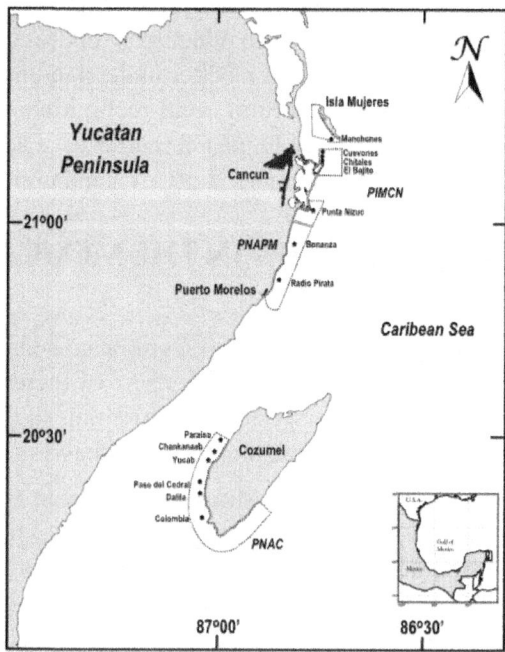

Figure 1: Sampling localities at the Marine National Parks of Isla Mujeres – Cancún –

Nizuc (PNIMCN), Arrecifes de Puerto Morelos (PNAPM) and Arrecifes de Cozumel (PNAC) during 2006 - 2007.

Pielou's index of evenness (J') which measures the proportion of the diversity observed in relation to the maximum expected diversity with values ranging from 0 to 1, where values close to 1 means that all species are equally abundant [18]. We used multivariate techniques in order to statistically support the comparisons between localities, both in the community structure of coral and fish. A similarity matrix was generated using the Bray-Curtis coefficient of similarity; this coefficient measures the similarity between two samples with values ranging from 0 to 1, where values close to 0 correspond to different samples and close to 1 correspond to equal samples. Using the similarity matrix we performed an Analysis of Similarity (ANOSIM) for assessing the differences in community structure between locations. In addition a cluster analysis classification was performed in order to detect if a group of samples have more similarity to others within a group [19]. In the case of nutrient concentrations we applied a variance analysis (ANOVA) to detect statistically significant differences between the parks studied.

COMMUNITY STRUCTURE OF CORALS AND BENTHIC COMPONENTS

The three National Marine Parks were established to conserve the natural ecosystems found within the area, but human activities have not been restricted entirely, and instead, human settlements have increased on the coast, using these habitats for recreation and tourism as well as for fishing within in the limits of the Parks, poaching included.

In order to determine the current status of the coral reefs of the Yucatan Peninsula, an analysis of benthic components of each of these marine parks was conducted. The characterization of benthic components of the reefs provides information about the current health status of the ecosystem. The percentage of coral cover, macroalgae, seagrass, and other invertebrates, is reported from the sampled localities and subsequently analyzed their spatial and temporal variability. During the months of November 2006 (beginning of the dry season) and May 2007 (beginning of wet season) we surveyed the benthic composition of the coral reefs of the three marine parks using the Line Intercept Transect (LIT) method [20]. We surveyed five reefs at the PNIMCN, two reefs at PNAPM, and six reefs at PNAC (Table 1).

There were a total of 40 species of coral identified in the study area with the Cozumel Reefs National Park consistently containing the highest number of species: 25 coral species found at Paso del Cedral reef and 19 at Chankanaab reef (which is its maximum for this locality but the minimum for the PNAC). In general, *Porites astreoides* and *Acropora palmata* are the most abundant species in the Parks of Northern Quintana Roo, i.e., in Isla Mujeres - Cancun - Nizuc and Puerto Morelos. In the reefs of Cozumel, the most abundant species are *Agaricia agaricites* and *Siderastrea siderea*. Of all the coral reefs studied, Chitales reef, located at Punta Cancun, is the one with the lesser number of species (only seven species) (Table 1).

As for the diversity indices, we see a decrease, both in diversity and in species richness from South to North along the Quintana Roo coast; Cozumel reefs have the highest richness index, being Dalila reef the one with the highest value of D '= 6.97. Also, Chankanaab reef together with 'Colombia' reef presented the highest diversity values of H' = 2.19 nits/Ind and H' = 2.18 nits/Ind, respectively. Diversity indices decrease towards the North, with theChitales reef having the lowest diversity value of H '= 1.60 nits/Ind. Based on the Pielou evenness index, the benthic community structure appears to be distributed evenly because the index values are relatively high; this is reinforced by the Simpson dominance index values obtained (Table 1) for Punta Nizuc reef (λ = 0.27) and for Yucab reef (λ = 0.19).

TABLE 1: Ecological Indices for the coral reefs of three marine national parks studied in northern Quintana Roo (PNIMCM: ParqueNacional Isla Mujeres,Cancún, Nizuc; PNAPM: ParqueNacionalArrecife de Puerto Morelos, PNAC: ParqueNacionalArrecifes de Cozumel)

	Reef Loc.	Coral (%)	Algae (%)	No. of sp.	Richness (D')	Eveness (J')	H' (nits/indv)	Dominance (λ)
PNIMCN	Manchones	27.00	48.00	13	3.64	0.69	1.78	0.24
	Cuevones	26.20	41.60	10	2.76	0.73	1.68	0.24
	Chitales	16.91	49.09	7	2.12	0.82	1.60	0.25
	El Bajito	16.17	50.17	17	5.75	0.73	2.07	0.21
	Punta Nizuc	17.40	58.00	14	4.55	0.71	1.88	0.27
PNAPM	Bonanza	12.40	41.20	12	4.37	0.77	1.93	0.21
	Radio Pirata	27.60	20.60	12	3.32	0.80	2.00	0.19
PNAC	Paraiso	11.32	24.51	14	5.40	0.77	2.03	0.18
	Chankanaab	15.69	39.31	17	5.84	0.77	2.19	0.15
	Yucab	9.31	17.01	15	6.50	0.73	1.96	0.19
	Paso del Cedral	18.33	21.18	18	5.93	0.72	2.09	0.20
	Dalila	18.40	32.57	21	6.95	0.69	2.09	0.22
	Colombia	22.22	21.88	16	4.85	0.79	2.18	0.14

With regard to the percentage coral cover and macroalage a trend can be observed contrary to the diversity indices and the number of species as the highest percentage of coral cover was observed in the reefs of northern Quintana Roo. Particularly Yucab reef at PNAC had the highest diversity and species richness but was the one with the lowest coral cover. In the case of algae cover there is also a decrease from North to South, as the National Reef Park Isla Mujeres- Cancun -Nizuc (PNIMCN) have the highest percentages in algal cover decreasing to 17.01% in Cozumel (Figure2). Consistently, always PNIMCN reefs have higher algal cover compared to coral. At the reefs of PNAC algal cover is consistently less than algal cover at PNPM and PNIMCN.

In order to simplify comparison of benthic component data obtained from the different reef localities, we formed a single component group called abiotc components by grouping bottom components such as rocks, sand, dead coral and other non-living components.

It is important to note that while Cozumel reefs have the highest species richness and diversity, and reduced algal cover, these reefs are characterized by a higher percentage of abiotic components (above 40%). In the case of Northern reefs it is the algae that contributes to the higher percentage of benthic components (Figure3). In the case of the reef sites of Punta Nizuc, Bonanza and Radio Pirata are characterized by the presence of seagrasses.

Applying the ANOSIM similarity analysis significant differences were detected between the various reefs studied (R = 0.497, p = 0.001), except for El Bajito reef, which in terms of the structure of benthic components it was not significantly different from Chitales reef (R = -0.006, P = 0.51), being the two reefs part of the same national park (PNIMCN). At the Cozumel Reefs National Park, the Paso del Cedral reef and Delilah reef, Paraiso reef and Chankana'ab reef are the only reefs that present no significant differences in the structure of benthic components (R = 0.047, P = 0.142; R = -0.039 p = 0.75). This can be confirmed by a cluster analysis, that based on a similarity matrix generated using the Bray-Curtis, can be seen that the diagram is clearly divided into two groups separating Cozumel reefs park (PNAC) from the parks of Puerto Morelos (PNAPM) and Isla Mujeres-Cancun-Nizuc (PNIMCN); and within this latter group, the separation between Puerto Morelos and PNIMCN can be clearly seen (Figure4).

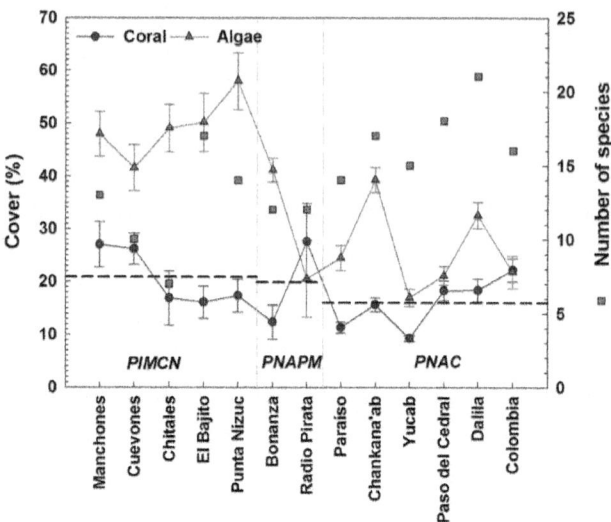

Figure 2: Coral and algal cover (percentage) at the Northern Quintana Roo National Reefs parks for the years 2006-07. The dashed line represents the average coral cover (percentage) for each park.

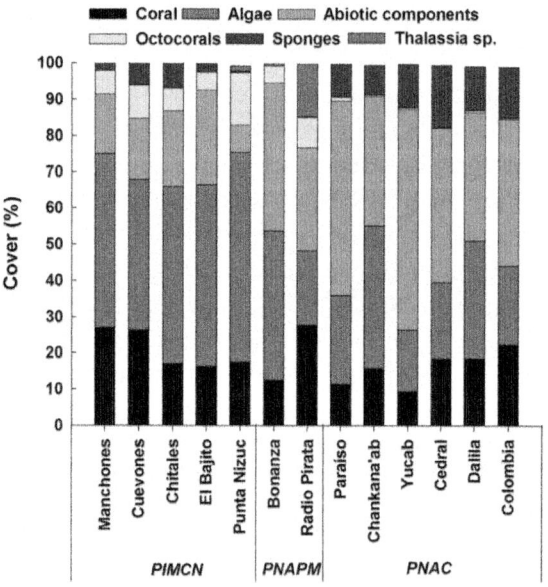

Figure 3: Benthic components of the coral reefs at three National Marine Parks in Northern Quintana Roo during 2006-2007.

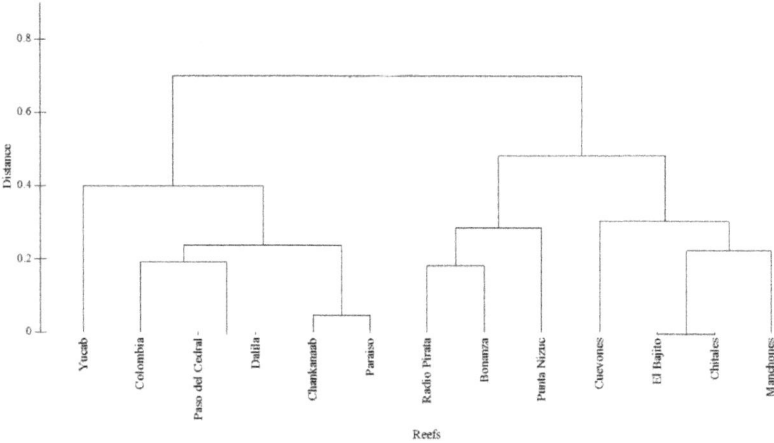

Figure 4: Cluster analysis of the coral reefs of Northern Quintana Roo (data for 2006-2007).

HYDROLOGY: NUTRIENT CONCENTRATIONS

The increase of anthropogenic nutrients is attributed to be a major cause of increased algal cover in coral reef ecosystems, based on the theory that in some reefs bottom-up ecological control in food chains regulates ecosystems. In order to try to better understand the processes that regulate ecosystem resilience we additionally characterized the hydrological conditions throughout the three marine parks studied here. Samples were taken from different stations distributed on the same reefs where benthic surveying was conducted (Figure 5a,b), additionally including the hydrologic (nutrient) characterization of the Nichupte Lagoon, bounded to the north by Punta Cancun and to the south by Punta Nizuc (Figure 5a). In order to characterize the concentration of nutrients, particularly from dissolved inorganic nitrogen (DIN) species that involve inorganic nitrogen: nitrate (NO_3^-), nitrite (NO_2^-) and ammonium (NH_4), water samples were collected at two different depths: at the water surface and near the substrate. Sampling was conducted in the months representing contrasting seasonal conditions (dry and rainy seasons) in the years 2006 and 2007.

The average concentration of DIN in the sampled coral reefs is shown in Table 2. By applying a factorial ANOVA considering the different climatic seasons and parks there were no significant differences in the concentration of DIN ($F = 0.422$, $p = 0.657$) but between the dry and rainy seasons ($F = 9.280$, $p = 0.003$). Consistently, one can observe a slight increase in the concentration of dissolved inorganic phosphate (DIP) in the three parks during the rainy season; this increase is mostly evident in the PNIMCN and PNAPM, because it almost doubled the average concentration measured during the dry season (Figure 6). As was expected for Nichupté Lagoon, it was characterized by a higher concentration of DIN with respect to the levels seen at the reef sites. In addition, Nichupté lagoon presents an opposite behavior to reef sites as a function of seasonality: while during the rainy season the DIN increases reefs, the concentration decreases in the lagoon, probably due to dilution effect (Table 2, Figure 6).

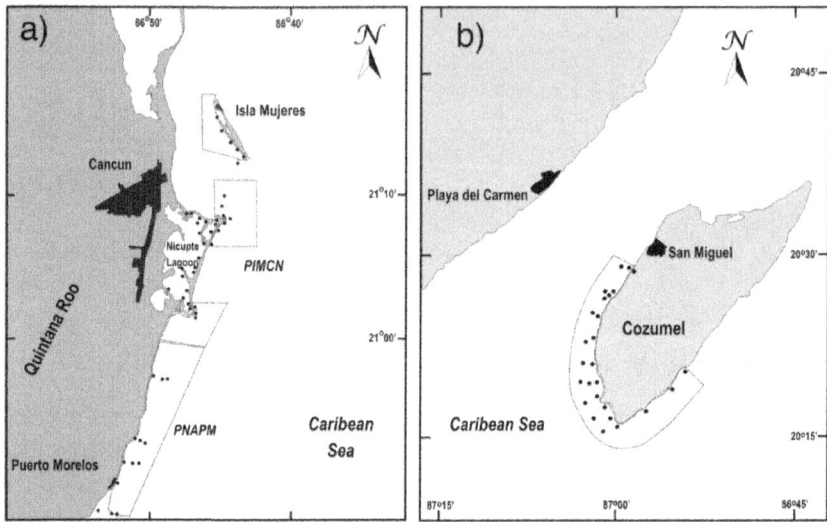

Figure 5: Sampling locations in the PNIMCN and PNAPM (a) and in the PNAC (b), used for characterizing the nutrient levels in the coral reefs of Northern Quintana Roo during 2006 – 2007.

TABLE 2: Average nutrient concentration in the Nichupté-Lagoon and the coral reefs of three National reef parks of Northern Quintana Roo during 2006 – 2007

Location	Season	DIN (μM)	Nitrate (μM)	Nitrite (μM)	Ammonium (μM)	Phosphate (μM)	Silicate (μM)
Nichupté-Lagoon	Dry	14.77	2.08	1.08	11.61	0.18	5.50
	Rainy	4.18	1.08	0.16	2.94	0.33	7.98
PNAC	Dry	1.26	0.60	0.09	0.58	0.26	6.48
	Rainy	1.96	0.54	0.09	1.33	0.29	4.58
PNAPM	Dry	0.91	0.25	0.09	0.57	0.24	3.15
	Rainy	2.15	0.53	0.04	1.58	0.16	1.85
PNIMCN	Dry	0.86	0.17	0.10	0.59	0.23	4.42
	Rainy	1.50	0.46	0.07	0.98	0.28	2.30

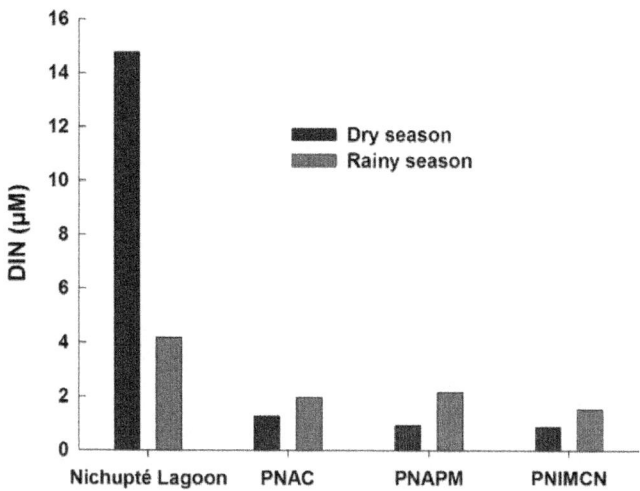

Figure 6: Average DIN concentration (μM) in in the Nichupté-Lagoon and the coral reefs of three National reef parks of Northern Quintana Roo (PNAC, PNIMCN, PNAPM) during 2006 – 2007 in two climatically contrasting seasons (rainy and dry).

Ammonium (NH_4) is the most important DIN species in the reef waters. Its concentration is consistently higher than nitrates and nitrites, except for the PNAC during the dry season in which ammonium and nitrate have approximately the same concentration (Figure 7). It was expected that ammonia were the most important DIN species, since the success of high productivity of coral reefs results from the high nutrient recycling that takes place within the ecosystem. Most of the new nitrogen entering the reef is through N2 fixation, so it is absorbed and converted to organic nitrogen which can be later consumed and passed through the food webs or returned to the system as ammonia by passing through the process of ammonification, so it is excreted in the urine of organisms [21].

During the dry season, there is no significant difference among the three species of DIN in the three parks, however the average concentration of nitrate is higher in the PNAC. During the rainy season nitrate concentration is high in all parks but ammonium is significantly higher, being PNAPM the park with the highest ammonium concentration (Table 2, Figure 7).

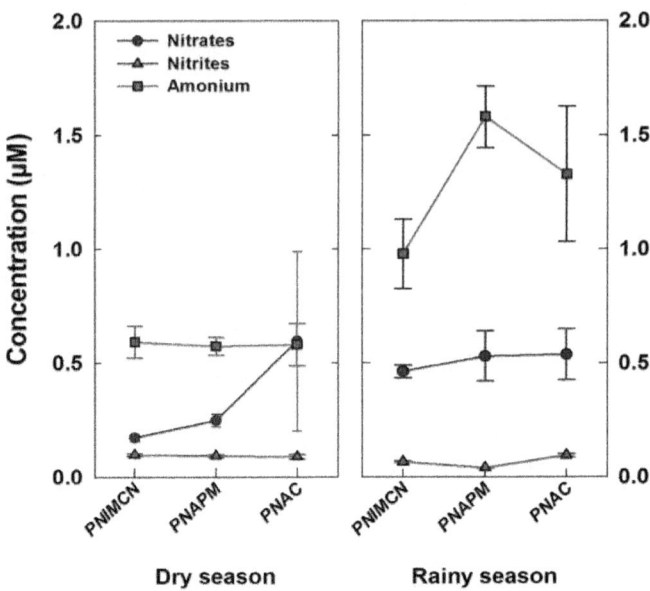

Figure 7: Average DIN concentration in the coral reefs of three National reef parks of Northern Quintana Roo (PNAC, PNIMCN, PNAPM) during 2006 – 2007 in two climatically contrasting seasons (rainy and dry). ANOVA statistical test (bars represent Std. Error).

The hydrographic information generated in this study indicates that there is no evidence of eutrophication in reef areas in the Mexican Caribbean. In general, nutrient concentrations are low, typical of the reef zones of the Wider Caribbean. The general average of DIN (1.3 ± 1.6 µM) in the Northern Quintana Roo reef parks is lower than that reported for the Florida reef tract (4.3 ± 7.4 µM; [22]) and within the observed range of the coral reefs of Tobago (1.6 ± 1.1 µM; [23]).

It has been shown that groundwater seepage into the coastal area of this region can supply significant amounts of nutrients to the water column [24-25]. However, a study in seagrass meadows of *Thalassia testudinum* in Puerto Morelos reef lagoon has shown that water seeping through the springs (locally called 'ojos') can enrich the water column --and seagrasses-- with phosphorus, but not with nitrogen [25]. These authors found that in *T. testudinum* meadows of Puerto Morelos, the pore water contains extremely low levels of nutrients ($1.2 - 3.42$ µM of ammonium and $1 - 1.5$ µM of phosphate) compared with the world average for seagrass meadows (~ 86 µM of ammonium and 12 µM of phosphate). This low DIN concentration in the water column of Puerto Morelos suggests that nitrogen could be limiting the growth of seagrass meadows. The nitrogen content in the tissue *T. testudinum* in the reef lagoon (% N> 1.8), however, is high enough to not show this limitation [25].

In contrast to the typical oligotrophic conditions that characterize coral reef areas, nutrient concentration inside the Nichupté Lagoon system is much higher. The weighted average concentration of DIN inside this lagoon system (14.7 ± 11.6 µM for winter and 4.18 ± 1.98 µM for summer) is between 2-7 times higher than the DIN average measured in Puerto Morelos, the coral reef area with the highest average DIN concentration (2.15 ± 0.84 µM) of our study area. These results, however, were expected because this lagoon system receives wastewater from the surrounding developments. After a few decades of continuous supply, there are now evident signs of eutrophication [25-26]. In spite of this situation, our results indicate that reef areas developing outside this lagoonal system are not affected, so far, in their hydrographic characteristics.

Favorably, the reef systems along the Mexican Caribbean coast still thrive under low nutrient concentrations. However, the low concentrations of DIN in the coastal waters and the evident overgrowth of macroalgae on the reefs studied suggest the existence of diffuse nitrogen sources fueling their growth. Nitrogen fixation could be a major source for these reefs (see further evidences of this in the isotopic section), and if this nitrogen source dispersed through the water column, it would raise the DIN up to 0.3 µM day^{-1}[25]. This assumption is reasonable, especially when considering that the nitrogen isotope values (δ^{15}N) in the tissues of macroalgae growing on these reefs (see below) are very close to the isotopic composition of atmospheric nitrogen ($\delta^{15}N_2 = 0$‰). This new nitrogen, however, may pass "undetected" in our monitoring sampling because it may be immediately assimilated by the macrophytes upon entering the coastal zone where the coral reefs develop. In this regard, the actual macrophyte biomass itself may be the best evidence of large nitrogen inputs into the otherwise oligotrophic environments that characterize coral reefs, where macrophytes' occurrence is commonly very scarce.

Isotope systematics of δ^{15}N in macroalgae

In order to differentiate potential sources of nitrogen to the reef zones we analyzed the nitrogen isotopic composition (δ^{15}N) of the tissue of several species of macroalgae collected from reefs studied. The δ^{15}N of macroalgae has been a widely used as tracer of nutrient dynamics [27] This approach has been applied mainly in areas where nutrient sources are diffuse or little obvious, but also in areas where sources are very different, such as nutrient inputs from sewage. However, the spatial extent of its influence is not clear [28-29].

The $\delta^{15}N$ values of macroalgae from the different reefs of the three national marine parks are shown in Figure 8. The more common genus found in the study area were *Dyctiota* spp. and *Halimeda* spp., followed by *Penicillus pyriformis*, *Ulotrix* spp. and the seagrass *Thalassia testudinum*. We grouped all species by location and compared the average-$\delta^{15}N$ (± 1SD) of macroalgae between locations and sampling period (Figure 8a for winter-2006, and Figure 8b for summer-2007). For reference, we have included in this figure the average nitrogen isotopic composition of nitrate ($\delta^{15}NO_3 = 4.37 ± 2.5$ ‰) for the three parks (Carriquiry unpublished data). The validity of this reference comparison rests on the assumption that nitrate is a major source of nitrogen for macroalgae.

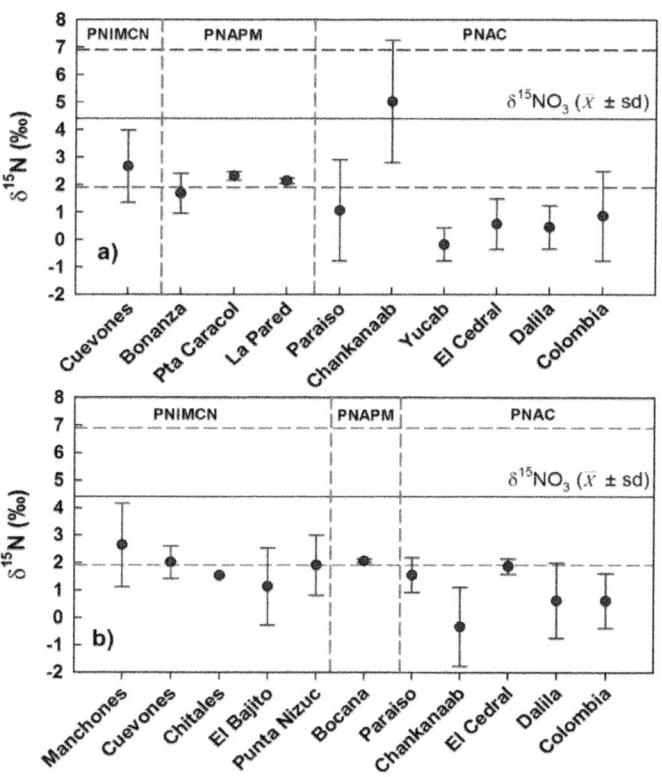

Figure 8: Average $\delta^{15}N$ composition (± 1SD) of macroalgae collected at each site during the winter of 2006 (a) and the summer of 2007 (b). The average isotopic composition of nitrate ($\delta^{15}NO_3$) in the coastal waters (Carriquiry, unpublished data) of the three coral reef national parks studied here is included as a thin horizontal line in each diagram.

A common feature for both collection periods is the low $\delta^{15}N$ values of macroalgae. During winter-2006, with the exception of Chankanaab reef at PNAC ($\delta^{15}N$ of 5.0 ± 2.2 ‰), the $\delta^{15}N$ of macroalgae is less than 3 ‰ and is markedly smaller than the isotopic value of nitrate ($\delta^{15}NO_3$) (Figure8a). As in the winter of 2006, the average $\delta^{15}N$ of macroalgae during the summer of 2007 varied from ~ 0.5 to 2.5 ‰, being always below the average $\delta^{15}NO_3$ (Figure 8b).

It is noteworthy to highlight the results obtained from Chankanaab reef (at PNAC) which shows the greatest contrast in the average $\delta^{15}N$ of macroalgae between sampling seasons: while the average $\delta^{15}N$ of macroalgae was the most positive of all sampling stations during the winter of 2006 (Figure 8a),during the summer of 2007 Chankanaab reef presented the most negative $\delta^{15}N$ values (-0.34 $\pm$ 1.43 ‰) of all the studied sites (Figure 8b). One factor that may explain this discrepancy is the difference in species composition between sampling sites. While in the winter of 2006 the $\delta^{15}N$ was measured in*Dyctiota* spp. and *Penicilluspyriformis*, during summer 2007 the species analyzed were*Lobophoravariegata*, *Dictyosphaera cavernous,Anadyomenestellata*and *Ulotrix* spp. The results for species indicates that the $\delta^{15}N$ of *Dyctiota* spp. and *P. pyriformis* were characterized by positive values in both sampling periods, while the species *L. variegata* always showed low $\delta^{15}N$ values, including *A. stellata* whose measured $\delta^{15}N$ was the most negative (-1.67 and -1.42 ‰ for Chankanaab and Dalila, respectively) in both sampling periods.

There are several alternatives to explain the low $\delta^{15}N$ values of macroalgae in this region. N_2 fixation in coral reefs is regarded as a major component of the nitrogen cycle that provides new nitrogen for these ecosystems. It has been estimated that fixed nitrogen can supply from one quarter to one half of the nitrogen requirements for the primary producers in these oligotrophic environments [30]. It has been shown in coral reef areas where nitrogen fixation is predominant that $\delta^{15}N$ value of macroalgae is close to 0 ‰, or even negative [31,32]. In agreement, the range of $\delta^{15}N$ values measured in the different species of macroalgae (<0 to <2.5 ‰) in our study, indicates that N_2 fixation may be playing an important role in fulfilling the macrophytes' nitrogen demand. This conclusion is also supported by [25] who found that the $\delta^{15}N$ values of 1.9 ‰ in the seagrass *Thalassia testudinum,* in Puerto Morelos lagoon, were the result of nitrogen fixation.

Alternatively, another possibility is the low concentration of dissolved inorganic nitrogen (DIN) in the studied reefs. Different studies have shown a positive relationship between the $\delta^{15}NO_3$ *versus* nitrate concentration [NO_3], both in temperate and tropical areas. It was found in a Massachusetts estuary that the $\delta^{15}NO_3$ values approach 0 ‰ when the nitrate concentration is reduced to levels <1 uM [33]. Similarly, other results obtained by [34]from the Mexican Caribbean reported that $\delta^{15}NO_3$ decreases linearly with the concentration of nitrate in the water column.Thus, consideringthe low DIN concentrations that characterizeour study area it would be expected that the $\delta^{15}N$-DIN available for macroalgae should be characterized by lower values.

Lastly, changes in the proportion of the different nitrogen species available for photosynthesis could explain the lower values in $\delta^{15}NO_3$. In our study, ammonia accounts for 60 to 76% of the DIN levels. This implies that ammonium, but not nitrate, could be the main source of nitrogen to seaweeds in the region. Hence, the $\delta^{15}N$ of macroalgae largely reflects the isotopic signature of ammonia. This hypothesis, although plausible, depends on the metabolic capacity of each species for using ammonium, depending on its availability in the environment.

THE FUNCTIONAL ROLE OF REEF FISH

Fishes are particularly recognized for their role as the main drivers of energy flow in coral reefs and can be separated into two major functional groups: the grazers or herbivores that regulate the abundance and community structure of algae; and corallivorous that selectively feed on coral tissue. Its importance, in addition to controlling the population of primary producers is that their feeding activity promotes biodiversity, since consumption of algae and coral leaves available space for colonization of new individuals or species [35]. Another important group are the predators as they play an important role in the ecosystem because they occupy the highest level in food webs, and from this position they regulate the organisms that are in the lower trophic levels. Additionally, they connect the dynamics of other communities and ecosystems that are apparently distinct as they often travel long distances. It has been suggested that the predator's ability to travel great distances in response to changes in prey abundance is important for maintaining stability of food webs [36]. Consequently, the loss of top predators could destabilize ecosystems through a chain reaction that eventually propagate down through the food web. With this in mind, the biomass of herbivorous and predatory fish have been calculated in order to determine whether there is spatial variability in these functional groups.

There is a close link between the fish community and benthic components, as any change in the structure of one of these communities has an effect on the structure of the other. [37] pointed out that in quantitative studies during the 80's reported high densities of predators, like sharks and groupers, associated with areas of high coral cover. Currently, the presence of large top predators is rare and is considered that the decrease in the abundance of this group has strongly affected the trophic flow patterns in reef communities [38]. In addition, the decrease in coral cover in coral reefs has been related to the decrease in the abundance and diversity of reef fishes [39].

In order to gain a better understanding of the role of fishes in our study area, we analyzed the community structure of reef fish sampled at PNAPM and PNAC; for PNIMCN we only had data for Manchones reef. By applying a similarity analysis (ANOSIM) to the abundance data, significant differences in the structure of the fish community were detected in these locations (R = 0.442, $p < 0.0001$). Paired tests showed significant differences between all locations, except Yucab and Paraiso Reefs ($p = 0.09$), and Yucab and Paso Cedral reefs ($p = 0.23$); also Dalila and Colombia do not show significant differences ($p = 0.20$). Using cluster analysis, based on a similarity matrix generated using the Bray-Curtis index, can be seen again the grouping of localities from each national park, leaving the Manchones reef alone as an isolated entity (Figure 9).

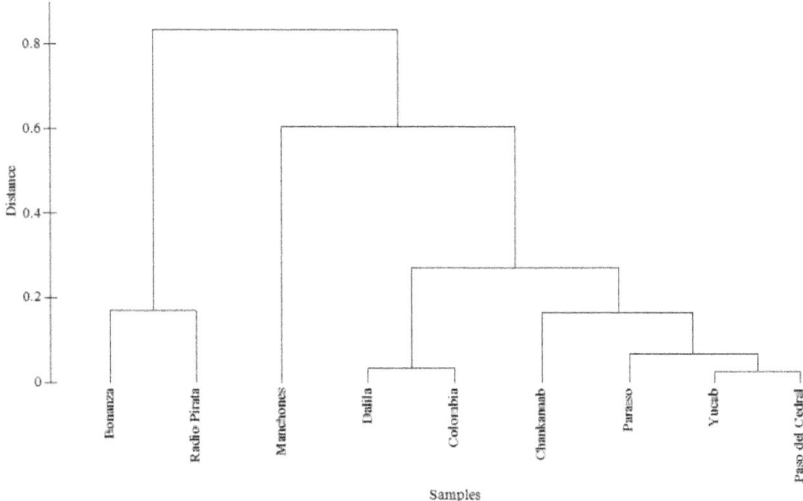

Figure 9: Cluster analysis of the reef fish community in northern Quintana Roo (2007 data).

The differences in the structure of the fish community, can be as well associated with structural differences in the benthic composition of the reefs, as the decline in coral cover reduces the structural complexity of the reef and therefore the space available for shelter and feeding [39] (Figure 10).

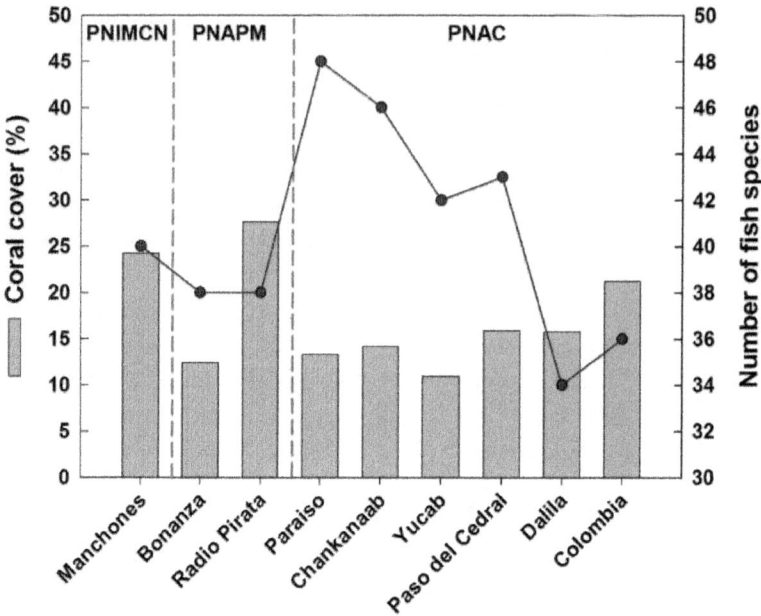

Figure 10: Trends in fish species richness and coral cover in the three national coral reef parks of Northern Quintana Roo, Mexico.

As to the relative abundance of trophic guilds, we can observe that higher predators generally have little relative abundance or simply not recorded during surveys, being Radio Pirate reef in Puerto Morelos, as well as Dalila and Colombia reefs in Cozumel were the only coral reefs where this trophic guild was recorded. The relative abundance of carnivores is up 50% Chankana'ab and Paso del Cedral reefs, however this trophic group also characterized by small fish whose dietary components include small invertebrates. Another group that has a high relative abundance of 60% and 50%, in Colombia and Dalila reefs respectively, are the herbivores (Figure 11).

Using the information of the size structure of three reefs (Manchones, Bonanza and Radio Pirata reefs) we calculated the relative biomass of the different trophic guilds using length – weight parameters for each species available at fish base [40]. This information allows assessing the degree of disturbance in the communities of each reef studied. In stable conditions or of low-disturbance, the dominant species of large size and longevity (K-strategists) are dominant in biomass and have low abundance. There are also present in the communities r-strategy species, opportunistic species with a short lifetime that are dominate in abundance but have low biomass contribution. When a community is disturbed, K-strategy species are usually not favored and opportunistic species increase in numbers and biomass [19].

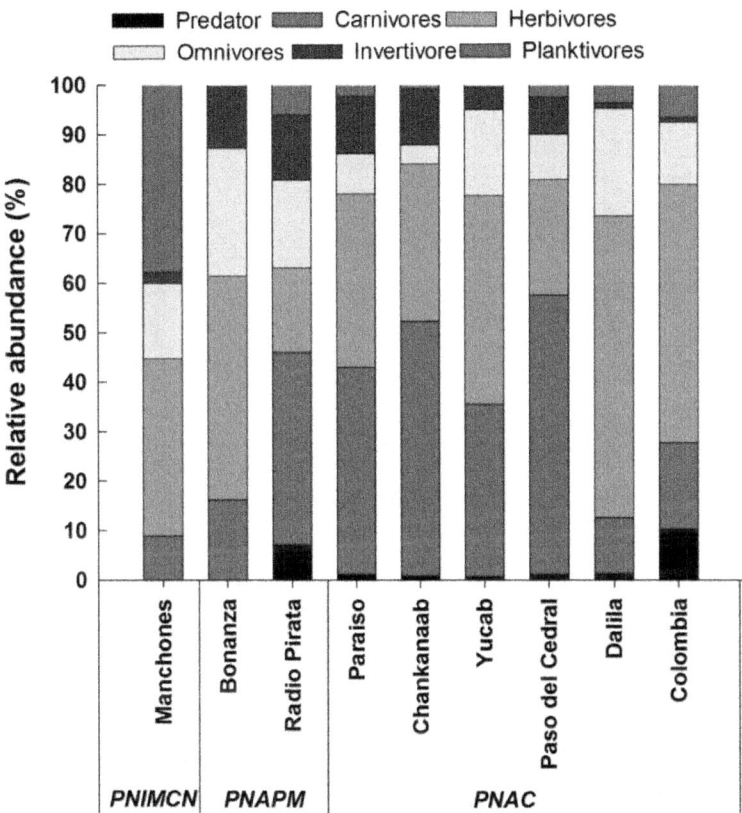

Figure 11: Relative abundance of the trophic guilds of the fish community associated with the coral reefs of northern Quintana Roo (2007 data).

For Radio Pirate and Bonanza reefs we can see that the relative biomass of carnivores is high (>50% of the biomass), however, the relative biomass of larger predators is virtually nonexistent. In the case of Manchones reefs, the herbivores contribute in greater proportion to the biomass (Figure 12). Such inverted biomass pyramids of fish have been reported in coral reefs characterized by low coral cover. To maintain these typesof biomass pyramids, however, a high primary production is required [37].

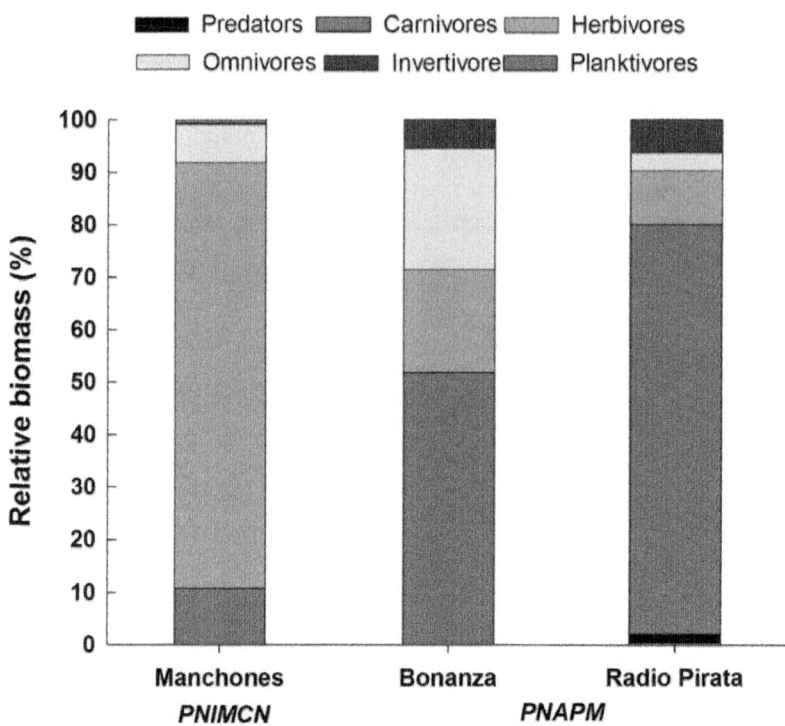

Figure 12: Relative biomass of trophic guilds in reef fish communities of northern Quintana Roo coral reefs (2007 data).

Additionally, a comparison analysis chart of abundance/biomass (Abundance / Biomass Comparison - ABC plots) was performed. This method presents a statistical test (W) that represents the abundance over biomass in a range of -1 to +1; if the statistical test generates a +1, biomass dominates over abundance and represents a system with no impact. On the contrary, when the result is -1, abundance dominates over biomass indicating that the system has been highly impacted. Values near zero indicate an intermediate disturbance [19].

In the case of reefs for which we had data on both, abundance and biomass, we obtained W = 0.146, indicating that the fish communities of these reefs are under a scenario of intermediate disturbance (Figure 13) that may result either from the degradation of their habitat or the effect of fishing and poaching that take place in these reef locations [41-43]. This analysis allow us to conclude that It is essential to conduct a monitoring program of the fish community structure, and their biomass, to better assess their status and how they contribute to the functioning of these ecosystems.

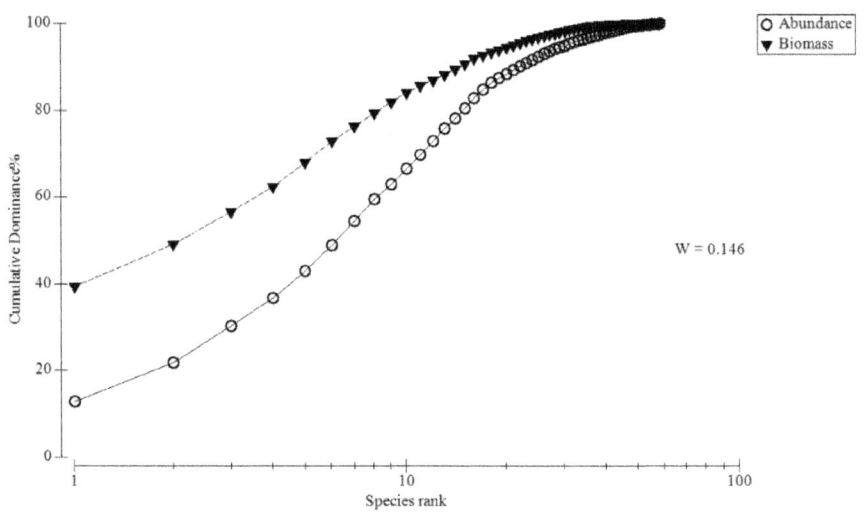

Figure 13: ABC-plot of reef fish for three coral reefs (Manchones, Bonanza and Pirate Radio) of northern Quintana Roo (2007 data).

CONCLUSION

The isotopic composition of algal tissue, along with the concentration of DIP and its various species, suggests that the low coral cover on reefs is caused by other factors rather than eutrophication of ecosystems. Plausibly, this may be an indication of reduced herbivorous-fish biomass, which is a key factor that regulates the abundance of macroalgae.

With regard to the fish community structure our results suggest an evident impact in the way biomass is distributed per trophic level, where large predators and consumers have significantly decreased. This type of inverted pyramids in fish biomass has been documented in other reef systems that show signs of anthropogenic disturbance, mainly by overfishing [37].

Fishing is the major environmental and economic problem facing most marine ecosystems, where pressure is exerted mainly on higher trophic levels in search of the largest fishes, and consecutively reducing the abundance and biomass of large predators, planktivorous and herbivorous fish, thus lowering the trophic level of the catches [2]. The ecological function of the fish is of great importance in the recovery and resilience of coral reefs and is likely to depend not only on food preferences of a trophic guild (e.g. herbivory), but also in the abundance and biomass of fish. Unfortunately, the limited information about fish populations prevent us to establish a reliable baseline for the coral reefs in the northern Mexican Caribbean describing an undisturbed ecosystem that can be used to compare with other coral reefs in the wider Caribbean. Most of fish studies mostly focus only on single fish taxon, which does not provide relevant information required to describe the conditions of the community. Thus, assessing the ecological effects of the loss of predators is difficult, but it is well documented that the decrease of fish diversity and of important functional groups (herbivores) brings significant structural impacts through food webs and benthic community, and ultimately in the functioning of coral reefs [8, 38].

The ecological symptoms of a coral reef with probabilities of collapsing (a phase-shift) are likely to include the loss of macro-fauna, reduced fish stocks, a change in the ecological role of herbivorous fish that are replaced by only one species of echinoid, destructive over-grazing and bioerosion due to food limitation suffered by sea urchins, and reduced coral recruitment. To be able to efficiently address the current global crisis of coral reefs, it is urgently needed to be able to generate information that allows us to assess the current status of these ecosystems. It is also required the active management of human activities that modify the essential ecological processes and functions of coral reef ecosystems.

NaN. Acknowledgements

We greatly thank CONABIO for financial support, and Mario Castañeda and Ma. Carmen Vázquez Rojas for their administrative support during the project's course. Likewise, we greatly thank the great logistic support provided by the authorities of the Marine Protected Areas studied here, especially Biol. Alfredo Arellano-Guillermo and Dr. Jaime González of SEMARNAT-CONANP, and the personnel from the Marine Park of Cancún (PAIMCN): Juan Carlos Huitrón, Alejandro Vega; from the Marine Park of Cozumel (PNAC): Biol. Ricardo Gómez-Lozano, Christopher Gonzalez, Nayeli Hernandez; and the Marine Park of Puerto Morelos (PNAPM): Biol. Daniela Guevara, Biol. Oscar Álvarez Gil. We are also very thankful to Pedro Castro and the students of the Geosciences Research Group of IIO-UABC for the great help in sample handling and processing.

REFERENCE

1. Gardner T.A., Côte I.M., GillJ.A., GrantA., WatkinsonA. R. Long-Term region-Wide declines in Caribbean Corals. Science2003; 301: 958 – 960.

2. Bellwood D.R., Hughes T.P.,Folke C., NyströmM. Confronting the coral reef crisis. Nature2004; 429: 827-833.

3. Connell J.H. Distrubance and recovery of coral assemblages. Coral Reefs 1997; 16: S101-s113.

4. Moberg F., Folke C. Ecological goods and services of coral reef ecosystems. Ecological Economics 1999; 29: 215-233.

5. Salvat B. Coral reefs – a challenging ecosystem for human societies. Global Environmental Change 1992; 2: 12-18.

6. Wilkinson B. Status of coral reefs of the world: 2004. Vol. 2. Australian Institute of Marine Science; 2004.

7. Wilkinson B. Status of coral reefs of the world: 2008. Global Coral Reef Monitoring Network and Reef an Rainforest Research Center; 2008.

8. Hughes T.P., Baird A.H.,Bellwood D.R.,Card M., ConnollyS.R.,FolkeC., GrosbergR., Hoegh-GuldbergO., JacksonJ.B.C., KleypasJ., LoughJ.M., MarshallP., NyströmM., PalumbiS.R., PandolfiJ.M., RosenB.,RoughgardenJ. Climate change, human impacts and the resilience of coral reefs. Science 2003; 301: 929-933.

9. Lapointe B.E. Nutrient thresholds for bottom-up control of macroalgal blooms on coral reefs in Jamaica and Southeast Florida. Limnology and Oceanography 1997; 42 (5 part 2): 119-1131.

10. Szmant A.M. Nutrient enrichment on coral reefs: is it a major cause of coral reef decline? Estuaries 2002; 25(4b): 743-766

11. Hughes T. P., GrahamN.A.J., JacksonJ.B.C., MumbyP.J.,SteneckR.S. Rising to the challenge of sustaining coral reef resilience. Trends in Ecology and Evolution2010; 25(11): 633-642.

12. Hoegh-Guldberg, O., MumbyP.J., HootenA.J., SteneckR.S., GreenfieldP., GomezE., HarvellC.D., SaleP.F., EdwardasA.J., CaldeiraK., KnowltonN., EakinC.M., Iglesias-PrietoR., MuthigaN., BradburyR.H., DubyA.,HatziolosM.E. Coral Reefs under rapid climate change and Ocean acidification. Science 2007; 318(5857): 1737-1742.

13. Ledlie M.H., Graham N.A.J., Bythell J.C., Wilson S.K., Jenings S., Polunin N.V.C., Hardcastle J. Phase shifts and the role of herbivory in the resilience of coral reefs. Coral Reefs 2007; 26: 641-653.

14. Done T.J. Phase shifts in coral reef communities and their ecological significance. Hydrobiologia 1992; 27: 121-132.

15. Nyström M., FolkeC.,MobergF. Coral reef disturbance and resilience in a human-dominated environment. TREE 2000; 15(10): 413-417.

16. Mora, C., AndrefouetS., CostelloM.J., KranenburgC., RolloA., VeronJ., GastonK.J., MyersR.A. Coral reefs and the global network of marine protected areas. Science 2006; 312: 1750-1751.

17. Hughes T.P. Community structure and diversity of coral reefs: the role of history. Ecology 1994; 70:275-279.

18. Magurran A.E. Measuring Biological Diversity.Blackwell Publishing; 2004.

19. Clarke K.R., Warwick R.M. Change in marine communities: an approach to statistical analysis and interpretation. 2a ed. PRIMER-E Pymouth; 2001.

20. English S., Wilkinson C. Baker V. (Eds) Survey Manual for Tropical Marine Resource. AEAN-Australia Marin Science Project: Living CoastalsResourse. AIMS, Townsville; 1994.

21. O'Neil J.M. Capone D.G. Nitrogen Cycling on Coral Reefs. Chapter X, In: Capone, D.G., Carpenter, E.J., Bronk, D.A. & Mulholland, M.R. (eds.) Nitrogen in the Marine Environment.Elselvier Press; 2008.p937-977.

22. LapointeB.E., Bedford B.J. Ecology and nutrition of invasive *Caulerpabrachypusf. parvifolia* blooms on coral reefs off southeast Florida, U.S.A., Harmful Algae 2010; 9:1-12.

23. Lapointe B.E., Langton R., Bedford B.J., Potts A.C., Day O, Hu Ch. Land-based nutrient enrichment of the Buccoo Reef Complex and fringing coral reefs of Tobago, West Indies. Marine Pollution Bulletin 2010; 60:334–343.

24. Herrera-Silveira J.A., Medina-Gomez I.,Colli R. Trophic status base on nutrient concentration scales and primary producers community of tropical coastal lagoons influenced by groundwater discharges. Hydrobiologia 2002; 475-476, 91-98.

25. Carruthers T.J.B., van Tussenbroek B.I. Dennison W.C. Influence of submarine springs and wastewater on nutrient dynamics of Caribbean seagrass meadows.Estuarine, Coastal and Shelf Science 2005; 64,191-199.

26. Merino M., Gonzalez A., Reyes E., Gallegos M., Czitrom S. Eutrophication in the lagoons of Cancun, Mexico. Science of the Total Environment1992; 126, 861 – 870.

27. LapointeB.E., Barile P.J., Littler M.M., Littler D.S., Macroalgal blooms on southeast Florida coral reefs II. Cross-shelf discrimination of nitrogen sources indicates widespread assimilation of sewage nitrogen. Harmful Algae 2005; 4, 1106-1122.

28. Gartner A., Larvery P.,Smit A.J. Use of δ15N signatures of different functional forms of macroalgae and filter-feeders to reveal temporal and spatial patterns in sewage dispersal. Marine Ecology Progress Series 2002; 235, 63-73.

29. Capone D. G., Dunham S.E., Horrigan S.G.,Duguay L.E.Microbial nitrogen transformations in unconsolidated coral reef sediments. Marine Ecological Progress Series1992; 80: 75-88.

30. Yamamuro M., Kayanne H.,Yamano H. $\delta15N$ of seagrass leaves for monitoring antrhropogenic nutrient increases in coral reef ecosystems. Marine Pollution Bulletin 2003; 46, 452-458.

31. France R.L. Estimating the assimilation of mangrove detritus by fiddler crabs in Laguna Joyuda, Puerto Rico, using dual stable isotopes. Journal of Tropical Ecology 1998; 14(4): 413-425.

32. Umezawa, Y., MiyajimaT., YamamuroM., KayanneH., KoikeI.Fine-scale mapping of land-derived nitrogen in coral reefs by $\delta15N$ in macroalgae. Limnology Oceanography 2002; 47(5): 1405-1416.

33. McClelland J.W.,Valiela I. Linking nitrogen in estuarine producers to land – derived sources. Limnology & Oceanography1998; 43, 577-585.

34. Mutchler, T., Dunton K.H., Townsend-Small A., Fredriksen S., Michael K.,Rasser M.K. Isotopic and elemental indicators of nutrient sources and status of coastal habitats in the Caribbean Sea, Yucatan Peninsula, Mexico. Estuarine, Coastal and Shelf Science2007; 74, 449-457.

35. Done, T.J., OgdenJ.C., WiebeW.J., RosenB.R. Biodiversity and Ecosystem Function of Coral Reefs. In: Mooney H.A., J.H. Cushman, E. Medina, O.E. Salas & E.-D. Schulze (Eds). Functional Roles of Biodiversity: A global Perspective; 1996. p393-429

36. Holt R.D. Predation and community organization in S.A. Levin, ed. The Princeton Guide to Ecology. Princeton university Press; 2009. p274-281

37. Sandin S.A. Smith J.E., De Martini E.E., Dinsdale E.A., Fonner S.D., Friedlander A.M., Konotchick T., Malay M., Maragos J.E., Obura D., Pantos O., Paulay G., Richie M., Rohwer F., Schroeder R.E., Walsh S., Jackson J.B.C., Knowlton N.,Sala E., Baselines and degradation of Coral Reefs in the Northern Line Islands. PLoS one 2008; 3(2): e1548.

38. Duffy J.E. Biodiversity loss, trophic skew and ecosystem functioning. Ecology Letters 2003; 6: 680-687.

39. Jones G.P., McCormick M.I., SrinivasanM., Eagle J.V. Coral decline threatens fish biodiversity in marine reserves. PNAS 2004; 101(21):8251-8253.

40. Froese R. PaulyD.Fishebase. World Wide Web electronic publication. http://www.fishbase.org. (accessed august 2012).

41. INE. Programa de Manejo del Parque Marino Nacional Costa Occidental de Isla Mujeres, Punta Cancún y Punta Nizuc. Instituto Nacional de

Ecología; 1998a.

42. INE. Programa de Manejo Parque Marino Nacional Arrecifes de Cozumel, Quintana Roo.Instituto Nacional de Ecología; 1998b.

43. INE. Programa de Manejo del Parque Nacional Arrecife de Puerto Morelos. Instituto Nacional de Ecología; 2000.

Chapter 11

COPING MECHANISMS OF PLANTS TO METAL CONTAMINATED SOIL

Melanie Mehes-Smith[1], Kabwe Nkongolo[2] and Ewa Cholewa[1]

[1]Department of Biology, Nipissing University, North Bay, Ontario, Canada
[2]Department of Biology, Laurentian University, Sudbury, Ontario, Canada

INTRODUCTION

Metals such as cobalt (Co), copper (Cu), iron (Fe) and nickel (Ni) are essential for normal plant growth and development since they contribute to the function of many enzymes and proteins. However, metals can potentially become toxic to plants when they are present at high levels in their bioavailable forms (Hall, 2002). Phytotoxic levels of one or more inorganic ions in soil can be found in various parts of the world. These toxic sites occurred through natural processes or by anthropogenic effects. Naturally toxic soils include saline, acidic and serpentine soils, while anthropogenic polluted soils occur through mining activities, aerial fallout, and the run-off from galvanized sources of electricity pylons or motorway verges polluted by vehicle exhaust fumes (Bradshaw, 1984). The biochemical effect of metals on plants varies and the excess metal usually results in oxidative damage which affects their phenotype (Kachout et al., 2009)

Plants colonizing metal-contaminated soils are classified as resistant and have adapted to this stressed environment. Heavy metal resistance can be achieved by avoidance and/or tolerance. Avoiders are plants that are able to protect themselves by preventing metal ions from entering their cellular cytoplasm, while tolerant plants are able to detoxify metal ions that have crossed the plasma membrane or internal organelle biomembranes (Millaleo et al., 2010). Based on strategies used by plants growing on metal-contaminated soils, Baker and Walker (1990) classified them into three categories; metal excluder, indicators and accumulators/hyperaccumulators. The excluder group includes the majority of plant species that limit the translocation of heavy metals and maintain low levels of contaminants in their aerial tissues over an extensive range of soil concentrations. Plants that are metal indicators accumulate metals in their harvestable biomass and these levels generally are reflective of the metal concentration in the soil. Metal accumulators/hyperaccumulators are plants that increase internal sequestration, translocation and accumulation of metals in their harvestable biomass to levels that far exceed those found in the soil (Mganga et al., 2011; Baker and Walker 1990). Plants can accumulate and cope with the effects of high internal metal concentrations by the upregulation of the antioxidant defense system. This system is activated in order to respond to the deleterious effects caused by reactive oxygen species (Solanki and Dhankhar 2011).

Coping strategies allow the establishment of plant communities on metal contaminated soils. This is possible since some plants have adapted to these hostile sites by evolving mechanisms to deal with the toxic effects of metals in soil on plants. There is a need of identifying plants that are able to deal with excess metal in soil. Without these plants, the lands would remain barren and unsustainable.

The importance of plants in the remediation of heavy metal polluted soil is discussed in details in the present chapter. A review of the current knowledge on metal resistance mechanisms, as well as the potential genes and their role in metal homeostasis in plants will be examined. Finally, the coping mechanisms used by plants growing under metal contamination will be discussed.

REMEDIATION OF HEAVY METAL CONTAMINATED SITES

Soils that are heavily contaminated by metals may pose health risks to humans and to other living organisms in an ecosystem. Current techniques used to remediate metal contaminated soils include excavation, chemical stabilization, soil washing or soil flushing, but these methods are costly and impractical. There is a need to develop effective, low-cost and sustainable methods for soil bioremediation. The revegetation of these sites appears to be the most suitable method for long term land reclamation since plants can improve nutrient soil conditions. This can lead to the establishment of a self-sustaining vegetative cover, which in turn can prevent soil erosion (Wei et al., 2005). Phytoremediation is an inexpensive and solar-driven approach that is performed in situ. It can be used to remove, stabilize and detoxify organic and inorganic pollutants including heavy metals from air, soil and liquid substrates (Salt et al., 1998). An example of a reclaimed metal contaminated site in the mining region of Northern Ontario (Canada) is illustrated in figure 1. Plant species selected for land reclamation should grow and spread fast and be able to establish an effective soil cover. It is therefore important to search for plants that have spontaneously colonized these disturbed sites. Moreover, heavy metal contaminated mining sites exhibit physiochemical characteristics that are not suitable for the vast majority of plant species; hence the colonization of these sites is slow. However, plants that are resistant to this toxic environment can easily spread since there is a lack of competitors. It has been demonstrated that annual species have an extensive adaptive capacity compared to perennial genotypes due to their long-term natural selection (Wei et al., 2005).

Phytoremediation is composed of five main subgroups: phytoextraction, phytovolatilisation phytostabilization, phytodegradation and rhizofiltration. Phytoextraction is a process by which plants extract metals from soil by accumulating them in their aerial biomass. These plants can be harvested and metals can be extracted from their tissues. Plants that accumulate metals in their aerial tissues have been involved in the phytoextraction of several metals including Cd, Cr, Cu, Hg, Pb, Ni Se and Zn (Yong and Ma 2002).

The accumulation of metals by plants is interesting from an environmental or agronomic point of view. In mining or industrial sites, as well as their surrounding areas, heavy metals are responsible for severe soil contamination. In these cases, accumulator plants could be used for phytoremediation as they are likely able to remove metals from soils (Salt et al., 1998; Salt et al., 1995). Since some heavy metals are also essential minerals that can be deficient in staple food crops, genetic determinants of hyperaccumulation could be utilized in biofortification to improve the nutritional value of these crops (Frérot et al., 2010; Cakmak, 2008; Jeong and Guerinot 2008; Mayer et al., 2008).

Figure 1: A metal contaminated site in Sudbury, Ontario, Canada; a) before remediation and b) after remediation (photos courtesy of Keith Winterhalder and David Pearson from Laurentian University).

These metal accumulator plants could also convert metals and release them in a volatile form. This process is known as phytovolatilization.

Phytostabilization on the other hand, is a method that uses plants with a low ability for metal uptake to stabilize the contaminated soil thus preventing erosion. This limits the metals from entering the food chain. Plants can also be utilized for phytodegradation since they can in combination with microorganisms degrade organic pollutants. Finally, rhizofiltration is a process by which plant roots absorb metals from waste streams (Pulford and Watson 2003; Dushenkov et al., 1995).

RESISTANCE

Levitt (1980) stated that heavy metal contaminated environments act as stress factors on plants, which causes physiological reaction change that reduces or inhibits plant vigor and growth. A plant showing injury or death due to metal stress is deemed sensitive to its environment. On the other hand, resistant plants can survive and reproduce under metal stress conditions (Ernst et al., 2008). In general, plants can achieve resistance to heavy metals by avoidance or tolerance.

AVOIDANCE

Avoidance occurs when plants restrict the uptake of metals within root tissue by several strategies. In environments where the soil metal contamination is heterogeneously distributed, plants can prevent metal uptake by exploring less contaminated soil. Another avoidance strategy involves mycorrhizal fungi, where they can extend their hyphae outside the plants rooting zone up to several tens of meters and transfer the necessary elements to the plant (Ernst, 2006; Baker, 1987). Also, these metal tolerant fungi can increase plant metal resistance by changing the metals speciation or by restricting the metal transfer into the plant (Ernst, 2006). Arines et al. (1989) found that mycorrhizal Trifolium pratense (red clover) plants growing in acid soils had lower levels of Mn in their roots and shoots as compared to the non-mycorrhizal plants. Plants can also restrict contaminant uptake in root tissues by immobilizing metals for example through root exudates in the rhizosphere. A role of root exudates is to chelate metals and stop their entry inside the cell. The cell wall has also been found to be involved in restricting metal uptake into the cell's cytoplasm (Mganga et al., 2011).

Tolerance

In the absence of avoidance strategies, some plants can grow and survive in soil contaminated with toxic levels of heavy metals which are otherwise lethal or detrimental for growth and survival of others genotypes of the same or of different species (Maestri and Marmiroli 2012). Plants exhibiting tolerance are internally protected from the stress of metals that have entered the cell's cytoplasm (Baker, 1987). Metallophytes (metal tolerant plant) can function normally even in the presence of higher plant-internal metal levels. Plants adapt to their environments by developing heritable tolerance mechanisms. Tolerance to specific metals has evolved independently several times in different species from local non tolerant ancestral plant populations (Schat et al., 2000). Plants can exhibit tolerance to metals that are present in surplus in the soil. Each metal is under control of specific genes.

According to Bradshaw (1991) most species are in a state of genostasis. It is the restriction of genetic variability which limits the evolution of the population/species. In the absence of avoidance pathways, metal contaminated soil acts as a selection force on a population, where only the plants with tolerant genotype can survive and reproduce. This leads to a bottleneck, where few individuals survive and reproduce. In turn, metal tolerant populations can evolve rapidly following a disturbance such as contamination of soil with heavy metals. Plant adaptation to these sites occurs in populations for which tolerance variability already exists prior to the contamination (Maestri and Marmiroli 2012; Baker, 1987). Genes for the tolerance of metals are pre-existing at a low frequency in non tolerant populations of certain plant species (Ernst, 2006; Macnair, 1987).

Variation In Tolerance And Accumulation Characteristics

Variation occurs between species, populations and clones for tolerance and accumulation of metals. Assunçãno et al. (2003) found differences in the degree of chlorosis and concentration of metal for Thlaspi caerulescens (currently named Noccaea caerulescens) populations when grown in hydroponic solutions containing various Ni, Cd and Zn concentrations. Visioli et al. (2010) found differences in growth, morphology and Ni accumulation capacity when the Ni hyperaccumulator T. caerulescens and the non-metal adapted T. caerluescens were exposed to different Ni concentrations in hydroponic solutions. Besnard et al. (2009) used cleaved amplified polymorphic sites (CAPs) and microsatellites to determine the genetic variation for T. caerulescens populations from metalliferous and non-metalliferous sites from Switzerland. They found a correlation between the level of heavy metals in soil and the variation at the target loci for the genes involved in encoding metal transporters. Basic et al. (2006) found similar results when they analyzed the genetic variation of different T. caerulescens population sampled from different soil types with single nucleotide polymorphism (SNPs) in target and non target genes. These results were also observed in Populus spp. Marmiroli et al. (2011) compared Populus clones and found variation in their capacity to accumulate Cd. This variation in Cd accumulation between clones was correlated with SNPs at some target genes. These results imply that gene flow is limited between individuals found on metal contaminated and those from uncontaminated sites, at least for the loci that are involved in the fitness of the individuals (Visioli et al., 2012). On metal-enriched soil, there is a strong selection of local offspring, which conserves the metal tolerant genotypes (Ernst, 2006).

Genetics Of Tolerance To Metals

Identifying genes involved in a specific adaptation is challenging. Metal tolerance and accumulation in plants are complex genetic systems. Plants have to modify their physiological processes in order to be able to survive in the environment in which they have germinated. In turn, the survival of a population to the contaminated environment is dependent on the inheritance of favourable traits. Tolerance mechanisms are heritable and variable, resulting from genes and gene products (Maestri and Marmiroli 2012). Variation in the evolution of metal tolerance exists over species, populations and clones (Baker, 1987). Some species do not show variation in tolerance and accumulations. In order to determine genes involved in metal tolerance and accumulation, segregating analyses were used, where parents with contrasting phenotypes were crossed to produce progeny. Studies have determined that in many species, metal tolerance and accumulation are genetically independent (Assunçãno et al., 2006). For example, in Arabidopsis halleri, Cd tolerance and accumulation segregated as independent traits while Cd and Zn tolerance and accumulation cosegretated. In this later species, two or more genes were proposed to be involved in Cd and Zn accumulation but only one gene for Cd and Zn tolerance (Bert et al., 2003; Bert et al., 2002). In T. caerulescens, no genes involved in Cd, Zn and Ni tolerance and accumulation cosegregated. This suggests that there is a high probability that the genetic and physiological mechanisms for these traits are distinct from each other (Yang et al., 2005a; Maestri et al., 2010; Richau and Schat 2009; Assunçãno et al., 2006; Zha et al., 2004). As a result, it is not possible to conclude that a plant with high levels of metals in aerial biomass is also metal tolerant. The concentration of metals in above ground tissue serves as an indication of the plant's potential metal tolerance (Frérot et al., 2010).

Several techniques have been used to isolate and identify genes involved in heavy metal tolerance in plants, one of which is the quantitative trait loci (QTL) mapping. QTL mapping is a powerful tool in examining complex adaptive traits and in determining the number of genes involved in a trait as well as the genes effects and their interactions (Willems et al., 2007). By mapping QTLs, it can be possible to identify or validate candidate genes involved in a complex trait such as metal tolerance and accumulation (Willems et al., 2007). Other techniques used to identify genes for metal tolerance and accumulation are functional complementation in yeast mutants defective in metal homeostasis. These methods use plant cDNA expression libraries, as well as the identification of hypothesized pathways based on sequence similarities with plant cDNA libraries and genomic sequences (Lal, 2010). Transcriptome analyses have also been used to reveal genes involved in hyperaccumulation by analysing the differences in expression profiles or regulation-level of hyperaccumulator and non hyperaccumulator plants (Colzi et al., 2011).

Few specific major genes have been found for Cd, Cu, Ni and Zn tolerance in Silene vulgaris by crossing plants from a metalliferous site with non tolerant plant from a nonmetalliferous site (Schat et al., 1996; Schat et al., 1993). Similar results were reported for Cu tolerance for Mimulus guttatus (Macnair, 1993) and Zn tolerance for Arabidipsis halleri (Bert et al., 2003). In S. vulgaris and M. guttatus, modifier genes (minor genes) were involved in Cu tolerance, thus increasing tolerance and enhancing the effect of the major gene(s) (Smith and Macnair 1998). Only two QTL were involved in Ni accumulation and tolerance in S. vulgaris (Bratteler, 2005).

Studies aiming at identifying associations between molecular markers and metal tolerance and accumulation trait have been performed using interspecific and intraspecific crosses. When a high Zn accumulating T. caerulescens parent was crossed with a low Zn accumulating parent, two major QTLs were found to be involved in the increased of Zn accumulation in root (Assunçãno et al., 2006). Deniau et al., (2006) performed QTL mapping for the hyperaccumulation of Zn and Cd in T. caerulescens. They found two QTLs responsible for Cd and two for Zn accumulation in roots. In addition, one QTL for Cd and three QTLs for Zn accumulation in shoot were characterized. Macnair et al. (1999) reported a major gene involved in Zn tolerance from the analysis of F2 progeny derived from a cross between A. halleri (tolerant parent) and Arabidopsis lyrata (sensitive parent). Willems et al. (2007) generated a backcross progeny from the interspecific cross between A. halleri (tolerant parent) and A. lyrata (sensitive parent) and identified three major additive QTLs involved in Zn tolerance in A. halleri. These QTLs were mapped to three different chromosomes (3, 4 and 6) and colocalized with genes that have been known to be involved in metal tolerance and accumulation. HMA4 (Heavy Metal ATPase 4) encodes a P-type ATPase pump localized at the plasmamembrane involved in loading Zn and Cd into the xylem. MTP1-A and MTP1-B are Metal Tolerance Protein- vacuolar transporters that are involved in Zn tolerance) (Gustin et al., 2009; Krämer, 2005). Three new QTLs were identified and mapped to chromosomes 4, 6 and 7 by Filatov et al. 2007, when F2 progenies from a similar interspecific cross were analyzed. Frérot et al. (2010) also found Zn accumulation to be polygenic using A. halleri X A. lyrata petraea progenies. They determined that Zn accumulation is controlled by two QTLs in low Zn concentration and three QTLs in high Zn concentration. Four of the five QTLs mapped for Zn accumulation in their study were also reported in previous studies using A. halleri X A. lyrata petraea progenies (Frérot et al., 2010; Filatov et al., 2007; Filatov et al., 2006).

Courbot et al. (2007), also using progeny from interspecific cross between A. halleri and A. lyarata determined three QTLs involved in Cd tolerance. A major QTL region was found to be common to Cd (Courbot et al 2007) and Zn (Willems et al 2007) tolerance and was colocalized with the HMA4 gene. Hanikenne et al (2008) identified the role of HMA4 using RNAi-mediated silencing. They reported that when the expression of HMA4 was down-regulated, less Zn was translocated from the root to the shoot. When this gene was expressed in A. thaliana, an increase in Zn translocation to aerial tissue was observed. This increase in Zn translocation in A. thaliana plants resulted in signs of Zn hypersensitivity. Therefore, the expression of AhHMA4 alone was not adequate for Zn detoxification. Additional genes are involved in the A. halleri Zn hyperaccumulation (Hanikenne and Nouet 2011; Frérot et al., 2010).

Using segregating progeny resulting from intraspecific crosses between a high Cd accumulating parent and a low Cd accumulating parent for Glycine max, Benitez et al. (2010) identified a major QTL in seeds that was named cd1. This gene was mapped on chromosome 9. Jegadeesan et al. (2010) also identified a major QTL, cda1, associated with Cd accumulation in seeds of G. max. These two major QTLs were mapped to the same region of chromosome 9 which suggested that cd1 and cda1 may be identical. Both QTLs were found to be a dominant major gene involved in the control of low Cd uptake. By analyzing the G. max genome, Benitez et al. (2010) revealed that the cd1 QTL is localized in the vicinity of the P1B-ATPase gene (designated as GmHMA1) and proposed that this gene is involved in the transport of Cd. Benitez et al. (2012) found a single-base substitution between two cultivars, Harosoy (high Cd content in seed Cd) and Fukuyutaka (low Cd content in seed) in this P1B–ATPase gene. This mutation resulted in an amino acid substitution (glycine in Fukuyutaka and glutamic acid in Harosoy) in GmHMA1a. Since the glycine residue at the amino acid substitution site was conserved in AtHMA3, AtHMA4, AtHMA6 and AtHMA7, it was suggested that the GmHMA1a from Fukuyutaka was the wild type, responsible for low Cd accumulation in seed (Benitez et al., 2012). A dominant major gene involved in the control of Cd uptake was also observed in wheat (Triticum aestivum) (Clarke et al., 1997) and oat (Avena sativa L.) (Tanhuanpää et al., 2007). QTL analyses have also been performed in radish (Raphanus sativus L.). Xu et al. (2012) found a major QTL and three minor QTLs responsible for Cd accumulation in radish roots which were mapped on linkage groups 1, 4, 6 and 9. Induri et al. (2012) identified major QTLs for Cd response in Populus by performing a pseudo-backcross pedigree of Populus trichocarpa Torr. & Gray and Populus deltoides Bart. These QTLs were mapped to two different linkage groups. They performed a whole-genome microarray study and they were able to identify nine Cd responsive genes, which included a metal transporter, putative transcription factor and an NHL repeat membrane-spanning protein. Additional candidate genes located in the QTL intervals included a glutathione-S-transferase and putative homolog of a glutamine cysteine ligase.

Several QTL studies on rice (Oryza sativa L.) have been conducted to determine the number of genes involved in metal accumulation and tolerance. Three putative QTLs involved in Cd accumulation have been found on chromosomes 3, 6 and 8 (Ishikawa et al., 2010; Ishikawa et al., 2005). Ueno et al. (2009) also identified another major QTL for Cd accumulation in O. sativa that was mapped on the short arm of chromosome 7. QTLs for the translocation of Cd from roots to sink regions were reported in O. sativa (Xu et al., 2012; Tezuka et al., 2010). Tezuka et al. (2010) revealed a major QTL (qCdT7), mapped to chromosome 7, which controlled the translocation of Cd from roots to shoots. This QTL explained 88% of the phenotypic variation indicating that low Cd accumulation was a dominant trait. Dufey et al. (2009), using recombinant inbred lines, identified in O. sativa 24 putative QTLs involved in Fe tolerance which were mapped to chromosomes 1, 2, 3, 4, 7 and 11. In addition, two QTLs, located on chromosomes 2 and 3, were involved in As concentration in shoots and in roots respectively.

In durum wheat (Triticum durum, L.), Cd accumulation is controlled by a major gene named Cdu1 and localized on chromosome 5BL (Knox et al., 2009; Clarke et al., 1997). Further, Ci et al. (2012) characterized 26 QTLs involved in Cd tolerance and accumulation in T. aestivum, where 16 were involved in Cd stress control, 8 for Cd tolerance and 2 for Cd accumulation in roots. In A. sativa L., a single QTL for Cd accumulation in grain has been reported (Tanhuanpää et al., 2007).

In wheat (T. aestivum L.), Mayowa and Miller (1991) reported QTLs involved in Cu tolerance and accumulation that were mapped to chromosomes 5A, 4D, 7A, 7B, 7D. Ganeva et al. (2003) also characterized QTLs for T. aestivum on chromosomes 1A, 1D, 3A, 3B, 4A and 7D. Bálint et al. (2003) identified QTLs associated with Cu tolerance located on T. aestivum chromosomes 3D, 5A, 5B, 5D, 6B and 7D. In addition, Bálint et al. (2007) also determined QTLs for Cu tolerance in T. aestivum. They reported one major QTL for Cu tolerance on chromosome 5D and minor QTLs on chromosomes 1A, 2D, 4A, 5B and 7D. A QTL affecting shoot Cu content under Cu stress conditions was mapped on chromosome 1BL and an additional QTL for Cu accumulation was found on chromosome 5AL. The role of these genes located on various chromosomes in these different studies suggests that Cu tolerance is a polygenic character, as well as the possibility of different gene expressions against distinct toxic Cu concentrations in different populations. The accumulation of Cu in the shoots is affected by different QTLs, suggesting a strong metal-specific uptake and/ or translocation. Bálint et al. (2007) reported a negative correlation between Cu tolerance and accumulation in the shoot indicating that the key tolerance mechanism in wheat could be the restriction of Cu uptake in the roots or the reduced translocation from root to shoot.

Categories Of Plants Growing On Metal Contaminated Soils

Baker and Walker (1990) categorized plants into three groups according to their strategy for coping with metal toxicity in soil; metal excluders, indicators and accumulators/ hyperaccumulators.

EXCLUDERS

The metal excluder strategy consist in limiting the amount of metals translocated from roots to shoots thus maintaining low levels of metal concentration in their aerial parts. Large amounts of metals in the roots of excluder species have been reported (Baker and Walker 1990). Examples of excluder species include Oenothera biennis, Commelina communis, Silene maritime, Agrostis stolonifera L., woody plants such as Salix, Populus and Pinus radiata (Maestri et al., 2010; Wei et al., 2005).

ACCUMULATORS/HYPERACCUMULATORS

Metal accumulators/hyperaccumulators are plants that can concentrate metals in their above-ground tissues to levels that exceed those in the soil or also to those in the non accumulating species found growing nearby with concentrations up to 100 times more than non hyperaccumulators (Salt et al., 1998). Accumulators/hyperaccumulators growing on metal contaminated environments can naturally accumulate higher levels of heavy metals in their shoots than in their roots (Kachout et al., 2009). Some plants can accumulate only a specific metal while others can accumulate multiple metals ((Mganga et al., 2011; Almås et al., 2009). Presently, at least 45 plant families comprising more than 400 species have been found to accumulate metals in their harvestable tissues, and the majority of them belong to the Brassicaceae family (Pal and Rai 2010). The best known genera from this family are Alyssum and Thlaspi. Thlaspi species can accumulate more than 3% of their shoots in Zn, 0.5% in Pb and 0.1% in Cd. A. halleri can also accumulate more than 1% of its above-ground biomass in Cd and Zn and Alyssum species can accumulate over 1% Ni in their harvestable parts (Di Baccio et al., 2011). There are variations among family, species and populations in the ability to accumulate metals. For example, Arabidipsis halleri can accumulate Cd and Zn in their harvestable parts where as A. thaliana is known to be a metal excluder and restricts metals in the roots. Betula spp. can accumulate Zn, while other trees species of the same family (Carpinus and Corylus) are unable to do so (Ernst, 2006; Ernst, 2004).

Indicators

Like accumulators, metal indicators accumulate metals in their aerial tissue, but the metal levels in the above ground tissue of these plants usually reflect the metal concentration in the surrounding environment (Baker and Walker 1990). If these plants continue to uptake metals, they will eventually die-off. These plants are of biological and ecological importance since they are pollution indicators and also, like accumulators, they absorb pollutants (Mganga et al., 2011).

Determination Of Excluders, Indicators And Hyperaccumulators Plants

A plant is classified as a hyperaccumulator when it meets four criteria including; a) when the level of heavy metal in the shoot divided by level of heavy metal in the root is greater than 1 (shoot/root quotient > 1); b) when the level of heavy metal in the shoot divided by total level of heavy metal in the soil is greater than 1 (extraction coefficient > 1) (Rotkittikhun et al., 2006; Harrison and Chirgawi 1989); c) when the plant takes up between 10 – 500 times more heavy metals than normal plants (uncontaminated plants - control plants) (Fifield and Haines 2000; Allen,1989); and d) more than 100mg/kg of cadmium, 1000g/kg of copper, lead, nickel, chromium; or more than 10000mg/kg of zinc (Mganga et al., 2011; Ernst, 2006; Brooks, 1998). An excluder is a plant that has high levels of heavy metals in the roots but with shoot/root quotients less than 1 (Boularbah et al., 2006). Finally, Baker and Walker (1990) classified a plant as an indicator when the levels of heavy metals within their tissues reflect those in the surrounding soil.

Physiological Mechanisms Of Metal Resistance

Resistant plants are able to grow on metal contaminated soil due to avoidance and/or tolerance strategies. Plant resistance to high levels of heavy metals in soils can result from either reduced uptake or once taken up, metals have to be transformed into a physiologically tolerable form.

Restriction Of Metal Uptake

The plasma membrane is the first structure of living cells exposed to heavy metals. The membrane functions as a barrier for the movement of heavy metals into cytoplasm. The restriction of metals at the plasma membrane limits the uptake and accumulation of metals by preventing their entry into the cytoplasm. This can be done by changing the ion binding capacity of the cell wall and/or decreasing the uptake of metal ions through modified ion channels, and/or by removing metals from cells with active efflux pumps and/or with root with root exudates (Tong et al., 2004).

THE CELL WALL

The cell wall and membrane interface could be a site of metal tolerance since a significant amount of metals has been reported to be accumulated there. Divalent and trivalent metal cations can bind plant cell walls because of the presence of functional groups such as –COOH, -OH and –SH. Pectins are polymers that contain carboxyl groups which enable the binding of divalent and trivalent heavy metals ions. In enriched heavy metal environments, some plants will increase the capacity of their cell wall to bind metals by increasing polysaccharides, such as pectins (Colzi et al., 2011; Pelloux et al., 2007). Konno et al. (2010; 2005) showed that the pectin in root cell walls was important in binding Cu in the fern, Lygodium japonicum, and in the moss, Scopelophila cataractae. The cell wall of Minuartia verna sp. hercynica growing on heavy metal contaminated medieval mine dumps has been found to have high concentrations of Fe, Cu, Zn and Pb (Solanki and Dhankhar 2011; Neumann et al., 1997). On the other hand, Colzi et al. (2012) found that a copper tolerant Silene paradoxa population restricted the accumulation of Cu in roots, when exposed to high Cu, by decreasing their pectin concentration in the cell wall and increasing pectin methylation thus preventing the binding of Cu.

Root Exudates

Resistant plants can also restrict the entry of metals by immobilizing them in the rhizosphere with root exudates outside the plasma membrane (Colzi et al., 2011). This has been reported in T. aestivum where the exudation of phytochelatins, citrate and malate may be responsible for Cu exclusion mechanisms in non accumulators (Yang et al., 2005b; Bálint et al., 2007). Hall (2002) also proposed a mechanism for Ni exclusion in plants involving Ni-chelating exudates which include histidine and citrate. In non hyperaccumulator plants, these Ni chelators accumulate in their root exudates which, in turn decreases Ni uptake. The copper exclusion could be due to its chelation with citrate and malate exudates in the rhizosphere of wheat roots. The restriction of Cu uptake in wheat by the efflux of these organic acids has been previously documented by Nian et al. (2002).

Chelation

The phytotoxic effect of free metal ions can be eliminated by their chelation by specific high-affinity ligands (Yong and Ma 2002). The chelation of metals allows for the restriction of metal uptake, the uptake of metal ions, sequestration and compartmentation, as well as xylem loading and transport within the plant. Baker et al. (2000) categorized these ligands according to the characteristic electron donor centers, which include sulfur donor ligands, oxygen donor ligands and nitrogen donor ligands.

Oxygen Donor Ligands

Organic acids such as malate, aconitate, malonate, oxalate, tartrate and citrate are involved in metal uptake restriction and dexotification in plants. These carboxylic acid anions form complexes with divalent and trivalent metal ions with high stability. They are involved in the restriction of metal entry into the cell, metal exclusion in the root cells, accumulation and transport within the plants. In wheat (T. aestivum), citrate and malate formed complexes with Cu in order to immobilize this metal in the rhizosphere thus preventing its entry into the cell (Yong and Ma 2002). Citrate was also involved in the hyperaccumulation of Ni in 17 New Caledonian plants and the amount of citrate produced was highly correlated with the accumulation of Ni (Lee et al., 1977). The accumulation of Zn in some plants is facilitated by the transport of malate-Zn complexes. Upon the Zn ions uptake into the cytoplasm, they are bound to malate, which serves as a carrier to transport the Zn ions to the vacuole. Once there, the Zn ions are complexed by a terminal acceptor and released from malate. The malate is then able to return to the cytoplasm and transport additional Zn ions to the vacuole (Yong and Ma 2002). Still and Williams (1980) reported that the transport of free Ni ions to root cells via membrane is restricted. However, when Ni is bound to organic compounds such as citric and malic acids, it can be transported across the plasma membrane (Yong and Ma 2002). In Zea mays, the production of organic acid is influenced by external aluminium ion concentration (Pintro et al., 1997). Also, in manganese tolerant T. aestivum cultivars, the production of malic, citric or aconitic acid was not induced when exposed to this metal but for the manganese sensitive cultivars, the organic acids concentration slightly increased (Burke et al., 1990).

Nitrogen Donor Ligands

This group consists of amino acids and their derivatives which have relatively high affinity for specific metals. Krämer et al., (1996) revealed histidine to be involved in the Ni tolerance and translocation of the hyperaccumulator plant Alyssum lesbiacum. The majority of Zn in roots of the Zn hyperaccumulator, T. caerulescens, was complexed with histidine (Salt et al., 1999). Studies have also shown histidine to be involved in the restriction of metal uptake. For example, plants chelate Ni with histidine in the rhizosphere which prevents the uptake of this metal (Wenzel et al., 2003).

SULFUR DONOR LIGANDS

In plants, sulfur donor ligands are composed of two classes of metal chelating ligands which are phytochelatins (PCs) and metallothioneins (MTs). Phytochelatins are small metal binding peptides synthesized from the tripeptide glutathione (γ-Glu-Cys)2-11-Gly) (Solanki and Dhankhar 2011; Hall, 2002). Since there is a γ-carboxamide linkage between glutamate and cysteine, PCs are not synthesized by translation of mRNA, but rather it is a product of an enzymatic reaction involving the enzyme PC synthase (Yong and Ma 2002). The production of PCs is positively correlated with metal accumulation in plant tissues (Pal and Rai 2010). PCs are produced in cells immediately after heavy metal exposure, including Cd, Pb, Zn, Ag, Hg, As and Cu as seen in Rubia tinctorum (Maitani et al., 1996). PC production can be induced in roots, shoots, and leaves as observed in Sedum alfredii when exposed to Cd (Pal and Rai 2010).

Several research groups concurrently and independently cloned and characterized genes encoding PC synthase. These genes were isolated from Arabidopsis thaliana, Schizosaccharomyces pombe, and T. aestivum, and were designated AtPSC1, SpPCS, and TaPCS1, respectively. They encoded 50-55kDa sequences with 40-50% similarity. The polypeptides were found to be active in the synthesis of PCs from glutathione (GSH) (Yong and Ma 2002). In cultured Silene cucubalis cells, the presence of heavy metals, such as Cd, Cu, Zn, Ag, Hg and Pb, induce the synthesis of PCs by PC synthase from the GSH like substance (Pal and Rai 2010). Gaudet et al. (2011) did a comparative analysis of two Populus nigra genotypes from contrasting environments. They determined that both genotypes responded differently to Cd stress. The southern genotype (Poli) was more tolerant than the northern genotype (58-861). This variation was due to different adaptation strategies to Cd stress. The thiol and PC content, which was associated with the glutathione S-transferase gene, was higher in the southern genotype as compared to the northern genotype, which under Cd stress, revealed differences in the use of phytochelatin pathway that might be related to the variation in their Cd tolerance.

The second class of sulfur donor ligands are metallotioneins (MTs). They are low molecular weight (4-14kDa), cysteine-rich, metal-binding proteins found in a wide range of organisms (animals, plants, eukaryotic microorganisms, and prokaryotes) (Huang and Wang 2010). Unlike PCs, they are encoded by structural genes (Yong and Ma 2002). They play essential roles in a variety of organisms including Cu, Cd and Hg detoxification by sequestration (Palmiter, 1998; Ecker et al., 1989), Zn homeostasis (Coyle et al., 2002) and also scavenging of reactive oxygen species (Wong et al., 2004). MTs have been divided into two classes based on their cysteine residue arrangements. Class I MTs are widespread in vertebrates and are composed of 20 highly conserved cysteine residues based on mammalian MTs. Class II MTs have slightly flexible cysteine arrangements and are found in plants, fungi and invertebrates. A third class includes phytochelatins (Chaturvedi et al., 2012). Based on the position and allocation of cysteine residues, class II plant MTs are additionally divided into four types (Cobbett and Goldsbrough 2002). Type 1 plant MT genes have been more highly expressed in roots compared to leaves while the reverse is observed for the expression of type 2 plant MT genes. Type 3 MT genes are highly expressed in ripening fruits or in leaves while the expression of type 4 plant MT gene is restricted to developing seeds (Sekhar et al., 2011; Cobbett and Goldsbrough 2002).

The expression of MT genes in plants subjected to metal stress has been studied. AtMT1 and AtMT2 genes showed increased expression levels when Arabidopsis plants were exposed to high levels of Cu and Cd (Sekhar et al., 2011). Van Hoof et al. (2001) reported that the copper tolerant S. vulgaris individuals showed higher SvMT2b expression in roots and shoots when exposed to high concentrations of copper compared to the copper sensitive plants. Huang and Wang (2009) reported an increase in BgMT2 mRNA expression in large-leafed mangrove plants (Bruguiera gymnorrhiza) when exposed to Zn, Cu or Pb. Similar results were described by Gonzalez-Mendoza et al. (2007) in black mangrove (Avicennia germinans) seedlings exposed to Cd or Cu, showing a significant increase in AvMT2. High levels of the CcMT1 transcripts were also observed in pigeon pea (Cajanus cajan L.) exposed to Cd and Cu (Sekhar et al., 2011). In general, there are variations between species in the expression of MTs to various metals. The up and down regulation of

MTs in response to metal stress is largely unknown in plants. The MT gene expression was shown to be strongly induced by Cu, Cd, Pb and Zn (Huang and Wang 2009; Gonzalez-Mendoza et al., 2007; van Hoof et al., 2001). MT gene expression is also influenced by other abiotic stressors including absciscic acid (ABA), drought, salinity, heat, cold light, wounding and senescence (Sekhar et al., 2011).

Mechanisms Involved In Internal Metal Tolerance

Metal Uptake

The uptake of metal from soil into roots is dependent on the bioavailabilty of the metal, as well as its mobility in the rhizosphere (Maestri et al., 2010). The bioavailability of various metals greatly varies. No correlation exists between the metal content in soils and in plants (Clemens, 2006). The bioavailability of metals in the rhizosphere is affected by the chemical environment. For example, in T. caerulescens, the chemical form of nitrogen influences the plants ability to uptake Cd and Zn (Maestri et al., 2010; Xie et al., 2009). Metals present in the rhizosphere of hyperaccumulators are more bioavailable than for those of non hyperaccumulators. Plants can render metals mobile in their rhizosphere by excreting root exudates, such as organic acids and phytosiderophores and by acidification with protons (Maestri et al., 2010; Marschner,1995). Bacteria in the soil also affect metal mobility and availability by lowering the pH, producing hormones, organic acids, antifungals, antibiotics and metal chelators which all enhance the root growth (Maestri et al., 2010; Wenzel et al., 2003). Higher amounts of bacteria were found in the rhizosphere of hyperaccumulators. Microorganisms found in the rhizosphere were linked to an increased uptake of Cd, Z, and Pb in Sedum alfredii and an enhanced root growth (Maestri et al., 2010; Xiong et al., 2008).

Metal Uptake Across The Plasma Membrane

The uptake of heavy metals in plants is mediated by a group of metal transporter families which consists of iron-responsive transport proteins (ZIP-IRT), the heavy metal-transporting P1B-type subfamily of P-type ATPases, the natural resistance associated macrophage proteins (NRAMP) and the cation diffusion facilitators (CDF) (Baxter et al., 2003). Transporters were originally identified for Fe2+ or Zn2+ homeostasis, but it was demonstrated that most transporters of essential metal ions can also carry non essential metals, such as Cd (Zhou et al., 2012). The uptake of non essential metals may be the result of their close chemical characteristics or metal ion size to essential metals. Some metal transporters, present in the plasma membrane of root cells, exhibit low substrate specificity which can lead to the accumulation of other metals in plants (Schaaf et al., 2006). For example, the non-functional metal Cd can be taken up via a Ca^{2+} transporter (Perfus-Barbeoch et al., 2002) or also via the Fe^{2+} transporter IRT1 (Korshunova et al., 1999). Plant tolerance to metal stress can be achieved with the modification of these transporter activities (Zhou et al., 2012). Plants can prevent the uptake of certain metals by down-regulating the expression of such transporters, as observed in S. vulgaris, where the tolerant plants restrict the uptake of Cu by the down-regulation of Cu-transporters (Assunçãno et al., 2003; Harmens et al., 1993). Since Fe and Ni belong to the group of transient metals and have similar chemical properties, Fe deficiency may be the result of Ni phytotoxicity. Ni competes with Fe in physiological and biochemical processes, and in turn roots, can uptake Ni by Fe transporters (Pandey and Sharma 2002).

Increased Zn uptake is driven by an overexpression of members of the ZIP family of transporters. Under Zn deficiency conditions, many members of the ZIP transporter family are overexpressed in non hyperaccumulator species, while in hyperaccumulators, they are independently expressed regardless of Zn supply (Verbruggen et al., 2009). Nishida et al. (2011) and Schaaf et al. (2006) showed that A. thaliana can increase the uptake of Ni in roots when Fe levels are low by the Iron-Regulated Transporter 1 (AtIRT1; member of Zrt/IRT-like ZIP family of transporters). AtIRT1 has a wide specificity for divalent heavy metals including Ni, Zn, Mg, Co and Cd and mediates the accumulation of such metals under Fe-deficient conditions. Nakanishi et al. (2006) reported that Cd was uptaken in yeast by two O. sativa Fe^{2+} transporters, OsIRT1 (Iron-Regulated Transporter 1) and OsIRT2.

The uptake of Ni of some Ni hyperaccumulator accessions of Thlaspi goesingense, Thlaspi japonicum and T. caerulescens has been reported to be inhibited in the presence of Zn. This demonstrated that Ni entered the cell via Zn uptake transporters, specifically the TcZNT1 transporter (Assunçãno et al., 2008). In Zn deficiency conditions, the expression of AtZIP4, the orthologue of TcZNT1 in A. thaliana, can be induced but when additional Ni was added, the expression was repressed. This suggested that Zn and Ni competed for their uptake via AtZIP4/TcZNT1 transporters (Hassan and Aarts 2011). In addition, in presence of high Zn concentration, the expression of ZNT1 was higher in Zn hyperaccumulator T. caerulescens roots than in the non hyperaccumulator Thlaspi arvense suggesting its involvement in the hyperaccumulator phenotype (Hassinen et al., 2007; Assunçãno et al., 2001; Assunçãno et al., 2001; Pence et al., 2000). Milner et al. (2012) also determined that NcZNT1, isolated from T. caerulescens, played a role in Zn uptake from the soil which was based on its high expression in root.

Heavy metal-transporting P1B-type transporters are also involved in metal-ion homeostasis and tolerance in plants by transporting essential and non essential heavy metals such as Cu, Zn, Cd, Pb across cell membrane. Transporters located at the plasma membrane function as efflux pumps by removing toxic metals form cytoplasm. They have also been found in membranes of intracellular organelles for compartmentalization of metals for sequestration in vacuoles, golgi or endoplasmic reticulum (Yang et al., 2005b). These ion pumps transport ions across a membrane by hydrolysing ATP (Benitez et al., 2012). Eight P1B-ATPases, AtHMA1–AtHMA8, have been reported in Arabidopsis (Baxter et al., 2003). AtHMA1, 2, 3, and 4 showed high similarity with $Zn2+/Co2+/Cd2+/Pb2+$ ATPases previously characterized in prokaryotes (Axelsen and Palmgren 2001). The AtHMA4 was located at the plasma membrane. The ectopic expression of AtHMA4 improved the growth of roots in the presence of toxic Zn, Cd and Co concentrations (Yang et al., 2005b). The heterologous expression of AtHMA4 enhanced Cd tolerance in yeast (Mills et al., 2003).

In addition, the gene Nramp encodes for another divalent metal transporter located at the plasma membrane. This transporter also removes toxic metals from the cytosol by efflux pumping. It has been reported to be expressed in roots of Arabidopsis and O. sativa (OsNramp1- expressed in rice roots where as OsNramp2 is expressed in leaves and OsNramp3 is expressed in both tissues). The OsNramp1 gene was found to be involved in the uptake of Mn, while the Nramp genes in Arabidopsis and rice were involved in the uptake of Cd, and other divalent metals (Yang et al., 2005b). The AtNRAMP1, 3, and 4 showed uptake of $Cd2+$ when they were expressed in the yeast Saccharomyces cerevisiae. In addition, $Cd2+$ hypersensitivity was observed in A. thaliana when AtNRAMP3 was overexpressed. This transporter was located in the vacuolar membrane where it is involved in the mobilization of metals from the vacuole (Clemens, 2006).

In bacteria and in some eukaryotes, Zn, Co and Cd are transported by the CDF transport proteins. Within the Arabidopsis genome, there are 12 nucleotide sequences that are predicted to encode members of CDF transporter family. However, these transporters might be involved in cation efflux out of the cytoplasm, by pumping ions out of the cytoplasm to the exterior of the cell or into intracellular compartments such as the vacuole (Yang et al., 2005b).

Plants can make metal ions more available for uptake by acidifying the rhizosphere and pumping protons via plasma membrane-localized proton pumps; and also by exuding low molecular weight (LMW) compounds that act as metal chelators (Clemens, 2006). The secretion of organic acids can render heavy metals mobile and enhance their absorption by plant roots. Krishnamurti et al. (1997) reported that when Cd was complexed with organic acids, it was readily available for transport across the membrane, while free Cd ions were restricted for uptake. Cieśliński et al. (1998) revealed a higher acetic acid and succinate in the rhizosphere of the T aestivum (Kyle) Cd accumulating genotype compared to the non accumulating (Arcola) wheat genotype. The Zn/Cd hyperaccumulating Sedum alfredii was able to extract high levels of Zn and Pb from its contaminated environment because of the release of root exudates (Li et al., 2005). In Alyssum, the Ni transport and accumulation was enhanced by secretion of histidine in the rhizosphere (Krämer et al., 1996).

Sequestration/Compartmentation

Some metal tolerant plants can accumulate large amounts of metals within the cell without exhibiting toxicity symptoms (Entry et al., 1999). These plants are able to store the surplus of accumulated metals where no sensitive metabolic activities occur such as organs or subcellular compartments (Ernst, 2006). This avoidance of metal poisoning involves the intracellular sequestration and apoplastic or vacuolar compartmentation of the toxic metal ions (Liu et al., 2007). Compartmentation of metals can also be found in the cells central vacuole. This was observed in the Zn resistant Deschampsia cespitosa where the excess Zn ion was removed from the cytoplasm and actively pumped into the vacuoles of root cells where as Zn sensitive plants had a much lower capacity to do so (Brookes et al., 1981).

Schaaf et al. (2006) determined that the transporter AtIREG2, located at the tonoplast, was involved in Ni detoxification in roots. AtREG2, confined to roots, prevents heavy metal translocation to shoots restricting metals to roots. This transporter counterbalances the low substrate specificity of transporter AtIRT1 and other iron transporters in iron deficient root cells. The AtIREG2 transporter, found in A. thaliana, was involved in the detoxification of Ni in roots under Fe deficiency conditions at pH 5 (Schaaf et al., 2006). The T. caerulescens ZTP1 gene was involved in the intracellular sequestration of Zn. The expression of the ZTP1 gene was higher in the roots and shoots of the Zn tolerant T. caerulescens compared to the non tolerant plant (Assunçáno et al., 2001).

Members of the CDF protein play a role in tolerance to various metals including Cd, Co, Mn, Ni and Zn by their sequestration into vacuoles (Montanini et al., 2007). Increased Zn tolerance and accumulation was reported in non accumulator A. thaliana when AtMTP1, PtdMTP1, AtMTP3 and TgMTP1 (members of the CDF family) were ectopically or heterologously expressed. This suggested that the function of these proteins was the creation of a sink of Zn in the vacuole of plant cells in instances of high intracellular Zn levels or as buffer in Zn deficiency situations (Hassan and Aarts 2011). Phytochelatins are also thought to be involved in the restriction of metals to the roots (Zenk, 1996). When Nicotiana tubacum seedlings were exposed to excess Cd, the level of phytochelatin increased (Vogelilange and Wagner 1990). The metal-phytochelatin complexes are formed when plants are exposed to high heavy metal concentrations. They are then sequestered into vacuoles for detoxification. A group of organic solute transporters actively transport phytochelatin-metal complexes into the plant's vacuole (Solanki and Dhankhar 2011; Salt and Rauser 1995). In the presence of excess Cu and Cd, phytochelatins form complexes with these metals in Zea mays and in turn reduce the root to shoot translocation (Galli et al., 1996). The synthesis of phytochelatins is catalyzed by the enzyme phytochelatin synthase (PCS), a constitutive enzyme which requires post-translational activation by heavy metals and/or metalloids that include Cd, Ag, Pb, Cu, Hg, Zn, Sn, As and Au (Solanki and Dhankhar 2011). Martínez et al. (2006) reported that the expression of a PCS gene isolated from T. aestivum improved the accumulation of Cd, Pb and Cu in Nicotiana glauca. The elevation of phytochelatin concentration in roots might reduce the root to shoot transport required for accumulation in shoots.

Root To Shoot Translocation

The translocation of metals to the aerial biomass can be an important biochemical process used by plants to remediate polluted areas. In some plants, the mobilization of metals from their roots to their above aerial organs can minimize the damage that could be exerted by these heavy metals on the root physiology and biochemistry (Zacchini et al., 2009). Excluders prevent or limit the translocation of toxic metals or essential metals from roots to shoots. On the other hand, accumulators/hyperaccumulators translocate metals from roots to shoots via the xylem with the transpiration stream. This is accomplished by increasing the uptake of metals in roots, and by reducing the sequestration of metals in the root.

The chelation of metals with ligands, such as organic acids, amino acids and thiols facilitates the movements of heavy metals from roots to shoots (Zacchini et al., 2009). The xylem cell wall has a high cation exchange capability, thus the movement of metal cations is severely retarded when the metals are not chelated by ligands. Organic acids are involved in the translocation of Cd in the species Brassica juncea (Salt et al., 1995).

The chelation of Ni to histidine is involved in the long distance translocation of Ni in the hyperaccumulator A. lesbiacum, where a 36-fold increase was reported in the histidine content of the xylem sap upon exposure to nickel (Solanki and Dhankhar 2011; Krämer et al., 1996). Richau et al. (2009) found that the Ni hyperaccumulator, T. caerulescens, had a higher free histidine concentration in roots compared to the non Ni hyperaccumulator T. arvense. Also, T. caerulescens had less Ni in root vacuoles than T. arvense because the histidine-Ni complexes were much less taken up by vacuoles than free Ni ions. Therefore, an increase in free histidine in roots inhibited the vacuolar sequestration of His-Ni in T. caerulescens compared to free Ni in T. arvense and also had enhanced histidine-mediated Ni xylem loading. The elevated free histidine in root cells appears to be involved in reduced vacuolar sequestration and enhanced xylem loading of Ni (Richau and Schat 2009). This was also the case for Zn and Cd for this hyperaccumulating species (Hassan and Aarts 2011). An increase in Ni accumulation was also observed in the Ni hyperaccumulator Sebertia acuminata where, when chelated to citrate, Ni was able to translocate to the shoot. In the absence of citrate, Ni was no longer accumulated in the aerial tissues (Lee et al., 1977).

The chelation of metals with nicotianamine (NA) also contributes to improved tolerance. Nicotianamine can chelate and transport divalent Ni, Cu and Zn (Takahashi et al., 2003; Pich et al., 2001; Ling et al., 1999). The nicotianamine synthase (NAS) enzyme is responsible for the synthesis of NA by trimerization of S-adenosylmethionine (Shojima et al., 1990). When exposed to high levels of Zn, Cd, and/or Ni, all four NAS genes were highly expressed in T. caerulescens compared to non hyperaccumulator A. thaliana (van de Mortel et al., 2006). In the presence of elevated Mn, Zn, Fe and Cu concentrations, Kim et al. (2005) reported an increased expression of the NAS gene, as well as NA levels for A. thaliana and N. tubacum. In addition, Pianelli et al. (2005) showed that the over-expression of the T. caerulescens NAS3 gene in the Ni excluder A. thaliana resulted in improved Ni tolerance and Ni accumulation in their aerial organs. An increase of Fe, Zn and Cu accumulation in O. sativa was associated with an overexpression of the NAS3 gene (Hassan and Aarts 2011; Kawachi et al., 2009).

Visioli et al. (2010) also showed that metallothioneins may be involved in the translocation of Ni in T. caerulescens. An increase in MT-1B in the individuals from the metal contaminated environment was observed when metallicolous T. caerulescens and non-metallicolous T. caerulescens individuals were grown in presence of high Ni concentrations, compared to non contaminated site. Additionally, Visioli et al. (2012) analyzed four T. caerulescens sub-population (MP1 to MP4) for their ability to accumulate and tolerate Ni. In four sub-populations analyzed, MP2p translocated the highest amount of Ni to the shoots. This sub-population also had the highest level of putative metallothionein protein (MT4C). Constitutively higher expressions of other MTs are also seen in the hyperaccumulators A. halleri, S. paradoxa and

S. vulgaris. Transporters are not only involved in the uptake of metals from the soil, but also in their transport out of the vacuole. These mobilized metals can then be translocated to aerial tissue. Visioli et al. (2012) subsequently found for sub-population MP2p, which exhibited the highest level of Ni translocation of the four sub-populations analyzed, significantly higher levels of the ABC27 transporter. This transporter is part of the ABC family of transporters which are involved in removing metals from the cytoplasm by pumping outside the cell wall, metals sequestered in vacuoles and other subcellular compartments (Visioli et al., 2012; Martinoia et al., 2002; Sanchez-Fernandez et al., 2001). Hassinen et al. (2007) showed that the AtMRP10 homolog, also part of the ABC family of transporters, had different expression in roots of two T. caerulescens populations with contrasting Zn tolerance and accumulation. In addition, the AtNramp3 transporter was also involved in the mobilization of vacuolar Cd back into the cytosol. This was oberved when AtNramp3 was overexpressed in A. thaliana. AtNramp3 was further hypothesized to play a role in the mobilization of Fe, Mn, and Zn in the vacuole (Clemens, 2006).

The passage of metal ions and/or metal ligand complexes from the cytosol of root cells into the vascular tissue requires their transport across the cell membrane. Transporters involved in this activity are the heavy metal transporting P-type ATPases (HMAs) (Clemens, 2006). The AtHMA2 and 4 are involved in translocation of Zn in A. thaliana. Stunted growth and chlorosis resulted in the hma2hma4 double mutant from inadequate Zn supply to the leaves. The two genes were expressed in vascular tissue which indicates their hypothesized function in xylem loading (Hussain et al., 2004). The AtHMA4 transporter was also involved in the transport of Cd^{2+} ions (Clemens, 2006). In T. caerulescens, the P-type ATPase, TcHMA4, was also involved in the translocation of Zn. When Zn and Cd levels were elevated or when Zn is deficient, the expression of TcHMA4 was induced in the roots. This transporter was involved in the xylem loading of Zn in plant roots (Hassinen et al., 2007; Papoyan and Kochian 2004). Milner et al. (2012) also determined that NcZNT1 in T. caerulescens was not only involved in Zn uptake from the soil but also could be involved in the long distance transport of Zn from root to shoot via the xylem.

Metal Storage

Metals have to undergo a xylem unloading process prior to their distribution and their detoxification in the shoot and their redistribution via the phloem (Schmidke and Stephan 1995). Once unloaded, the metals are either taken up into surrounding cells and are symplastically transported through the leaf tissues or they are apoplastically distributed over the leaf (Hassan and

Aarts 2011; Marschner, 1995). NA is important in the chelation of metals for their symplastic transport through the leaf. This occurs through the Yellow Stripe Like proteins (YLS) (Hassan and Aarts 2011; DiDonato et al., 2004). In the hyperaccumulator T. caerulescens, three YSL genes (TcYLS3, TcYSL5 and TcYSL7) were highly expressed in shoots around vascular tissues. This high level of expression was not observed in the excluder plant A. thaliana orthologues (Hassan and Aarts 2011; Gendre et al., 2007). For the TcYSL3, it was suggested that its function was to unload Ni-NA complexes from the xylem into leaf cells and to distribute it to storage cells. Using yeast complementation and uptake measurement studies, it was determined that TcYSL3 was also a Fe/Ni-NA influx transporter. Considering that YSL proteins have a role in the transport of Fe-NA complexes, it was proposed that they might also be involved in the hyperaccumulation of Fe-NA in some plants (Hassan and Aarts 2011; Curie et al., 2009).

The sequestration of excess essential and non essential metals is localized in various parts of the aerial tissue, such as trichomes, leaf epidermal cell vacuole and mesophyll vacuole. Broadhurst et al. (2004) grew five Alyssum hyperaccumulator species/ecotypes on Ni-enriched soil and determined that the majority of hyperaccumulated Ni was stored in either leaf epidermal cell vacuoles or in the basal section of stellate trichomes. They also found that the metal concentration in the basal part of the trichome was 15% to 20% of dry weight. This was among the highest metal concentrations reported in healthy vascular plant tissues. In A. halleri, the majority of Zn ions were stored in the vacuoles of mesophyll cells, while for T. caerulescens, most Zn ions were located in the vacuoles of epidermal cells (Verbruggen et al., 2009). The transport of metals through the phloem sap is less documented. The sole molecule identified as a phloem metal transporter is nicotianamine which is involved in the transport of Fe, Cu, Zn and Mn (Stephan et al., 1994).

Antioxidative Defence Involved In Metal Tolerance

In environments, where metals are present in toxic levels, the elevated activities of antioxidant enzymes and non-enzymatic constituents are important in the plant tolerance to stress. Metal tolerance may be enhanced by the plant's antioxidant resistant mechanisms. There is an indication that the alleviation of oxidative damage and increased resistance to stresses in the environment is often correlated with an effective antioxidative system. The minimization of damage due to oxidative stress is a universal feature of plants defense responses (Kachout et al., 2009). The detrimental effect of heavy metals in plants is due to the production of ROS and induction of oxidative stress. Oxidative stress is expressed by the increase levels of reactive oxygen species such as singlet oxygen (1O2), superoxide radical (O−2), hydrogen peroxide (H2O2) and hydroxyl radical (OH−) (Salin, 1988). ROS are strong oxidizing agents that lead to oxidative damage to biomolecules, for instance lipids and proteins and can eventually result in cell death (Gunes et al., 2006). It is shown that plant tolerance to metals is correlated with a rise in antioxidants and activity of radical scavenging enzymes (Kachout et al., 2009). Plants respond to oxidative stress by activating antioxidative defence mechanisms which involve enzymatic and non-enzymatic antioxidants. The enzymatic components include superoxide dismutase (SOD), catalase (CAT), ascorbate peroxidase (APX) and enzymes of ascorbate glutathione cycle whereas the non-enzymatic antioxidants include ascorbate and glutathione and atocoperol (Solanki and Dhankhar 2011; Kachout et al., 2009). These antioxidants are responsible for elimination and destruction of the reactive oxygen species (Solanki and Dhankhar 2011).

Oxidative damage could result when the balance between the detoxification of the ROS products and the antioxidative system is altered (Kachout et al., 2009). The tolerance of deleterious environmental stresses, such as heavy metals, is correlated with the increased capacity to scavenge or detoxify activated oxygen species (Kachout et al., 2009). Boominathan and Doran (2003a,b) determined the role of antioxidative metabolism of heavy metal tolerance in T. caerulescens. They determined that superior antioxidant defenses, mainly catalase activity, may have an important role in the hyperaccumulator phenotype of T. caerulescens. Kachout et al (2010) determined the effects of Cu, Ni, Pb and Zn on the antioxidative defense systems of Atriplex plants. They found that when the plants were exposed to different levels of metals, their dry matter production and shoot height decreased. Of the antioxidant enzymes, metal toxicity only diminished the levels of superoxide dismutase (SOD) and probably ascorbate peroxidase (APX) but increased the activity of catalase (CAT) and glutathione reductase (GR). The plants showed an intermediate level of tolerance to the metal stress conditions imposed. The antioxidative activity may be of fundamental significance for the Atriplex plants in their response against environmental stress.

Problems Associated With Plant Metal Tolerance

Soils enriched with metals are demanding on tolerant and accumulator plants. The costs associated with their adaptation to these sites are related to energy and resources allocations. When a metal tolerant or accumulator plant is growing in a metal contaminated soil, there is an increase in cost because the organism has to spend energy to counter the effects of the metals (Maestri et al., 2010). Slow growth and low reproduction are the main characteristics of plants growing on metal enriched soils (Ernst, 2006; Ernst et al., 2000). Haldane

(1954) stated that costs are associated with the natural selection of new alleles. More energy and resources are required for the maintenance of the tolerance mechanisms at the cellular level. It has been demonstrated that tolerant plants have increased synthesis of complexing molecules in the cytosol. For example, metallothioneins and phytochelatins for the detoxification of metals such as As, Cd, and Cu. ATP are also needed for the active transport of metals across the plasma membrane and tonoplast. The synthesis of these agents withdraws N, S and energy from the primary metabolism (Ernst, 2006; Verkleij et al., 1998). Energy is also required for the translocation of metals from root to shoot as well as for their allocation to various tissues and cell types. The reduced biomass of metal tolerant plants compared to their non metal tolerant ancestors might also be the result of less than favourable environmental conditions such as low water and nutrient supply. The diminished biomass and seed production might be the result of all costs associated with their survival to these metal contaminated sites, such as adaptation and environmental constraints (Ernst, 2006). Plants have an advantages growing on metal contaminated soil. As previously mentioned, there is a lack of competitive species on these sites. With high metal accumulation of metals in their aerial tissues, the "elemental hypothesis" speculates that hyperaccumulators can deter predators such as herbivores from feeding on them (Maestri et al., 2010; Vesk and Reichman 2009). However, some insects feed on hyperaccumulator plants and in turn accumulate the metals in their tissue which then aid in their defence against predators (Maestri et al., 2010). This contradiction may explain why there is a mix of excluders, accumulator and hyperaccumulators growing on metal contaminated sites. Another advantage of hyperaccumulation is the elimination of competitive plants by further contaminating the surrounding soil by shedding their metal contaminated leaves (Maestri et al., 2010).

Effects Of Metals On Plant Population Diversity And Structure

Elevated accumulations of metals in soil and vegetation have been documented within short distances of the smelters compared to control sites (Nkongolo et al., 2008; Gratton et al., 2000). Several authors have reported differences in genetic structure of plants growing in contaminated areas (Vandeligt et al., 2011; Nkongolo et al., 2008; Scholz and Bergmann 1984). Enzymatic studies of Norway spruce (Picea abies) revealed genetic differences between groups of sensitive trees in polluted areas (Scholz and Bergmann 1984). It has been demonstrated that the evolution of heavy metal tolerant ecotypes occurs at an unexpectedly rapid rate (Wu et al., 1975) and that despite founder effect and selection, in several cases, the recently established tolerant populations maintain a high level of variation and appear to be at least as variable as non tolerant populations. Observations of higher heterozygosity in tolerant plants of European beech (Fagus sylvatica) in Germany (Muller and Starck 1985), scots pine (Pinus sylvestris) in Germany and Great Britain (Geburek et al., 1987), trembling aspen (Populus tremuloides) and red maple (Acer rubrum) in the United States (Berrang et al., 1986) have been reported. Several studies, however, have reported the detection of bottleneck effects (Nordal et al., 1999; Vekemans and Lefebre 1997; Mejnartowicz, 1983). Mejnatowicz (1983) presented evidence of loss of genes and heterozygosity in tolerant Scots pines. The frequent lack of a bottleneck effect has been explained by different hypotheses: successive colonization events, a high number of tolerant plants in the primary populations, pollen flow from the neighboring populations, environmental heterogeneity and human disturbance (Nkongolo et al., 2007).

Molecular analyses of several conifer and hardwood species clearly indicated that the exposure to metals for more than 30 years has no effect on genetic structure and diversity of early generations of Picea mariana, P. glauca, Pinus banksiana, P. rubens, P. strobus, and several hardwood populations in Northern Ontario (Narendrula et al., 2012; Nkongolo et al., 2012; Dobrzeniecka et al., 2011; Vandeligt et al., 2011). This lack of association between the level of genetic variation and metal content can be attributed to the long life span of these tree species. Table 1 shows similar level of genetic variabilities in pine populations growing in metal contaminated sites for more than 30 years compared to control in Northern Ontario, Canada. This is in contrast to data observed in herbaceous species such as D. cespitosa where a high level of metal accumulation reduced significantly the level of genetic variation (Table 2) (Nkongolo et al., 2007). Metals impose severe stress on plants, especially in the rooting zone, which has led to the evolution of metal resistant ecotypes in several herbaceous species like D. cespitosa (Cox and Hutchinson 1980).

TABLE 1: Genetic variability parameters of Pinus banksiana populations growing in the Sudbury, Ontario (Canada) area based on ISSR data

Populations	P (%)	h	I	Ne	Na
Vale site 1 (metal contaminated)	31.25	0.1120	0.1653	1.2035	1.3125
Vale site 2 (metal contaminated	31.25	0.1171	0.1727	1.2061	1.3125
Xtrata 2 (metal contaminated)	27.08	0.0995	0.1467	1.1758	1.2708
Xtrata 3 (metal contaminated)	20.83	0.0630	0.0982	1.1004	1.2083
Vale Tailing (metal contaminated)	35.42	0.0977	0.1552	1.1514	1.3542
Temagami site (control)	29.17	0.0818	0.1284	1.1310	1.2917
Low Water Lake (control)	31.25	0.0812	0.1297	1.1256	1.3125
Mean	31.63	0.1001	0.1528	1.1679	1.3163

P represents percentage of polymorphic loci; h, Nei's gene diversity; I, Shannon's information index; Ne, effective number of alleles; Na, observed number of alleles.

Table 2: Genetic variability within Deschampsia cespitosa populations from Northern Ontario generated with ISSR primers.

Region	Site	Polymorphism per site (%)	Mean polymorphism per region (%)
Sudbury (moderately contaminated)	Coniston	72	74 (Sudbury)
	Xtrata	92	
	Copper Cliff	67	
	Walden	65	
Cobalt (highly contaminated)	Cobalt-3	48	46 (Cobalt)
	Cobalt-4	46	
	Cobalt-5	44	
Manitoulin (control)	Little Current	70	69 (Manitoulin)
	Mississagi Lighthouse	68	

P represents percentage of polymorphic loci; Sudbury and Cobalt regions were moderately and highly contaminated with metals, respectively. Manitoulin Island region was not contaminated with metals and was used as a control region.

CONCLUSION

Plants play an essential role in the remediation of metal enriched soils. Coping mechanisms developed by some group of plants growing on metal contaminated soil facilitate the establishment of sustainable ecosystems in areas that would otherwise remain barren. A number of studies have been completed to explain the complex mechanisms involved in tolerance genotypes, and also the biological variability in their environmental adaptation. Depending on the circumstances, metal excluders or hyperaccumulators may be used to remediate polluted soil. Excluders may be useful for soil stabilization by preventing wind and water erosion and also by limiting the entry of heavy metals in the food chain. Metals can also be extracted by hyperaccumulators but since majority of these plants have low biomass, the extraction of metals from soil is very slow. The remediation of these sites using this technique may take up to hundreds of years. With genetic engineering, it may be possible to design the ideal plant prototype for the remediation of metal contamination in different environments. Many genes and mechanisms have been identified to have a role in tolerance and hyperaccumulation of metals. However, there is still a need for a better understanding of the mechanisms such as characterization of promoters of genes controlling metal tolerance and hyperaccumulation. This new knowledge would significantly contribute to a better understanding of the regulation and expression of different genes in hyperaccumulators. It is essential to mimic this regulation and expression of genes in high biomass non hyperaccumulators in order to obtain the hyperaccumulator phenotype.

NAN. ACKNOWLEDGEMENTS

We express our appreciation to the Natural Sciences and Engineering Research Council of Canada (NSERC) for a Postdoctoral Fellowship to M. Mehes-Smith.

REFERENCES

1. S. E Allen, 1989Chemical analysis of ecological materials. Blackwell Scientific Publications, Oxford. 368

2. Å. R Almås, C Kweyunga, and M. L. K Manoko, 2009Investigation of trace metal concentrations in soil, sediments and waters in the vicinity of "Geita Gold Mine" and "North Mara Gold Mine" in North West Tanzania (with Barrick Response). IPM-report.

3. J Arines, A Vilarino, and M Sainz, 1989Effect of Different Inocula of Vesicular-Arbuscular Mycorrhizal Fungi on Manganese Content and Concentration in Red-Clover (Trifolium pratense L) Plants. New Phytol. 112215219

4. A. G. L Assunçãno, W. M Bookum, H. J. M Nelissen, R Vooijs, H Schat, and W. H. O Ernst, 2003Differential metal-specific tolerance and accumulation patterns among Thlaspi caerulescens populations originating from different soil types. New Phytol. 159411419

5. A Assunçãno, B Pieper, J Vromans, P Lindhout, M Aarts, and H Schat, 2006Construction of a genetic linkage map of Thlaspi caerulescens and quantitative trait loci analysis of zinc accumulation. New Phytol. 1702132

6. A Assunçãno, P Martins, S De Folter, R Vooijs, H Schat, and M Aarts, 2001Elevated expression of metal transporter genes in three accessions of the metal hyperaccumulator Thlaspi caerulescens. Plant Cell and Environment 24217226

7. A. G. L Assunçãno, P Bleeker, ten Bookum, W.M., Vooijs, R. and Schat, H., 2008Intraspecific variation of metal preference patterns for hyperaccumulation in Thlaspi caerulescens: evidence from binary metal exposures. Plant Soil 303289299

8. K Axelsen, and M Palmgren, 2001Inventory of the superfamily of P-type ion pumps in Arabidopsis. Plant Physiol. 126696706

9. A. J. M Baker, and P. L Walker, 1990Ecophysiology of metal uptake by tolerant plants: Heavy metal uptake by tolerant plants. In: Shaw, A.J. (Editors). Evolutionary Aspects. CRC, Boca Raton.

10. A. J. M Baker, S. P Mcgrath, R. D Reeves, and J. A. C Smiith, 2000Metal hyperaccumulator plants: A review of the ecolgy and physiology of a biological resource for phytoremediation of metal-polluted soil. In: Terry, N. and Banuelos, G. (Editors). Phytoremediation of contaminated soil and water. CRC Press LLC, Boca Ratton. 85108

11. A. J. M Baker, 1987Metal tolerance. New Phytologist 106 (Suppl s1): 93-111.

12. A. F Bálint, G Kovacs, A Börner, G Galiba, and J Sutka, 2003Substitution analysis of seedling stage copper tolerance in wheat. Acta agronimica hungarica 51397404

13. A. F Bálint, M. S Röder, R Hell, G Galiba, and A Börner, 2007Mapping of QTLs affecting copper tolerance and the Cu, Fe, Mn and Zn contents in the shoots of wheat seedlings. Biol. Plant. 51129134

14. N Basic, N Salamin, C Keller, N Galland, and G Besnard, 2006Cadmium hyperaccumulation and genetic differentiation of Thlaspi caerulescens populations. Biochem. Syst. Ecol. 34667677

15. I Baxter, J Tchieu, M Sussman, M Boutry, M Palmgren, M Gribskov, J Harper, and K Axelsen, 2003Genomic comparison of P-type ATPase ion pumps in Arabidopsis and rice. Plant Physiol. 132618628

16. E. R Benitez, M Hajika, and R Takahashi, 2012Single-base substitution in 1Bgene is associated with a major QTL for seed cadmium concentration in soybean. J. Hered. 103: 278-286.

17. E. R Benitez, M Hajika, T Yamada, K Takahashi, N Oki, N Yamada, T Nakamura, and K Kanamaru, 2010A Major QTL Controlling Seed Cadmium Accumulation in Soybean. Crop Sci. 5017281734

18. P Berrang, D. F Karnosky, R. A Mickler, and J. P Bennett, 1986Natural selection for ozone tolerance in Populus tremuloides. Canadian Journal of Forest Research 1612141216

19. V Bert, I Bonnin, P Saumitou-laprade, P De Laguerie, and D Petit, 2002Do Arabidopsis halleri from nonmetallicolous populations accumulate zinc and cadmium more effectively than those from metallicolous populations? New Phytol. 1554757

20. V Bert, P Meerts, P Saumitou-laprade, P Salis, W Gruber, and N

Verbruggen, 2003Genetic basis of Cd tolerance and hyperaccumulation in Arabidopsis halleri. Plant Soil 249918

21. G Besnard, N Basic, P Christin, D Savova-bianchi, and N Galland, 2009Thlaspi caerulescens (Brassicaceae) population genetics in western Switzerland: is the genetic structure affected by natural variation of soil heavy metal concentrations? New Phytol. 181974984

22. R Boominathan, and P Doran, 2003bCadmium tolerance and antioxidative defenses in hairy roots of the cadmium hyperaccumulator, Thlaspi caerulescens. Biotechnol. Bioeng. 83158167

23. R Boominathan, and P Doran, 2003aOrganic acid complexation, heavy metal distribution and the effect of ATPase inhibition in hairy roots of hyperaccumulator plant species. J. Biotechnol. 101131146

24. A Boularbah, C Schwartz, G Bitton, W Aboudrar, A Ouhammou, and J. L Morel, 2006Heavy metal contamination from mining sites in South Morocco: 2. Assessment of metal accumulation and toxicity in plants. Chemosphere 63811817

25. A. D Bradshaw, 1991Genostasis and the Limits to Evolution. Philosophical Transactions of the Royal Society of London Series B-Biological Sciences 333289305

26. A. D Bradshaw, 1984Adaptation of plants to soils containing toxic metals- a test for conceit. Origins and development of adaptation, CIBA Foundation Symposium 102, Pitman, London, 419

27. M. G Bratteler, 2005Genetic architecture of traits associated with serpentine adaptation of Silene vulgaris (Caryophyllaceae). Eidgenoessische Technische Hochschule Zuerich (Switzerland), Switzerland. 91

28. C Broadhurst, R Chaney, J Angle, T Maugel, E Erbe, and C Murphy, 2004Simultaneous hyperaccumulation of nickel, manganese, and calcium in Alyssum leaf trichomes. Environ. Sci. Technol. 3857975802

29. A Brookes, J. C Collins, and D. A Thurman, 1981The mechanism of zinc tolerance in grasses. J. Plant Nutr. 3695705

30. R. R Brooks, 1998Plant that accumulate heavy metals. CAB International, Wallingford.

31. D. G Burke, K Watkins, and B. J Scott, 1990Manganese toxicity effects on visible symptoms, yield, manganese levels, and organic-acid levels in tolerant and sensitive wheat cultivars. Crop Sci. 30275280

32. I Cakmak, 2008Enrichment of cereal grains with zinc: Agronomic or genetic biofortification? Plant Soil 302117

33. A. K Chaturvedi, A Mishra, V Tiwari, and B Jha, 2012Cloning and

transcript analysis of type 2 metallothionein gene (SbMT-2) from extreme halophyte Salicornia brachiata and its heterologous expression in E. coli. Gene 499280287

34.   D Ci, D Jiang, S Li, B Wollenweber, T Dai, and W Cao, 2012Identification of quantitative trait loci for cadmium tolerance and accumulation in wheat. Acta Physiol Plant 34191202

35.   G Cieslinski, K. C. J Van Rees, A. M Szmigielska, G. S. R Krishnamurti, and P. M Huang, 1998Low-molecular-weight organic acids in rhizosphere soils of durum wheat and their effect on cadmium bioaccumulation. Plant Soil 203109117

36.   J Clarke, D Leisle, and G Kopytko, 1997Inheritance of cadmium concentration in five durum wheat crosses. Crop Sci. 3717221726

37.   S Clemens, 2006Toxic metal accumulation, responses to exposure and mechanisms of tolerance in plants. Biochimie 8817071719

38.   C Cobbett, and P Goldsbrough, 2002Phytochelatins and metallothioneins: Roles in heavy metal detoxification and homeostasis. Annual Review of Plant Biology 53159182

39.   I Colzi, S Doumett, Del Bubba, M., Fornaini, J., Arnetoli, M., Gabbrielli, R. and Gonnelli, C., 2011On the role of the cell wall in the phenomenon of copper tolerance in Silene paradoxa L. Environ. Exp. Bot. 727783

40.   I Colzi, M Arnetoli, A Gallo, S Doumett, Del Bubba, M., Pignattelli, S., Gabbrielli, R. and Gonnelli, C., 2012Copper tolerance strategies involving the root cell wall pectins in Silene paradoxa L. Environ. Exp. Bot. 789198

41.   M Courbot, G Willems, P Motte, S Arvidsson, N Roosens, P Saumitou-laprade, and N Verbruggen, 2007A Major Quantitative Trait Locus for Cadmium Tolerance in Arabidopsis halleri Colocalizes with HMA4, a Gene Encoding a Heavy Metal ATPase. Plant physiology 144: 1065.

42.   R. M Cox, and T. C Hutchinson, 1980Multiple metal tolerances in the grass Deschampsia cespitosa (L.) Beauv. from the Sudbury smelting area. New Phytologist 84631647

43.   P Coyle, J Philcox, L Carey, and A Rofe, 2002Metallothionein: The multipurpose protein. Cellular and Molecular Life Sciences 59627647

44.   C Curie, G Cassin, D Couch, F Divol, K Higuchi, Le Jean, M., Misson, J., Schikora, A., Czernic, P. and Mari, S., 2009Metal movement within the plant: contribution of nicotianamine and yellow stripe 1-like transporters. Annals of Botany 103111

45.   A. X Deniau, B Pieper, Ten Bookum, W.M., Lindhout, P., Aarts, M.G.M.

and Schat, H., 2006QTL analysis of cadmium and zinc accumulation in the heavy metal hyperaccumulator Thlaspi caerulescens. Theor. Appl. Genet. 113907920

46. Di BaccioD., Galla, G., Bracci, T., Andreucci, A., Barcaccia, G., Tognetti, R., Sebastiani, L. and Moshelion, M., 2011Transcriptome analyses of Populus × euramericana clone I-214 leaves exposed to excess zinc. Tree Physiol. 3112931308

47. DiDonatoR., Roberts, L., Sanderson, T., Eisley, R. and Walker, E., 2004Arabidopsis Yellow Stripe-Like2 (YSL2): a metal-regulated gene encoding a plasma membrane transporter of nicotianamine-metal complexes. Plant Journal 39403414

48. S Dobrzeniecka, K. K Nkongolo, P Michael, M Mehes-smith, and P Beckett, 2011Genetic Analysis of Black Spruce (Picea mariana) Populations from Dry and Wet Areas of a Metal-Contaminated Region in Ontario (Canada). Water Air and Soil Pollution 215117125

49. I Dufey, P Hakizimana, X Draye, S Lutts, and P Bertin, 2009QTL mapping for biomass and physiological parameters linked to resistance mechanisms to ferrous iron toxicity in rice. Euphytica 167143160

50. V Dushenkov, P Kumar, H Motto, and I Raskin, 1995Rhizofiltration- the use of Plants to Remove Heavy-Metals from Aqueous Streams. Environ. Sci. Technol. 2912391245

51. D. J Ecker, T. R Butt, and S. T Crooke, 1989Yeast Metallothionein- Gene-Function and Regulation by Metal-Ions. Met. Ions Biol. Syst. 25147169

52. J Entry, L Watrud, and M Reeves, 1999Accumulation of Cs-137 and Sr-90 from contaminated soil by three grass species inoculated with mycorrhizal fungi. Environmental Pollution 104449457

53. W. H. O Ernst, 2004Vegetation, organic matter and soil quality. In: Doelman, P. and Eijsackers, H.J.P. (Editors). Vital soil : Function, value, and properties. Amsterdam, Elsevier, 2004, 4198

54. W. H. O Ernst, 2006Evolution of metal tolerance in higher plants. Forest Snow and Landscape Research 80251274

55. W Ernst, H Nelissen, and Ten Bookum, W., 2000Combination toxicology of metal-enriched soils: physiological responses of a Zn- and Cd-resistant ecotype of Silene vulgaris on polymetallic soils. Environ. Exp. Bot. 435571

56. W. H. O Ernst, G Krauss, J. A. C Verkleij, and D Wesenberg, 2008Interaction of heavy metals with the sulphur metabolism in angiosperms from an ecological point of view. Plant Cell and Environment 31123143

57. F. W Fifield, and P. J Haines, 2000Environmental Analytical Chemistry. Wiley-Blackwell, 512

58. V Filatov, J Dowdle, N Smirnoff, B Ford-lloyd, H. J Newbury, and M Macnair, 2006Comparison of gene expression in segregating families identifies genes and genomic regions involved in a novel adaptation, zinc hyperaccumulation. Molecular Ecology 15: 3059.

59. V Filatov, J Dowdle, N Smirnoff, Ford Lloyd, B., Newbury, H.J. and Macnair, M.R., 2007A quantitative trait loci analysis of zinc hyperaccumulation in Arabidopsis halleri. New Phytologist 174: 590.

60. H Frérot, M Faucon, G Willems, C Godé, A Courseaux, A Darracq, N Verbruggen, and P Saumitou-laprade, 2010Genetic architecture of zinc hyperaccumulation in Arabidopsis halleri : the essential role of QTL x environment interactions. New Phytol. 187355367

61. U Galli, H Schuepp, and C Brunold, 1996Thiols in cadmium- and copper-treated maize (Zea mays L). Planta 198139143

62. G Ganeva, S Landjeva, and M Merakchijska, 2003Effects of chromosome substitutions on copper toxicity tolerance in wheat seedlings. Biol. Plant. 47621623

63. M Gaudet, F Pietrini, I Beritognolo, V Iori, M Zacchini, A Massacci, G. S Mugnozza, M Sabatti, and R Tognetti, 2011Intraspecific variation of physiological and molecular response to cadmium stress in Populus nigra L. Tree Physiol. 3113091318

64. T. H Geburek, F Scholz, W Knabe, and A Vornweg, 1987Genetic studies by isozyme gene loci on tolerance and sensitivity in an air polluted Pinus sylvestris field trial. Silvae Genetica 364953

65. D Gendre, P Czernic, G Conejero, K Pianelli, J Briat, M Lebrun, and S Mari, 2007TcYSL3, a member of the YSL gene family from the hyper-accumulator Thlaspi caerulescens, encodes a nicotianamine-Ni/Fe transporter. Plant Journal 49115

66. D Gonzalez-mendoza, A. Q Moreno, and O Zapata-perez, 2007Coordinated responses of phytochelatin synthase and metallothionein genes in black mangrove, Avicennia germinans, exposed to cadmium and copper. Aquatic Toxicology 83306314

67. W Gratton, K Nkongolo, and G Spiers, 2000Heavy metal accumulation in soil and jack pine (Pinus banksiana) needles in Sudbury, Ontario, Canada. Bull. Environ. Contam. Toxicol. 64550557

68. A Gunes, G Soylemezoglu, A Inal, E. G Bagci, S Coban, and O Sahin, 2006Antioxidant and stomatal responses of grapevine (Vitis vinifera L.)

to boron toxicity. Scientia Horticulturae 110279284

69. J. L Gustin, M. E Loureiro, D Kim, G Na, M Tikhonova, and D. E Salt, 2009MTP1-dependent Zn sequestration into shoot vacuole's suggests dual roles in Zn tolerance and accumulation in Zn hyperaccumulating plants. Plant Journal 5711161127

70. J. B. S Haldane, 1954The biochemistry of genetics. Allen & Unwin, London. 144

71. J Hall, 2002Cellular mechanisms for heavy metal detoxification and tolerance. J. Exp. Bot. 53111

72. M Hanikenne, and C Nouet, 2011Metal hyperaccumulation and hypertolerance: a model for plant evolutionary genomics. Curr. Opin. Plant Biol. 14252259

73. H Harmens, N. G. C. P. B Gusmao, P. R Denhartog, J. A. C Verkleij, and W. H. O Ernst, 1993Uptake and transport of zinc in zinc-sensitive and zinc-tolerant Silene vulgaris. J. Plant Physiol. 141309315

74. R. M Harrison, and M. B Chirgawi, 1989The assessment of air and soil as contributors of some trace-metals to vegetable plants.1. Use of a filtered air growth cabinet. Sci. Total Environ. 831334

75. Z Hassan, and M. G. M Aarts, 2011Opportunities and feasibilities for biotechnological improvement of Zn, Cd or Ni tolerance and accumulation in plants. Environ. Exp. Bot. 725363

76. V. H Hassinen, A. I Tervahauta, P Halimaa, M Plessl, S Peraniemi, H Schat, M. G. M Aarts, K Servomaa, and S. O Karenlampi, 2007Isolation of Zn-responsive genes from two accessions of the hyperaccumulator plant Thlaspi caerulescens. Planta 225977989

77. G Huang, and Y Wang, 2010Expression and characterization analysis of type 2 metallothionein from grey mangrove species (Avicennia marina) in response to metal stress. Aquatic Toxicology 999692

78. G Huang, and Y Wang, 2009Expression analysis of type 2 metallothionein gene in mangrove species (Bruguiera gymnorrhiza) under heavy metal stress. Chemosphere 7710261029

79. D Hussain, M Haydon, Y Wang, E Wong, S Sherson, J Young, J Camakaris, J Harper, and C Cobbett, 2004P-type ATPase heavy metal transporters with roles in essential zinc homeostasis in Arabidopsis. Plant Cell 1613271339

80. B. R Induri, D. R Ellis, G. T Slavov, T Yin, X Zhang, W Muchero, G. A Tuskan, and DiFazio, S.2012Identification of quantitative trait loci and candidate genes for cadmium tolerance in Populus. Tree Physiol. 32:

626-638.

81. S Ishikawa, N Ae, and M Yano, 2005Chromosomal regions with quantitative trait loci controlling cadmium concentration in brown rice (Oryza sativa). New Phytol. 168345350

82. S Ishikawa, T Abe, M Kuramata, M Yamaguchi, T Ando, T Yamamoto, and M Yano, 2010A major quantitative trait locus for increasing cadmium-specific concentration in rice grain is located on the short arm of chromosome 7. J. Exp. Bot. 61923934

83. S Jegadeesan, K Yu, V Poysa, E Gawalko, M. J Morrison, C Shi, and E Cober, 2010Mapping and validation of simple sequence repeat markers linked to a major gene controlling seed cadmium accumulation in soybean [Glycine max (L.) Merr]. Theor. Appl. Genet. 121:.

84. J Jeong, and M. L Guerinot, 2008Biofortified and bioavailable: The gold standard, for plant-based diets. Proc. Natl. Acad. Sci. U. S. A. 10517771778

85. S. S Kachout, Ben Mansoura, A., Leclerc, J.C., Mechergui, R., Rejeb, M.N. and Ouerghi, Z., 2009Effects of heavy metals on antioxidant activities of Atriplex hortensis and A. rosea. J. Food Agric. Environ. 7938945

86. M Kawachi, Y Kobae, H Mori, R Tomioka, Y Lee, and M Maeshima, 2009A Mutant Strain Arabidopsis thaliana that Lacks Vacuolar Membrane Zinc Transporter MTP1 Revealed the Latent Tolerance to Excessive Zinc. Plant and Cell Physiology 5011561170

87. S Kim, M Takahashi, K Higuchi, K Tsunoda, H Nakanishi, E Yoshimura, S Mori, and N Nishizawa, 2005Increased nicotianamine biosynthesis confers enhanced tolerance of high levels of metals, in particular nickel, to plants. Plant and Cell Physiology 4618091818

88. R. E Knox, C. J Pozniak, F. R Clarke, J. M Clarke, S Houshmand, and A. K Singh, 2009Chromosomal location of the cadmium uptake gene (Cdu1) in durum wheat. Genome 52741747

89. H Konno, T Nakato, S Nakashima, and K Katoh, 2005Lygodium japonicum fern accumulates copper in the cell wall pectin. J. Exp. Bot. 5619231931

90. H Konno, S Nakashima, and K Katoh, 2010Metal-tolerant moss Scopelophila cataractae accumulates copper in the cell wall pectin of the protonema. J. Plant Physiol. 167358364

91. Y Korshunova, D Eide, W Clark, M Guerinot, and H Pakrasi, 1999The IRT1 protein from Arabidopsis thaliana is a metal transporter with a

broad substrate range. Plant Mol. Biol. 403744

92. U Krämer, 2005MTP1 mops up excess zinc in Arabidopsis cells. Trends Plant Sci. 10313315

93. U Krämer, J Cotter-howells, J Charnock, A Baker, and J Smith, 1996Free histidine as a metal chelator in plants that accumulate nickel. Nature 379635638

94. G Krishnamurti, G Cieslinski, P Huang, and K Vanrees, 1997Kinetics of cadmium release from soils as influenced by organic acids: Implication in cadmium availability. J. Environ. Qual. 26271277

95. N Lal, 2010Molecular mechanisms and genetic basis of heavy metal toxicity and tolerance in plants. In: Ashraf, M., Ozturk, M. and Ahmad, M.S.A. (Editors). Plant Adaptation and Phytoremediation. Springer Netherlands, 3558

96. J Lee, R. D Reeves, R. R Brooks, and T Jaffré, 1977Isolation and identification of a citrato-complex of nickel from nickel-accumulating plants. Phytochemistry 1615031505

97. J Levitt, 1980Responses of Plants to Environmental Stresses: Water, radiation, salt, and other stresses. Academic Press, New York. 607

98. T Li, X Yang, X Jin, Z He, P Stoffella, and Q Hu, 2005Root responses and metal accumulation in two contrasting ecotypes of Sedum alfredii Hance under lead and zinc toxic stress. Journal of Environmental Science and Health Part A-Toxic/hazardous Substances & Environmental Engineering 4010811096

99. H Ling, G Koch, H Baumlein, and M Ganal, 1999Map-based cloning of chloronerva, a gene involved in iron uptake of higher plants encoding nicotianamine synthase. Proc. Natl. Acad. Sci. U. S. A. 9670987103

100. C. P Liu, Z. G Shen, and X. D Li, 2007Accumulation and detoxification of cadmium in Brassica pekinensis and B. chinensis. Biol. Plant. 51116120

101. M. R Macnair, V Bert, S. B Huitson, P Saumitou-laprade, and D Petit, 1999Zinc tolerance and hyperaccumulation are genetically independent characters. Proceedings of the Royal Society of London. Series B, Biological Sciences, 26621752179

102. M. R Macnair, 1993The genetics of metal tolerance in vascular plants. New Phytologist 124541559

103. M. R Macnair, 1987Heavy metal tolerance in plants: a model evolutionary system. Trends in Ecology and Evolution 2354359

104. E Maestri, and M Marmiroli, 2012Genetic and molecular aspects of metal tolerance and hyperaccumulation. In: Gupta, D.K. and Sandalio, L.M.

(Editors). Metal toxicity in plants: Perception, signaling and remediation. Sprigner-Verlag Berlin Heidelberg, 4163

105. E Maestri, M Marmiroli, G Visioli, and N Marmiroli, 2010Metal tolerance and hyperaccumulation: Costs and trade-offs between traits and environment. Environ. Exp. Bot. 68113

106. T Maitani, H Kubota, K Sato, and T Yamada, 1996The composition of metals bound to class III metallothionein (phytochelatin and its desglycyl peptide) induced by various metals in root cultures of Rubia tinctorum. Plant Physiol. 11011451150

107. M Marmiroli, F Pietrini, E Maestri, M Zacchini, N Marmiroli, A Massacci, and R Tognetti, 2011Growth, physiological and molecular traits in Salicaceae trees investigated for phytoremediation of heavy metals and organics. Tree Physiol. 3113191334

108. H Marschner, 1995Mineral nutrition of higher plants. Edited by Horst Marschner. second edition. London; San Diego, Academic Press. 899

109. M Martínez, P Bernal, C Almela, D Velez, P Garcia-agustin, R Serrano, and J Navarro-avino, 2006An engineered plant that accumulates higher levels of heavy metals than Thlaspi caerulescens, with yields of 100 times more biomass in mine soils. Chemosphere 64478485

110. E Martinoia, M Klein, M Geisler, L Bovet, C Forestier, U Kolukisaoglu, B Muller-rober, and B Schulz, 2002Multifunctionality of plant ABC transporters- more than just detoxifiers. Planta 214345355

111. J. E Mayer, W. H Pfeiffer, and P Beyer, 2008Biofortified crops to alleviate micronutrient malnutrition. Curr. Opin. Plant Biol. 11166170

112. N. M Mayowa, and T. E Miller, 1991The Genetics of Tolerance to High Mineral Concentrations in the Tribe Triticeae- a Review and Update. Euphytica 57175185

113. L Mejnartowicz, 1983Changes in genetic structure of Scotch pine (Pinus sylvestris L.) population affected by industrial emission of fluoride and sulphur dioxide. Genetica Polonica 244150

114. N Mganga, M. L. K Manoko, and Z. K Rulangaranga, 2011Classification of plants according to their heavy metal content around north mara gold mine, Tanzania: Implication for phytoremediation. Tanzania Journal of Science 37109119

115. R Millaleo, M Reyes- Diaz, A. G Ivanov, M. L Mora, and M Alberdi, 2010Manganese as essential and toxic element for plants: Transport, accumulation and resistance mechanisms. Journal of Soil Science and Plant Nutrition 10470481

116. R Mills, G Krijger, P Baccarini, J Hall, and L Williams, 2003Functional expression of AtHMA4, a 1BATPase of the Zn/Co/Cd/Pb subclass. Plant Journal 35: 164-176.

117. M. J Milner, E Craft, N Yamaji, E Koyama, J. F Ma, and L. V Kochian, 2012Characterization of the high affinity Zn transporter from Noccaea caerulescens, NcZNT1, and dissection of its promoter for its role in Zn uptake and hyperaccumulation. New Phytol. 195113123

118. B Montanini, D Blaudez, S Jeandroz, D Sanders, and M Chalot, 2007Phylogenetic and functional analysis of the Cation Diffusion Facilitator (CDF) family: improved signature and prediction of substrate specificity. BMC Genomics 8: 107.

119. R Narendrula, K. K Nkongolo, P Beckett, and G Spiers, 2012Total and bioavailable metals in two contrasting mining regions (Sudbury in Canada and Lubumbashi in DR-Congo): relation to genetic variation in plant populations. Chemistry and Ecology In press.

120. K Neumann, DrogeLaser, W., Kohne, S. and Broer, I., 1997Heat treatment results in a loss of transgene-encoded activities in several tobacco lines. Plant Physiol. 115939947

121. H Nian, Z Yang, S Ahn, Z Cheng, and H Matsumoto, 2002A comparative study on the aluminium- and copper-induced organic acid exudation from wheat roots. Physiol. Plantarum 116328335

122. S Nishida, C Tsuzuki, A Kato, A Aisu, J Yoshida, and T Mizuno, 2011AtIRT1, the primary iron uptake transporter in the root, mediates excess nickel accumulation in Arabidopsis thaliana. Plant and Cell Physiology 5214331442

123. K. K Nkongolo, M Mehes, A Deck, and P Michael, 2007Metal content in soil and genetic variation in Deschampsia cespitosa populations from Northern Ontario (Canada): application of ISSR markers. European Journal of Genetics Toxicology March: 138

124. K. K Nkongolo, R Narendrula, M Mehes-smith, S Dobrzeniecka, K Vandeligt, M Ranger, and P Beckett, 2012Genetic Sustainability of fragmented conifer populations from stressed areas in Northern Ontario (Canada): Application of molecular markers. In: Blanco, J.A. and Lo, Y.L. (Editors). Forest Ecosystems- More than just trees. Intech publisher, 315336

125. K. K Nkongolo, A Vaillancourt, S Dobrzeniecka, M Mehes, and P Beckett, 2008Metal content in soil and black spruce (Picea mariana) trees in the Sudbury region (Ontario, Canada): Low concentration of arsenic, cadmium, and nickel detected near smelter sources. Bull. Environ.

Contam. Toxicol. 80107111

126. I Nordal, K. B Haraldsen, A Ergon, and A. B Eriksen, 1999Copper resistance and genetic diversity in Lychnis alpina (Caryophyllaceae) populations on mining sites. Folia Geobotanica 34471481

127. R Pal, and J. P. N Rai, 2010Phytochelatins: peptides involved in heavy metal detoxification. Appl. Biochem. Biotechnol. 160945963

128. R Palmiter, 1998The elusive function of metallothioneins. Proc. Natl. Acad. Sci. U. S. A. 9584288430

129. N Pandey, and C Sharma, 2002Effect of heavy metals Co2+, Ni2+ and Cd2+ on growth and metabolism of cabbage. Plant Science 163753758

130. A Papoyan, and L Kochian, 2004Identification of Thlaspi caerulescens genes that may be involved in heavy metal hyperaccumulation and tolerance. Characterization of a novel heavy metal transporting ATPase. Plant Physiol. 13638143823

131. J Pelloux, C Rusterucci, and E. J Mellerowicz, 2007New insights into pectin methylesterase structure and function. Trends Plant Sci. 12267277

132. N Pence, P Larsen, S Ebbs, D Letham, M Lasat, D Garvin, D Eide, and L Kochian, 2000The molecular physiology of heavy metal transport in the Zn/Cd hyperaccumulator Thlaspi caerulescens. Proc. Natl. Acad. Sci. U. S. A. 9749564960

133. L Perfus-barbeoch, N Leonhardt, A Vavasseur, and C Forestier, 2002Heavy metal toxicity: cadmium permeates through calcium channels and disturbs the plant water status. Plant Journal 32539548

134. K Pianelli, S Mari, L Marques, M Lebrun, and P Czernic, 2005Nicotianamine over-accumulation confers resistance to nickel in Arabidopsis thaliana. Transgenic Res. 14739748

135. A Pich, R Manteuffel, S Hillmer, G Scholz, and W Schmidt, 2001Fe homeostasis in plant cells: Does nicotianamine play multiple roles in the regulation of cytoplasmic Fe concentration? Planta 213967976

136. J Pintro, J Barloy, and P Fallavier, 1997Effects of low aluminum activity in nutrient solutions on the organic acid concentrations in maize plants. J. Plant Nutr. 20601611

137. I. D Pulford, and C Watson, 2003Phytoremediation of heavy metal-contaminated land by trees- a review. Environ. Int. 29529540

138. K. H Richau, and H Schat, 2009Intraspecific variation of nickel and zinc accumulation and tolerance in the hyperaccumulator Thlaspi caerulescens. Plant Soil 314253262

139. P Rotkittikhun, M Kruatrachue, R Chaiyarat, C Ngernsansaruay, P

Pokethitiyook, A Paijitprapaporn, and A. J. M Baker, 2006Uptake and accumulation of lead by plants from the Bo Ngam lead mine area in Thailand. Environmental Pollution 144681688

140. M. L Salin, 1988Toxic oxygen species and protective systems of the chloroplast. Physiol. Plantarum 72681689

141. D. E Salt, and W. E Rauser, 1995Mgatp-dependent transport of phytochelatins across the tonoplast of oat roots. Plant Physiol. 10712931301

142. D. E Salt, R. D Smith, and I Raskin, 1998Phytoremediation. Annu. Rev. Plant Physiol. Plant Mol. Biol. 49643668

143. D. E Salt, R. C Prince, I. J Pickering, and I Raskin, 1995Mechanisms of cadmium mobility and accumulation in indian mustard. Plant Physiol. 10914271433

144. D Salt, R Prince, A Baker, I Raskin, and I Pickering, 1999Zinc ligands in the metal hyperaccumulator Thlaspi caerulescens as determined using X-ray absorption spectroscopy. Environ. Sci. Technol. 33713717

145. R Sanchez-fernandez, T Davies, J Coleman, and P Rea, 2001The Arabidopsis thaliana ABC protein superfamily, a complete inventory. J. Biol. Chem. 2763023130244

146. G Schaaf, A Honsbein, A. R Meda, S Kirchner, D Wipf, and N Von Wiren, 2006AtIREG2 encodes a tonoplast transport protein involved in iron-dependent nickel detoxification in Arabidopsis thaliana roots. J. Biol. Chem. 2812553225540

147. H Schat, L Mercè, and R Bernhard, 2000Metal-specific patterns of tolerance, uptake and transport of heavy metals in hyperaccumulating and nonhyperaccumulating metallophytes. In: Terry, N. and Banuelos, G. (Editors). Phytoremediation of Contaminated Soil and Water. CRC Press, Boca Raton. 171188

148. H Schat, E Kuiper, W Tenbookum, and R Vooijs, 1993A General-model, for the genetic-control of copper tolerance in Silene Vulgaris- evidence from crosses between plants from different tolerant populations. Heredity 70142147

149. H Schat, R Vooijs, and E Kuiper, 1996Identical major gene loci for heavy metal tolerances that have independently evolved in different local populations and subspecies of Silene vulgaris. Evolution 5018881895

150. I Schmidke, and U. W Stephan, 1995Transport of metal micronutrients in the phloem of castor bean (Ricinus communis) seedlings. Physiol. Plantarum 95147153

151. F Scholz, and F Bergmann, 1984Selection pressure by air-pollution as studied by isozyme-gene-systems in norway spruce exposed to sulfur-dioxide. Silvae Genet. 33238241

152. K Sekhar, B Priyanka, V. D Reddy, and K. V Rao, 2011Metallothionein 1 (CcMT1) of pigeonpea (Cajanus cajan L.) confers enhanced tolerance to copper and cadmium in Escherichia coli and Arabidopsis thaliana. Environ. Exp. Bot. 72131139

153. S Shojima, N. K Nishizawa, S Fushiya, S Nozoe, T Irifune, and S Moris, 1990Biosynthesis of phytosiderophores- invitro biosynthesis of 2'-deoxymugineic acid from L-methionine and nicotianamine. Plant Physiol. 9314971503

154. S Smith, and M Macnair, 1998Hypostatic modifiers cause variation in degree of copper tolerance in Mimulus guttatus. Heredity 80760768

155. R Solanki, and R Dhankhar, 2011Biochemical changes and adaptive strategies of plants under heavy metal stress. Biologia 66195204

156. U. W Stephan, I Schmidke, and A Pich, 1994Phloem translocation of Fe, Cu, Mn, and Zn in ricinus seedlings in relation to the concentrations of nicotianamine, an endogenous chelator of divalent metal-ions, in different seedling parts. Plant Soil 165181188

157. E. R Still, and R. J. P Williams, 1980Potential methods for selective accumulation of nickel (II) ions by plants. J. Inorg. Biochem. 133540

158. M Takahashi, Y Terada, I Nakai, H Nakanishi, E Yoshimura, S Mori, and N Nishizawa, 2003Role of nicotianamine in the intracellular delivery of metals and plant reproductive development. Plant Cell 1512631280

159. P Tanhuanpää, R Kalendar, A. H Schulman, and E Kiviharju, 2007A major gene for grain cadmium accumulation in oat (Avena sativa L.). Genome 50588594

160. K Tezuka, H Miyadate, K Katou, I Kodama, S Matsumoto, T Kawamoto, S Masaki, H Satoh, M Yamaguchi, K Sakurai, H Takahashi, N Satoh-nagasawa, A Watanabe, T Fujimura, and H Akagi, 2010A single recessive gene controls cadmium translocation in the cadmium hyperaccumulating rice cultivar Cho-Ko-Koku. Theor. Appl. Genet. 12011751182

161. Y Tong, R Kneer, and Y Zhu, 2004Vacuolar compartmentalization: a second-generation approach to engineering plants for phytoremediation. Trends Plant Sci. 979

162. D Ueno, E Koyama, I Kono, T Ando, M Yano, and J. F Ma, 2009Identification of a novel major quantitative trait locus controlling distribution of cd between roots and shoots in rice. Plant and Cell

Physiology 5022232233

163. van de MortelJ.E., Almar Villanueva, L., Schat, H., Kwekkeboom, J., Coughlan, S., Moerland, P.D., van Themaat, E.V.L., Koornneef, M. and Aarts, M.G.M., 2006Large expression differences in genes for iron and zinc homeostasis, stress response, and lignin biosynthesis distinguish roots of Arabidopsis thaliana and the related metal hyperaccumulator Thlaspi caerulescens. Plant Physiol. 14211271147

164. N. A Van Hoof, V. H Hassinen, H. W Hakvoort, K. F Ballintijn, H Schat, J. A Verkleij, W. H Ernst, S. O Karenlampi, and A. I Tervahauta, 2001Enhanced copper tolerance in Silene vulgaris (Moench) Garcke populations from copper mines is associated with increased transcript levels of a 2b-type metallothionein gene. Plant Physiol. 12615191526

165. K. K Vandeligt, K. K Nkongolo, M Mehes, and P Beckett, 2011Genetic analysis of Pinus banksiana and Pinus resinosa populations from stressed sites contaminated with metals in Northern Ontario (Canada). Chem. Ecol. 27369380

166. X Vekemans, and C Lefebre, 1997The evolution of heavy metal tolerant populations in Armenia maritime: evidence from allozyme variation and reproductive barriers. Journal of Evolutionary Biology 10175191

167. N Verbruggen, C Hermans, and H Schat, 2009Molecular mechanisms of metal hyperaccumulation in plants. New Phytol. 182781781

168. J Verkleij, P Koevoets, M Blake-kalff, and A Chardonnens, 1998Evidence for an important role of the tonoplast in the mechanism of naturally selected zinc tolerance in Silene vulgaris. J. Plant Physiol. 153188191

169. P Vesk, and S Reichman, 2009Hyperaccumulators and herbivores- a bayesian meta-analysis of feeding choice trials. J. Chem. Ecol. 35289296

170. G Visioli, S Vincenzi, M Marmiroli, and N Marmiroli, 2012Correlation between phenotype and proteome in the Ni hyperaccumulator Noccaea caerulescens subsp. caerulescens. Environmental and Experimental Botany 77156164

171. G Visioli, A Pirondini, A Malcevschi, and N Marmiroli, 2010Comparison of protein variations in Thlaspi Caerulescens populations from metalliferous and non-metalliferous soils. Int. J. Phytoremediation 12805819

172. R Vogelilange, and G. J Wagner, 1990Subcellular-localization of cadmium and cadmium-binding peptides in tobacco-leaves- implication of a transport function for cadmium-binding peptides. Plant Physiol. 9210861093

173. S Wei, Q Zhou, and X Wang, 2005Identification of weed plants excluding the uptake of heavy metals. Environ. Int. 31829834

174. W Wenzel, M Bunkowski, M Puschenreiter, and O Horak, 2003Rhizosphere characteristics of indigenously growing nickel hyperaccumulator and excluder plants on serpentine soil. Environmental Pollution 123131138

175. G Willems, D. B Dräger, M Courbot, C Godé, N Verbruggen, and P Saumitou-laprade, 2007The Genetic Basis of Zinc Tolerance in the metallophyte Arabidopsis halleri ssp. halleri (Brassicaceae): An Analysis of Quantitative Trait Loci. Genetics 176659674

176. H Wong, T Sakamoto, T Kawasaki, K Umemura, and K Shimamoto, 2004Down-regulation of metallothionein, a reactive oxygen scavenger, by the small GTPase OsRac1 in rice. Plant Physiol. 13514471456

177. L Wu, A. D Bradshaw, and D. A Thurman, 1975The potential for evolution of heavy metal tolerance in plants. Heredity 34165187

178. H. L Xie, R. F Jiang, F. S Zhang, S. P Mcgrath, and F. J Zhao, 2009Effect of nitrogen form on the rhizosphere dynamics and uptake of cadmium and zinc by the hyperaccumulator Thlaspi caerulescens. Plant Soil 318205215

179. J Xiong, Z He, D Liu, Q Mahmood, and X Yang, 2008The role of bacteria in the heavy metals removal and growth of Sedum alfredii Hance in an aqueous medium. Chemosphere 70489494

180. L Xu, L Wang, Y Gong, W Dai, Y Wang, X Zhu, T Wen, and L Liu, 2012Genetic linkage map construction and QTL mapping of cadmium accumulation in radish (Raphanus sativus L.). Theoretical and Applied Genetics 125659670

181. X Yang, Y Feng, Z He, and P Stoffella, 2005bMolecular mechanisms of heavy metal hyperaccumulation and phytoremediation. Journal of Trace Elements in Medicine and Biology 18339353

182. X Yang, X Jin, Y Feng, and E Islam, 2005aMolecular mechanisms and genetic basis of heavy metal tolerance/hyperaccumulation in plants. Journal of Integrative Plant Biology 4710251035

183. C Yong, and L. Q Ma, 2002Metal tolerance, accumulation, and detoxification in plants with emphasis on arsenic in terrestrial plants. In: Biogeochemistry of Environmentally Important Trace Elements. American Chemical Society, 95114

184. M Zacchini, F Pietrini, Scarascia Mugnozza, G., Iori, V., Pietrosanti, L. and Massacci, A., 2009Metal tolerance, accumulation and translocation in poplar and willow clones treated with cadmium in hydroponics. Water Air Soil Pollut. 1972334

185. M. H Zenk, 1996Heavy metal detoxification in higher plants- a review. Gene 1792130

186. H Zha, R Jiang, F Zhao, R Vooijs, H Schat, J Barker, and S Mcgrath, 2004Co-segregation analysis of cadmium and zinc accumulation in Thlaspi caerulescens interecotypic crosses. New Phytol. 163299312

187. Z. S Zhou, H. Q Zeng, Z. P Liu, and Z. M Yang, 2012Genome-wide identification of Medicago truncatula microRNAs and their targets reveals their differential regulation by heavy metal. Plant Cell and Environment 358699

CITATION

CHAPTER 1

Noelle Aarts and Anne Marike Lokhorst (2012). The Role of Government in Environmental Land Use Planning: Towards an Integral Perspective, Environmental Land Use Planning, Dr. Seth Appiah-Opoku (Ed.), ISBN: 978-953-51-0832-0, DOI: 10.5772/50684.

CHAPTER 2

R. Canales-Pastrana and M. Paredes, "Phytoremediation Dynamic Model as an Assessment Tool in the Environmental Management," Open Journal of Applied Sciences, Vol. 3 No. 2, 2013, pp. 208-217. doi: 10.4236/ojapps.2013.32028.

CHAPTER 3

Wolfgang Römer (2013). Environmental Change and Geomorphic Response in Humid Tropical Mountains, Environmental Change and Sustainability, Dr. Steven Silvern (Ed.), ISBN: 978-953-51-1094-1,

CHAPTER 4

Humberto Barbosa, Carolien Tote, Lakshmi Kumar and Yazidhi Bamutaze (2013). Harnessing Earth Observation and Satellite Information for Monitoring Desertification, Drought and Agricultural Activities in Developing Countries, Environmental Change and Sustainability, Dr. Steven Silvern (Ed.), ISBN: 978-953-51-1094-1, DOI: 10.5772/55499.

CHAPTER 5

christopher kipkoech saina, daniel kipkosgei murgor and florence a.c murgor; climate change and food security; doi:10.1038/climate.2007.58.

CHAPTER 6

Paulo Roberto Ferreira Carneiro and Marcelo Gomes Miguez (2012). A Flood Control Approach Integrated with a Sustainable Land Use Planning in Metropolitan Regions, Environmental Land Use Planning, Dr. Seth Appiah-Opoku (Ed.), ISBN: 978-953-51-0832-0, DOI: 10.5772/50573.

CHAPTER 7

Dirk Loehr (2012). The Role of Tradable Planning Permits in Environmental Land Use Planning: A Stocktake of the German Discussion, Environmental Land Use Planning, Dr. Seth Appiah-Opoku (Ed.), ISBN: 978-953-51-0832-0, DOI: 10.5772/50469.

CHAPTER 8

Christine Majale-Liyala (2013). Policy Arrangement for Waste Management in East Africa's Urban Centres, Environmental Change and Sustainability, Dr. Steven Silvern (Ed.), ISBN: 978-953-51-1094-1, InTech, DOI: 10.5772/54382

CHAPTER 9

Jennifer Koch, Florian Wimmer, Rüdiger Schaldach and Janina Onigkeit (2012). An Integrated Land-Use System Model for the Jordan River Region, Environmental Land Use Planning, Dr. Seth Appiah-Opoku (Ed.), ISBN: 978-953-51-0832-0, DOI: 10.5772/51247.

CHAPTER 10

José D. Carriquiry, Linda M. Barranco-Servin, Julio A. Villaescusa,.Victor F. Camacho-Ibar, Hector Reyes-Bonilla and Amílcar L. Cupul-.Magaña (2013). Conservation and Sustainability of Mexican Caribbean Coral Reefs and the Threats of a Human-Induced Phase-Shift, Environmental Change and Sustainability, Dr. Steven Silvern (Ed.), ISBN: 978-953-51-1094-1, DOI: 10.5772/54339.

CHAPTER 11

Melanie Mehes-Smith, Kabwe Nkongolo and Ewa Cholewa (2013). Coping Mechanisms of Plants to Metal Contaminated Soil, Environmental Change and Sustainability, Dr. Steven Silvern (Ed.), ISBN: 978-953-51-1094-1, InTech, DOI: 10.5772/55124.

INDEX